Germany's Nature

Dear Dad,
   Here's some more exciting reading for your travels.
   Merry Christmas,
         Tom

Germany's Nature

Cultural Landscapes and Environmental History

THOMAS LEKAN
THOMAS ZELLER
EDITORS

RUTGERS UNIVERSITY PRESS
New Brunswick, New Jersey, and London

Library of Congress Cataloging-in-Publication Data

Germany's nature : cultural landscapes and environmental history / Thomas Lekan and Thomas Zeller, editors.
   p. cm.
   Includes bibliographical references and index.
   ISBN-13: 978-0-8135-3667-5 (hardcover : alk. paper)
  1. Forests and forestry—Germany—History. 2. Nature conservation—Germany—History. 3. Landscape protection—Germany—History. I. Lekan, Thomas M. II. Zeller, Thomas, 1966–

SD195.G76 2005
333.72'0943—dc22

2005002578

A British Cataloging-in-Publication record for this book is available from the British Library.

This collection copyright © 2005 by Rutgers, The State University
Individual chapters copyright © 2005 in the names of their authors
All rights reserved
No part of this book may be reproduced or utilized in any form or by any means, electronic or mechanical, or by any information storage and retrieval system, without written permission from the publisher. Please contact Rutgers University Press, 100 Joyce Kilmer Avenue, Piscataway, NJ 08854–8099. The only exception to this prohibition is "fair use" as defined by U.S. copyright law.

Manufactured in the United States of America

Contents

	Acknowledgments *vii*	
Introduction	The Landscape of German Environmental History THOMAS LEKAN AND THOMAS ZELLER	*1*

Part I   Seeing Like a State: Water, Forests, and Power

Chapter 1	Germany as a Focus of European "Particularities" in Environmental History JOACHIM RADKAU	*17*
Chapter 2	Conviction and Constraint: Hydraulic Engineers and Agricultural Amelioration Projects in Nineteenth-Century Prussia RITA GUDERMANN	*33*
Chapter 3	A Sylvan People: Wilhelmine Forestry and the Forest as a Symbol of Germandom MICHAEL IMORT	*55*
Chapter 4	Forestry and the German Imperial Imagination: Conflicts over Forest Use in German East Africa THADDEUS SUNSERI	*81*

Part II   The Cultural Landscapes of Home

Chapter 5	Organic Machines: Cars, Drivers, and Nature from Imperial to Nazi Germany RUDY KOSHAR	*111*
Chapter 6	Biology—*Heimat*—Family: Nature and Gender in German Natural History Museums around 1900 SUSANNE KÖSTERING	*140*

Part III The Politics of Conservation

Chapter 7	Indication and Identification: On the History of Bird Protection in Germany, 1800–1918 FRIEDEMANN SCHMOLL	161
Chapter 8	Protecting Nature between Democracy and Dictatorship: The Changing Ideology of the Bourgeois Conservationist Movement, 1925–1935 JOHN ALEXANDER WILLIAMS	183
Chapter 9	Protecting Nature in a Divided Nation: Conservation in the Two Germanys, 1945–1972 SANDRA CHANEY	207

Notes on Editors and Contributors   245
Index   249

Acknowledgments

WE ARE INDEBTED to the German Historical Institute in Washington, D.C., and its director, Christof Mauch, and the Department of History, University of Maryland, and its chair, Gary Gerstle, for their generous support of this publication. Our editor at Rutgers University Press, Audra Wolfe, has expertly guided us through the publication of the volume and we are grateful to her as well. Finally, we thank Michael Wettengel, the erstwhile archivist at the Bundesarchiv in Koblenz, for introducing us to one another in 1994.

Thomas Lekan
Thomas Zeller

Germany's Nature

Introduction

The Landscape of German Environmental History

THOMAS LEKAN
THOMAS ZELLER

THERE IS SOMETHING odd about the state of environmental history in Germany. When Americans think of a Western country with a strong environmental record, Germany will be among the top contenders. The Rhine River is cleaner than it has been in decades, recycling is considered a civic duty, and German manufacturers of pollution-control technology are exporting their goods all over the globe. And yet, the very country that regularly elects Green politicians to its highest offices is relatively weak when it comes to scholarship in environmental history. Currently, at best only a handful of professorships in this country of some 80 million people and more than 110 research universities is specifically dedicated to environmental history. On the other hand, the United States, which many Germans are quick to associate with profligate consumption of natural resources and reckless attitudes toward the environment, is home to probably the most vigorous academic society for environmental history with lively annual meetings and its own journal, dozens of professorships in environmental history, and a widespread interest in the field. The contrast between Germany and the United States in this respect could hardly be sharper. While many historians in Germany have been practicing environmental history for years, their work has not entered the historiographical mainstream, let alone the wider public's consciousness. When a German publisher recently decided on a title for the translation of John McNeill's comprehensive history of the twentieth-century environment, it settled on the English buzzword "Blue Planet" rather than a German heading, thus indicating how foreign the concept of environmental history still is to a wide audience.[1] One wonders whether there might be deeper cultural reasons for these diverging historiographies, whether historians of the United States are temperamentally better suited to think about the environmental dimension of history. "Time is the basic dimension of history, but the basic dimension of the American imagination is space," as Richard Hofstadter once remarked.[2]

While we would like to leave it to future historians of the discipline to analyze the reasons for this gulf between politics and scholarship, we hope that present scholars and students of German history in both the United States and Germany will use *Germany's Nature* to enrich their understanding of the tangled relationships between humans and the environment in this country. The volume as a whole does not aim to be comprehensive; rather, the contributions deal with previously neglected aspects or new interpretations of existing themes within German environmental history in the nineteenth and twentieth centuries. Our goal is to offer innovative approaches to environmental-historical analysis and to spearhead intellectual engagement in German environmental history beyond the Green movement and the recent environmental past, to uncover the enduring yet ever-changing cultural patterns, social institutions, and geographic factors that have sustained Germany's Nature over the past two hundred years.

The articles in this volume are informed by renewed interest in the concept and practice of landscape in Germany among historians, geographers, art historians, and other social researchers.[3] This landscape tradition envisioned the country's natural landscapes—from the Black Forest to the Lüneburger Heath, from the Bavarian Alps to the Rhine River valley—as objects of environmental concern, cultural inspiration, and national identity.[4] At the birth of German nationhood during the Napoleonic Wars, the thinker Johann Gottfried Herder (to cite one example), whose view of history and interest in German antiquity provided important ideas for the Romantics, extolled untouched nature as a divine presence and a source of Germanic character. The National Socialist regime later perverted Herder's cosmopolitan outlook by envisioning the German landscape as a font of primordial racial energies; their slogan "Blood and Soil" proclaimed a close symbiosis between ethnic Germans and their living space, or *Lebensraum*. More recently, German Greens have decried the death of Germany's forests (*Waldsterben*) due to acid rain as a consequence of postwar material success, the dark underside of the so-called economic miracle. German environmental practices have also had a decisive impact on international conservation efforts and Green movements worldwide. Wise-use advocates such as Gifford Pinchot, for example, at times saw German scientific forestry as a model of careful resource management, while Ralph Nader and other eco-minded American citizens have recently drawn inspiration from the German Green Party's electoral successes in building an equivalent party in the United States. Modern German history has taken place in specific physical and symbolic topographies that shaped and, in turn, were shaped by the country's economic, political, and cultural systems, yet German environmental historians have rarely examined the connection between landscape perception and environmental change. The contributions in *Germany's Nature* offer such an analysis, offering one of the first collections of essays in English devoted to German environmental and landscape history.[5]

The essays presented here are united by their attempt to understand the ideological underpinnings and environmental significance of the country's diverse cultural landscape, or *Kulturlandschaft*, a concept first coined by the conservative folklorist and social theorist Wilhelm Heinrich Riehl in the mid-nineteenth century. Following Herder, Riehl argued that there was an inextricable link between a "people" and a particular landscape; geographers such as Friedrich Ratzel and Karl Haushofer, as well as Nazi theorists such as Alfred Rosenberg and Paul Schultze-Naumburg, later used Riehl's ideas to argue that each nation-state was an organism that required a particular living space to survive.[6] Before becoming tainted by its association with social Darwinism and National Socialism, however, the cultural landscape ideal emphasized regional diversity and vernacular landscapes as the foundations of German culture and formed the basis of German land use planning and environmental stewardship well into the twentieth century.[7] Unlike American environmental culture, which is still dominated by debates about wilderness preservation and the retention of untouched spaces, German landscape perception has long recognized human impacts as part of the "natural" order.

As the historian Mark Cioc noted in a recent review of German environmental history, "the impact of human activity over the past two millennia in Central Europe has been so conspicuous" that German environmental historians have always assumed that the object of their analyses is a cultivated "cultural landscape."[8] This pastoral ideal also shaped the goals and perception of Germany's environmentalist traditions. Whereas American wilderness advocates such as John Muir and Robert Marshall valued sublime natural spaces devoid of human influence, Germany's early twentieth-century nature conservation (*Naturschutz*) and homeland protection (*Heimatschutz*) movements envisioned the ideal environment as an anthropogenic terrain that blended the natural, cultivated, and built environments in an aesthetically harmonious whole. In the German environmental imagination, this cultural landscape, the result of centuries-long interaction between human communities and their natural surroundings, was the spatial and symbolic glue that anchored each German to a particular locality despite the dislocations of political and social modernization. The landscape was the foundation of homeland, or

Heimat, signifying a uniquely and often regionally defined German sense of place.[9]

As the essays in *Germany's Nature* demonstrate, environmental debates in Germany generally have centered on the best way to harmonize human priorities and the organic order, rather than glorify wilderness as a place to escape from industrial society altogether. In the early nineteenth century, for example, the Baden engineer Johann Gottfried Tulla, who was responsible for canalizing and straightening the Rhine, argued that human beings could improve upon nature, preventing floods, eliminating disease, creating farmlands, and ensuring active commerce by harnessing the power of water for utilitarian ends.[10] In the decades before World War I, on the other hand, nature conservationists and homeland protectionists attempted to temper this faith in technological progress by setting aside local "natural monuments," including historic oak trees, scenic agricultural fields, and indigenous species, as symbols of a natural Heimat that was disappearing under the onslaught of new factories, mines, roads, power lines, and apartment buildings.[11] The concept of *Heimat*, loosely translated as "homeland" and a signifier of regional diversity and a sense of place, became an aesthetically charged and socially conservative pathway to broader ecological understanding. Between the discourses of improvement and preservation were a variety of other voices—including engineers, landscape architects, regional planners, state officials, hikers, and agrarian reformers—who argued that it was possible to bridge the differences between nature and culture by creating a modern culture that remained rooted to the soil. The ideal of the cultural landscape offered a precarious middle ground, rather than a sharp dichotomy, between wilderness and civilization within Germany's rapidly industrializing and urbanizing society.

The essays in *Germany's Nature* weave narratives of human-environmental interaction that explore the methodological possibilities, ideological contours, and environmental consequences of the cultural landscape in modern German society. The contributions examine the dynamic interplay between representations of nature, political and social development, and ecological change in a country that, on the whole, not only industrialized and urbanized later and more rapidly than its counterparts in Britain, France, and Belgium, but also reinvented itself politically in the form of six different political regimes between 1871 and 1990. The selections in *Germany's Nature* use a variety of sites, including forests, waterways, roadway networks, even natural history museums, to explore Germans' contradictory perceptions, transformations, and reactions to their natural landscape. Each essay is driven by a methodological commitment to the idea that both culture and nature have authored the German landscape. As Simon Schama has noted aptly in *Landscape and Memory*,

"Landscapes are culture before they are nature; constructs of the imagination projected onto wood and rock."[12] Yet these essays show that cultural projections have the power to reshape the human *habitus* in manifold and often unexpected ways. While geography and ecological change have delimited the range of economic, social, political, and cultural possibilities in Germany, culture itself has acted as a dynamic environmental agent that has shaped Germans' perceptions of the natural world and their relationship to it.

The contributions demonstrate that many of the large-scale processes of Germany's modern historical development, such as nation-state building, class formation, gender construction, technological development, and urbanization, cannot be understood without reference to environmental history. Though several of the contributions analyze the cultural narratives and economic prerogatives that have sustained what William Rollins has called the "hegemony of environmental exploitation" in modern Germany, none of them offers an easy, overarching moral imperative regarding the natural environment.[13] The landscapes that emerge from these essays are instead sites of ideological debate and social contestation. Landscape changes that one social group views as environmental improvement can destroy the material and ecological foundations of another group. Furthermore, the definition of nature has fluctuated between political regimes and among different constituencies, often complicating Germans' ability to formulate an ecological remedy to grave environmental concerns. The Central European landscape also defies any attempt to write a history of human encroachment on pristine ecosystems. As Hansjörg Küster reminds us, even the most canonical sites of sublime nature in Central Europe are products of human modification. The Lüneburger Heath, Germany's first national park, acquired its characteristic vegetation due to agricultural usage, rather than ecological succession alone, while man-made landscapes—rural hedgerows, rock quarries, railway corridors—provided the ecological niches for birds and other animals that many environmentalists now see as "natural," indigenous species.[14] Given these parameters, writing the history of *Germany's Nature* involves analyzing successive anthropogenic regimes defined by shifting environmental, economic, political, and cultural priorities.

For all its conceptual ambiguity and perhaps because of it, landscape offers one of the best tools for conceptualizing and narrating the messy, dynamic interaction between these different elements. It enables scholars to move beyond simple dichotomies between use and abuse, materialism and ideology, representation and reality. Today's landscape scholars avoid the essentializing tendencies of previous approaches, thus making room for open-ended, richer accounts.[15] For them, landscape is a "way of seeing," a frame for representing, experiencing, and manipulating nature in a particular way.[16] Wilhelm

Heinrich Riehl already noted in 1850 that each century not only had its own world-view (*Weltanschauung*), but also its particular mode of landscape view (*Landschaftsanschauung*).[17] Riehl's nineteenth-century notion of landscape as a potentially beautiful artifact, for example, broadened itself in the twentieth century to encompass scientific insights into the ecological functions of landscapes.[18] Landscapes are thus physically and intellectually malleable meeting places for culture, ecology, and technology.

Making the cultural landscape a useful category of analysis for environmental history is not without its conceptual and ethical challenges, especially in Germany, where the cultural landscape is still burdened by its association with Romantic aestheticism, völkisch nationalism, and Nazi racism. Practitioners of German environmental history, or *Umweltgeschichte*, for example, have often focused their efforts on studying the causes and effects of pollution and urban infrastructures, topics that once received short shrift in mainstream American environmental history.[19] As Joachim Radkau has correctly pointed out, German environmental historians had been studying urban pollution for years while their American colleagues were still debating whether or not this was a suitable topic for environmental history.[20]

While the effects of industrial effluents and urbanization are undoubtedly critical environmental issues in the heartland of Europe's coal, steel, chemical, pharmaceutical, and automobile industries, scholars in the home country of the cultural landscape paid insufficient attention to studying changes in that landscape. The focus on air, water, soil, and noise pollution often relegated landscape to the realm of bourgeois aestheticism—at best an idealized alternative to industrial blight, at worst a bucolic veil that led middle-class Germans to turn a blind eye to the plight of manual laborers who suffered as a result of toxic exposure and unsafe working conditions.[21] We do not see the analyses of pollution and of landscape as competing approaches, but rather as complementary fields of environmental history and present *Germany's Nature* in this spirit. Additionally, the papers in this volume aim to contribute to the recent rapprochement between environmental history and the history of technology, a union that has produced some of the most exciting scholarship in these fields. Rather than viewing environmental changes as mere by-products of inexorable technological transformations, many scholars now conceive of science, technology, and the environment as a continuum conditioned by cultural and social negotiation. Local farmers and environmentalists, for example, have often protested plans for hydroelectric dams and reservoirs, challenging scientists' and engineers' claim to exclusive expert knowledge. The modern landscapes that form *Germany's Nature* are thus products of social contestation, not of technological determinism. David Nye's recent dictum that technology is not alien to nature, but integral to it, could be hardly more appropriate than for the German case.[22]

The omission of cultural landscape analysis within German environmental history also reflected landscape's association with reactionary tendencies in German culture. In the 1960s and 1970s, for example, the work of cultural historians such as George Mosse, Fritz Stern, and Klaus Bergmann portrayed Romantic landscape ideals and the discourses of *Heimat* as harbingers of anti-Enlightenment, antimodern, and protofascist tendencies in German society and culture that sought to turn back the clock on political democratization and industrial modernization.[23] Echoing the notion of Germany's special path, or *Sonderweg*, into the modern world, these cultural historians disparaged landscape pastoralism as völkisch, agrarian-romantic, and culturally pessimistic. They saw such idealization of nature as a product of Germany's broader failure to establish political liberalism and cultural modernity in tandem with rapid industrial development, which left "feudalized" and authoritarian cultural structures intact well into the twentieth century. The last two decades have seen a gradual shift away from this view. Since the 1980s, many historians have been offering evidence that Imperial Germany and even the Third Reich were as dynamic and modern, if not as politically liberal, as other Western European societies.[24]

A rising interest in social history, growing environmental awareness, and greater appreciation of the "modernity" of the Third Reich have recast cultural landscape studies in light of these new methodological tendencies and political concerns in the 1980s and 1990s. In the work of scholars such as Celia Applegate and Alon Confino, for example, Heimat appreciation, regionalism, and everyday nature study emerged as agents of political and cultural modernization in Germany, rather than atavistic nostalgia. Other scholars, such as William Rollins, Jost Hermand, and Andreas Knaut, have explored the environmental ethic that underpinned the Heimat ideal. These writers have challenged the view that *Heimatschutz*, or Heimat protection and nature conservation movements sought to turn back the clock on German modernization or provided any direct ideological link to Nazism.[25] As Thomas Rohkrämer has argued, Wilhelmine and Weimar environmentalists were part of disparate reform tendencies that sought an "other" modernity, one that was more socially and environmentally sustainable, rather than a rejection of technological civilization altogether.[26] The essays in *Germany's Nature* are indebted to these new interpretive tendencies, as they attempt to avoid pigeonholing environmental ideas in terms of black-and-white distinctions between progressive and reactionary. In Germany, environmental ideas have influenced individuals from across the political spectrum, from anarcho-socialists to ecofascists, from green idealists to nationalist conservatives.[27]

One way to break free from the overly traveled special environmental path is to analyze German environmental traditions in a comparative perspective, as Joachim Radkau offers in his essay for the volume, "Germany as a Focus of European 'Particularities' in Environmental History." Radkau dispels the myth that Germans are culturally "closer to nature" than other European or Asian peoples, arguing that institutions—not mentalities—are decisive in how a particular society relates to its natural environment. He asserts that Germany's premodern pattern of petty states and free cities and the country's lack of colonies before the late nineteenth century created decentralized natural resource decision-making that was decisive in maintaining sustainable land use policies. Local elites' interest in retaining their woodland resources evolved into scientific forestry in Central Europe, ensuring a steady supply of timber and livestock feeding grounds for the area's burgeoning population in the eighteenth and nineteenth centuries.

Radkau's critique of a national way of thinking as an explanatory tool for measuring German environmental "peculiarities" across societies suggests the need for envisioning culture in provocative and more sophisticated ways, as the other contributions in part 1, "Seeing Like a State: Water, Forests, and Power," attempt to do. These essays apply James C. Scott's landmark analysis of the unintended consequences of large-scale social and spatial engineering to hydraulic engineering and scientific forestry, conservation practices that were supposed to improve society and consolidate state control over natural resources.[28] In her essay "Conviction and Constraint: Hydraulic Engineers and Agricultural Amelioration Projects in Nineteenth-Century Prussia," Rita Gudermann studies the politics and environmental effects of water rectification and land improvement strategies in Northern Germany's lowland plains. She shows that engineers and civil servants in charge of these projects believed that controlling nature would resolve social problems, yet encountered considerable resistance from small farmers who recognized the threat of land amelioration and river rectification projects to traditional extensive agriculture.

Nationalists often have deemed the forest to be the quintessential landscape of German nature, symbolic and ecological indicator of broader trends in German society and culture. In "A Sylvan People: Wilhelmine Forestry and the Forest as a Symbol of Germandom," Michael Imort argues that nationalist German foresters around 1900 believed that Germans needed to recapture a common sylvan identity to create a unified nation state. Many foresters at this time longed for the robust woodlands of Romantic lore. They disavowed scientific forestry techniques in favor of forests deemed to be "traditional"—mixed species, unevenly aged, and selectively cut, rather than regimented in monocultural tree farms harvested by clear cutting. The legal adoption of such sustainable forestry or *Dauerwald* techniques in the Nazi era, Imort suggests, actually hindered the further development of ecological forestry after World War II until green foresters rediscovered this technique in the 1990s. In his essay "Forestry and the German Imperial Imagination: Conflicts over Forest Use in German East Africa," Thaddeus Sunseri extends the discussion of forest conservation to the imperial state. He challenges the belief that conservation was one of the "gifts" of European civilization to the African continent by showing how forestry reserves served the state's fiscal and social control needs, rather than ecological imperatives. Such efforts proved disastrous in the long run by disrupting the balance between human hunters and wild game that pre-existed colonial administration, spurring native resistance and declining economic prospects. Taken together, these three essays demonstrate that the rise of the modern state has brought an entirely new ecological regime in Central Europe and beyond, one that has produced unintended and often catastrophic consequences for human communities and the natural environment alike.

In part 2, "The Cultural Landscapes of Home," the authors analyze the ideological aims that underpinned the cultural landscape ideal, examining the relationship between technology, environmental perception, and broader sociopolitical changes in late nineteenth and early twentieth-century Germany. Rudy Koshar casts the automobile as an "organic machine" in his essay "Organic Machines: Cars, Drivers, and Nature from Imperial to Nazi Germany." Koshar shows that cars were originally conceived of as tourist vehicles that would bring Germans closer to nature, rather than purveyors of environmental pollution and rampant consumerism. The object of automobile drivers' gaze was a cultural landscape that could nonetheless be naturalized by urban, bourgeois observers who perceived peasants, farmsteads, forests, and mountains as part of a preindustrial (and thus "natural") topography. The power of technology to shape environmental perception is also the theme of Susanne Köstering's contribution "Biology—*Heimat*—Family: Nature and Gender in German Natural History Museums around 1900." Köstering shows that natural history museums at this time began to develop new technologies of perception—particularly proto-ecological exhibits and dermoplastic animal families—that replaced taxonomically organized and taxodermically prepared specimens and displays. This transformation of museum culture coincided with larger societal changes that sought to compensate for vanishing natural landscapes and posited the national Heimat and the nuclear middle-class family as natural foundations for society. Nationalizing nature and naturalizing the nation thus went hand in hand.

The history of nature conservation and Heimat protection is a growing area of historical interest, as scholars have abandoned older models of agrarian romanticism and cultural pessimism. Instead, they emphasize German landscape preservationists' differentiated response to modernity, blossoming ecological awareness, and multivalent signification of nature. While Wilhelmine nature conservation and present-day Greens have received generous scholarly attention, the authors in part 3, "The Politics of Conservation," demonstrate the importance of alternative conservation groups and the diffusion of environmental ideas between 1880 and 1970.

In his "Indication and Identification: On the History of Bird Protection in Germany, 1800–1918," Friedemann Schmoll studies bird protection in Imperial Germany, the most avant-garde of conservation organizations, on two levels. First, he understands birds as objects of communication and emotional identification for scientific as well as nonacademic audiences. Also, birds became indicators of broader ecological connections, hastening the development of nonutilitarian, nonanthropocentric attitudes toward nature among large sections of German society. In contrast to accounts of German nature conservation that stress the special German path, John Williams finds a significant discontinuity between nature conservation movements of the Wilhelmine–early Weimar era and those of the late Weimar–Nazi periods. Williams locates the ideological shift away from nineteenth-century Romanticism and toward a technocratic and racist conception of nature conservation in the period between 1925 and 1935 in his essay "Protecting Nature between Democracy and Dictatorship: The Changing Ideology of the Bourgeois Conservationist Movement, 1925–1935." Williams shows that social anxieties about working-class radicalism and disaffected youth prompted conservationists to develop landscape planning strategies, known as *Landschaftspflege*, which promised to bring "order" to society and nature alike. Sandra Chaney offers the first systematic comparison of West and East German conservation practices in her essay "Protecting Nature in a Divided Nation: Conservation in the Two Germanys, 1945–1972." Chaney argues that, despite radically different political conditions, nature conservationists in both countries found it necessary to embrace utilitarian and pragmatic rationales for environmental protection due to the drive for economic development on both sides of the wall. Although East German conservationists touted the advantages of socialism for large-scale landscape planning, Chaney argues, the opportunity for democratic mobilization in West Germany was far more effective in making ecological issues an object of state and public concern. The essays by Schmoll,

Williams, and Chaney help scholars to understand the ideological contours and long-term vitality of environmentalism in German political culture more broadly. They demonstrate that environmental debate is often driven as much by social anxieties, cultural experimentation, and ideological contestation as by ostensibly "objective" physical conditions or scientifically defined ecological problems.

Taken together, these essays in *Germany's Nature* present a rich array of case studies showcasing the pivotal role of environmental history in modern German society and the need to analyze cultural landscapes in the writing of environmental history. We hope that this volume will contribute to and expand ongoing scholarly discussions in environmental history and German history, engage more specific debates on the roles of landscapes, technology, forests, and environmentalism in the modern era, and offer a critique of the essential "German" love for nature. If these papers give at least a partial answer to the question what is peculiar and what is ordinary about Germany's Nature, they will have succeeded.

Notes

1. John McNeill, *Blue Planet: Die Geschichte der Umwelt im 20. Jahrhundert* (Frankfurt am Main: Campus, 2003). The title of the original book is *Something New under the Sun: An Environmental History of the Twentieth-Century World* (New York: W. W. Norton, 2000).
2. Richard Hofstadter, *The Progressive Historians: Turner, Beard, Parrington* (New York: Alfred A. Knopf, 1968), 5.
3. On the rebirth of landscape studies in Germany, see the review essay by Norbert Fischer, "Der neue Blick auf die Landschaft: Die Geschichte der Landschaft im Schnittpunkt von Sozial-, Geistes-, und Umweltgeschichte," *Archiv für Sozialgeschichte* 36 (1996): 434–442, and Thomas Zeller, *Straße, Bahn, Panorama: Verkehrswege und Landschaftsveränderung in Deutschland von 1930 bis 1990* (Frankfurt am Main and New York: Campus, 2002), 21–40.
4. On the relationship between landscape and German national identity, see Bernd Weyergraf, *Die Deutschen und ihr Wald*, Ausstellung der Akademie der Künste vom 20. September bis 15. November 1987 (Berlin: Akademie, 1987); Jost Hermand, *Grüne Utopien in Deutschland: Zur Geschichte des ökologischen Bewußtseins* (Frankfurt am Main: Fischer, 1991); Rudy Koshar, *Germany's Transient Pasts: Preservation and National Memory in the Twentieth Century* (Chapel Hill: University of North Carolina Press, 1998); Thomas Lekan, *Imagining the Nation in Nature: Landscape Preservation and German Identity* (Cambridge, Mass.: Harvard University Press, 2004).
5. Recent collections do address the cultural dimensions of Green ideas and German environmentalism, however, such as Colin Riordan, ed., *Green Thought in German Culture: Historical and Contemporary Perspectives* (Cardiff: University of Wales Press, 1997), and Axel Goodbody, ed., *The Culture of German Environmentalism: Anxieties, Visions, Realities* (New York and Oxford: Berghahn, 2002). The volume *Nature in German History*, ed. Christof Mauch (New York and Oxford: Berghahn, 2004) was published after work on this volume was finished.

6. Wilhelm Heinrich Riehl, *Die Naturgeschichte des Volkes als Grundlage einer deutschen Social-Politik*, vol. 1, *Land und Leute*, 5th ed. (Stuttgart: J. G. Cotta, 1861), trans. David Diephouse as *A Natural History of the German People* (Lewiston, N.Y.: Edwin Mellen, 1990), 49. On Riehl, see also Celia Applegate, *A Nation of Provincials: The German Idea of Heimat* (Berkeley: University of California Press, 1990), 34–41; Jasper von Altenbockum, *Wilhelm Heinrich Riehl 1823–1897: Sozialwissenschaft zwischen Kulturgeschichte und Ethnographie* (Cologne: Böhlau, 1994); David Thomas Murphy, *The Heroic Earth: Geopolitical Thought in Weimar Germany, 1918–1933* (Kent, Ohio: Kent State University Press, 1997).
7. Colin Riordan, "Green Ideas in Germany: A Historical Survey," in *Green Thought*, 3; Franz-Josef Brüggemeier, Mark Cioc, and Thomas Zeller, eds., *How Green Were the Nazis? Nature, Environment, and Nation in the Third Reich* (Athens: Ohio University Press, 2005).
8. Mark Cioc, "The Impact of the Coal Age on the German Environment: A Review of the Literature," *Environment and History* 4, no. 1 (February 1998): 105–124, here 106. The cultural historian William Rollins briefly discusses this "wilderness hangup" in American environmental culture in *A Greener Vision of Home: Cultural Politics and Environmental Reform in the German Heimatschutz Movement* (Ann Arbor: University of Michigan Press, 1997), 267–273, a topic that deserves further discussion in light of the debates spurred by William Cronon's recent critique of wilderness in the American environmental imagination. See William Cronon, "The Trouble with Wilderness, or Getting Back to the Wrong Nature," in *Uncommon Ground: Rethinking the Human Place in Nature* (New York: W. W. Norton, 1996), 69–90, and J. Baird Callicott and Michael P. Nelson, eds., *The Great Wilderness Debate* (Athens: University of Georgia Press, 1998). On German environmentalism from a comparative perspective, see Raymond Dominick, "The Roots of the Green Movement in the United States and West Germany," *Environmental Review* 12, no. 3 (Fall 1988): 1–30 and Christoph Spehr, *Die Jagd nach Natur. Zur historischen Entwicklung des gesellschaftlichen Naturverhältnisses in den USA, Deutschland, Großbritannien und Italien am Beispiel von Wildnutzung, Artenschutz und Jagd* (Frankfurt am Main: IKO, 1994).
9. The extensive *Heimat* literature includes Celia Applegate, *A Nation of Provincials*; Alon Confino, *The Nation as a Local Metaphor: Württemberg, Imperial Germany, and National Memory* (Chapel Hill: University of North Carolina Press, 1997); Elizabeth Boa and Rachel Palfreyman, *Heimat—A German Dream: Regional Loyalties and National Identity in German Culture, 1890–1990* (Oxford: Oxford University Press, 2000).
10. On Tulla's "rectification" of the Rhine River, see Mark Cioc, *The Rhine: An Eco-Biography* (Seattle: University of Washington Press, 2002); Christoph Bernhardt, "The Correction of the Upper Rhine in the Nineteenth Century: Modernizing Society and State by Large-Scale Water Engineering," 183–202; and Dieter Schott, "Remodeling 'Father Rhine': The Case of Mannheim 1825–1914," in *Water, Culture, and Politics in Germany and the American West*, ed. Susan C. Anderson and Bruce H. Tabb (New York: Peter Lang, 2001), 203–226.
11. Recent works that survey German nature conservation, landscape preservation, or environmentalism in a broad-ranging manner include Rollins, *A Greener Vision of Home*; Lekan, *Imagining the Nation in Nature*; Raymond Dominick, *The Environmental Movement in Germany: Prophets and Pioneers, 1871–1971* (Bloomington: Indiana University Press, 1992); Michael Wettengel, "Staat und Naturschutz, 1906–1945: Zur Geschichte der Staatlichen Stelle für Naturdenkmalpflege in Preussen und der Reichsstelle für Naturschutz," *Historische Zeitschrift* 257, no. 2 (October

1993): 355–399; and Franz-Josef Brüggemeier, *Tschernobyl, 26. April 1986. Die ökologische Herausforderung* (Munich: Deutscher Taschenbuch Verlag, 1998).
12. See Simon Schama, *Landscape and Memory* (New York: Alfred A. Knopf, 1995), 61.
13. Rollins, *Greener Vision of Home*, 25.
14. Hansjörg Küster, *Geschichte der Landschaft in Mitteleuropa. Von der Eiszeit bis zur Gegenwart* (Munich: C. H. Beck, 1995).
15. Gerhard Böhme, *Natürlich Natur—Über Natur im Zeitalter ihrer technischen Reproduzierbarkeit* (Frankfurt am Main: Suhrkamp, 1992); Trevor J. Barnes and James Duncan, *Writing Worlds: Discourse, Text and Metaphor in the Representation of the Landscape* (London: Routledge, 1992); Anne Spirn, *The Language of Landscape* (New Haven: Yale University Press, 1998); Kenneth Olwig, *Landscape, Nature, and the Body Politic: From Britain's Renaissance to America's New World* (Madison: University of Wisconsin Press, 2002); Stefan Kaufmann, ed., *Ordnungen der Landschaft. Natur und Raum technisch und symbolisch entwerfen* (Würzburg: Ergon, 2002); W.J.T. Mitchell, *Landscape and Power*, 2nd ed. (Chicago: University of Chicago Press, 2002).
16. Denis E. Cosgrove, *Social Formation and Symbolic Landscape*, 2nd ed. (Madison: University of Wisconsin Press, 1998).
17. Wilhelm Heinrich Riehl, "Das landschaftliche Auge" (1850), in *Culturstudien aus drei Jahrhunderten* (Stuttgart: J. G. Cotta, 1859), 57–79, here 57.
18. Riehl, "Das landschaftliche Auge," 67; Werner Konold, ed., *Naturlandschaft-Kulturlandschaft: Die Veränderung der Landschaften nach der Nutzbarmachung durch den Menschen* (Landsberg am Lech: ecomed, 1996).
19. On the historiography of *Umweltgeschichte* and its focus on the impact of industrial technologies and pollution, see Franz-Josef Brüggemeier and Thomas Rommelspacher, eds., *Besiegte Natur. Geschichte der Umwelt im 19. und 20. Jahrhundert*, 2nd ed. (Munich: C. H. Beck, 1989); Werner Abelshauser, ed., *Umweltgeschichte. Umweltverträgliches Wirtschaften in historischer Perspektive* (Göttingen: Vandenhoeck und Ruprecht, 1994); Günter Bayerl, Norman Fuchsloch, and Torsten Meyer, eds., *Umweltgeschichte—Methoden, Themen, Potentiale* (Münster and New York: Waxmann, 1996).
20. Joachim Radkau, *Natur und Macht: Eine Weltgeschichte der Umwelt* (Munich: C. H. Beck, 2000). This tendency is certainly related to the adoption of environmental history by German historians of technology in the early 1980s.
21. For a critique of such bourgeois aestheticism, see Arne Andersen, "Heimatschutz. Die bürgerliche Naturschutzbewegung," in *Besiegte Natur: Geschichte der Umwelt im 19. und 20. Jahrhundert*, ed. Franz-Josef Brüggemeier and Thomas Rommelspacher (Munich: C. H. Beck, 1989), 156–157. The relegation of landscape to the realm of bourgeois aestheticism is by no means a particular German phenomenon. In his landmark 1988 essay "Doing Environmental History," the dean of American environmental history, Donald Worster, does not even include landscape within a subfield devoted to the "role and place of nature in human life." See Donald Worster, "Doing Environmental History," in Donald Worster, ed., *The Ends of the Earth: Perspectives on Modern Environmental History* (New York: Cambridge University Press, 1988), 289–307.
22. Jeffrey K. Stine and Joel A. Tarr, "At the Intersection of Histories: Technology and the Environment," *Technology and Culture* 39 (1998): 601–640; David E. Nye, "Technologies of Landscape," in David E. Nye, ed., *Technologies of Landscape: From Reaping to Recycling* (Amherst: University of Massachusetts Press, 1999), 3–17, here 10.

23. George Mosse, *The Crisis of German Ideology: Intellectual Origins of the Third Reich* (New York: Grosset and Dunlap, 1964); Klaus Bergmann, *Agrarromantik und Großstadtfeindschaft* (Meisenheim am Glan: Anton Hain, 1970); Gert Gröning and Joachim Wolschke-Bulmahn, *Die Liebe zur Landschaft: Natur in Bewegung. Zur Bedeutung natur—und freiraumorientierter Bewegungen der ersten Hälfte des 20. Jahrhunderts für die Entwicklung der Freiraumplanung* (Munich: Minerva, 1986).
24. The classic critique of the *Sonderweg* is David Blackbourn and Geoff Eley, *The Peculiarities of German History: Bourgeois Society and Politics in Nineteenth-Century Germany* (Oxford: Oxford University Press, 1984).
25. Rollins, *Greener Vision*; Jost Hermand, *Mit den Bäumen sterben die Menschen: zur Kulturgeschichte der Ökologie* (Cologne: Böhlau, 1993); Andreas Knaut, *Zurück zur Natur! Die Wurzeln der Ökologiebewegung* (Greven: Kilda-Verlag, 1993). Other works in this vein include Karl Ditt, *Raum und Volkstum: Die Kulturpolitik des Provinzialverbandes Westfalen, 1923–1945* (Münster: Aschendorff, 1988) and John Williams, "'The Chords of the German Soul are Tuned to Nature': The Movement to Preserve the Natural *Heimat* from the *Kaiserreich* to the Third Reich," *Central European History* 29, no. 3 (1996): 339–384.
26. Thomas Rohkrämer, *Eine andere Moderne? Zivilisationskritik, Natur und Technik in Deutschland, 1880–1933* (Paderborn: Schöningh, 1999). See also Jeffrey Herf, *Reactionary Modernism: Technology, Culture and Politics in Weimar and the Third Reich* (Cambridge: Cambridge University Press, 1984).
27. Ulrich Linse, *Ökopax und Anarchie: Eine Geschichte der ökologischen Bewegungen in Deutschland* (Munich: Deutscher Taschenbuch Verlag, 1986).
28. James C. Scott, *Seeing Like a State: How Certain Schemes to Improve the Human Condition Have Failed* (New Haven: Yale University Press, 1998).

Part I

Seeing Like a State
Water, Forests, and Power

Chapter 1

Germany as a Focus of European "Particularities" in Environmental History

JOACHIM RADKAU

IN HER AUTOBIOGRAPHY, the German writer Ricarda Huch (1864–1947) recalls a telling episode about national styles of nature appreciation. Huch kept many pets for her daughter, whose father was Italian, including a dog, a cat, a raven, a squirrel, a parrot, and rabbits, assuming that the child would love them as much as the mother did. "My daughter, however, looked at these animals, which should have been brothers and sisters for her, with a reserved coolness which she had inherited from her Italian father."[1] To be sure, Huch was far from being a Germanic racist; on the contrary, she is still remembered for her courageous retirement from the Prussian Academy of Arts in 1933 in protest over the Nazi takeover. Nevertheless, when it came to her daughter, she expressed a conviction about Germanic feeling for nature that was popular at that time, especially among animal lovers: that the Germanic peoples had a natural feeling of fraternity with animals that other nationalities lacked.

This metaphor of Germanic love for nature found expression in several laws and popular movements in the modern era. For example, one of the first laws that the Nazis enforced after their seizure of power made vivisection illegal, even threatening offenders with concentration camps; the law was the apogee of a long-standing animal rights movement whose supporters included the composer Richard Wagner, the idol of many German chauvinists.[2] The bird protection movement, which had its highest number of adherents among British and German women in the decades before World War I, sometimes displayed racist tendencies. Bird lovers directed their wrath against French and Italian bird hunting: Germanic love of animals was pitted against Latin insensitivity toward animals.[3] But in former times, bird hunting had been widespread in German-speaking countries, too.

In recent decades, many German environmentalists have believed that love of the woods and opposition against nuclear technology are typically German traits, while the French are characterized as being indifferent to environmental issues. But in reality the difference between national attitudes and policies toward the natural environment are not as sharp as has often been assumed.[4] As I suggest in this essay, the most important differences in national environmental traditions lie at the level of institutions, rather than being deeply rooted in national mentalities. In the case of the Franco-German environmental divide, for example, local hearings were an element of licensing procedures in the Federal Republic of Germany, but not in the centralized French nuclear economy. Protest movements had a better chance to succeed in Germany than in France; legal structures, rather than a cultural affinity for nature, best explain the relative popularity and success of the German antinuclear campaign, not cultural traits.

Therefore, with regard to Germany's Nature—the theme of this volume—it is important to distinguish between the history of the imagination, on the one hand, and the history of the real, effective relations between man and nature, on the other. To be sure, for the dyed-in-the-wool adherents of constructivism there exists only the first history: the history of the words, the conceptions, the discourses. But even David Blackbourn, who is well versed in postmodern approaches to environmental history, warned environmental historians against going too far on the path of constructivism when he cautioned them at the Nineteenth International Congress of Historical Sciences in Oslo in 2000, "Not all places are imagined." The environmental historian should retain a sense of material realities, even in a volume devoted to the environmental history of the German cultural landscape.

In light of these conceptual challenges, an institutional approach to environmental history can combine discursive history with material history to a certain degree. In this context, I conceive of the term "institution" in a broad sense, as it is understood by the institutional school of economics, comprising not only administrative bodies and established organizations, but also rules and customs fixed in a stable way for a long time. My basic philosophy is very simple: it is not ideas and individual actions that are of decisive importance in the evolution of the human use of nature, but rather the durable patterns of everyday collective behavior and the institutions that generate and maintain those patterns. I presume that on this institutional level a distinctive European as well as German relationship between society and nature can be recognized clearly and concretely, even if some questions about the extent of national environmental differences will remain. Over the centuries, the management of environmental problems in Europe has been deeply influenced by old European traditions of *Verrechtlichung*—regulation by law—dating back to Roman times. Under the polycratic conditions of Europe, where different sources of law always existed side by side, the law has frequently been a subject of discussion, controversy, and competition between an array of governing bodies. Law was not simply imposed by an omnipotent ruler, but was conceived as something an individual could use to fight, even against power elites.

German-speaking Central Europe was the best example of such European polycracy, a many-colored patchwork of different political territories loosely confederated as part of the Holy Roman Empire for many centuries. Such competing jurisdictions left space for a variety of conflicts over natural resources. For over five hundred years the forest communities (*Markgenossenschaften*) of the Tyrolean peasants, for example, instituted legal proceedings against the Habsburg government in order to preserve their right to regulate the forests (*Jus regulandi silvas*), and in the end, in 1847, they won.[5] The dispersion of territorial sovereignty made it impossible for Vienna to impose its will in outlying areas. I have the impression that it would be difficult to find similar cases in other regions of the world; based on my comparative reading of European environmental history, it appears that even the idea of such a challenge to central authority does not exist. The basic conditions of European and German environmental history have been deeply imprinted by this longtime process of decentralized regulation by law.

To be sure, this legal regulation of forests and other natural resources in Central Europe has not been a complete success story, especially not in environmental history. In fact, historical actors often pay a high price if they fight over resources through litigation, rather than through informal channels. The price is especially high in those areas where resources are managed by a state bureaucracy with an extensive administrative apparatus, rather than by the ones who are directly affected by resource management decisions. As Max Weber has already shown, the main difference between the Prussian-German model of state growth and those of England and America—France is more typical of Germany in this case—has been a distinctly greater role for bureaucratic officials, at least until the early twentieth century. Without any doubt, this traditional difference between Prussian-German and Anglo-American modernization has affected how historical actors dealt with environmental problems. One might expect Germans to have acted more slowly, but more systematically, in resolving resource conflicts than their counterparts in the United States, even though today there is no shortage of environmental bureaucracy in the United States.

It thus remains an open question whether centralized bureaucracies regulated environmental matters more effectively than local officials. In the course of the present European unification process, for example, many observers have placed high hopes in common European environmental policies. But we often need to be reminded that truly effective environmental policies are, as a rule, best achieved within relatively small nations such as Denmark or the Netherlands. An effective consensus is reached in the easiest manner where communication is not too complicated. The opinion that problems will be solved by integration might turn out to be a fundamental error. Indeed, in past times European polycentrism had considerable advantages in regard to the handling of environmental issues. The peculiar European process of regulation by law is conditioned by this polycentrism; when only one single authority exists, the law cannot be discussed and nothing can be obtained by individual or local engagement.

Even more important is another point: truly effective environmental management can only be achieved by institutions that are not too far removed from the site where action is demanded. Forest and water management—the two classic areas of governmental intervention in environmental matters—both present numerous historical examples of how advantageous it was to tackle problems locally, not from a faraway capital. Even if the Chinese emperor had been determined to protect the forests, he would have been unable to do so effectively because an appropriate forest policy can be organized only on a regional level, not on the level of a huge empire. In this regard, a comparison between China and Japan is instructive. Even though a high estimation of woodlands does not seem to be more deeply rooted in the Japanese cultural tradition than in the Chinese tradition, for strictly practical reasons, Japanese institutions had to start a forest protection policy in the course of the eighteenth century, one that achieved remarkable effectiveness.[6] The Chinese, on the other hand, were less successful at regulating their forests, a product of the relative ease of regulating woodlands in a small island empire faced with growing scarcity, rather than in a far-flung empire plagued by bureaucratic inertia.

In Europe, a comparison between France and Germany reveals a similar advantage to smaller-scale environmental management. Under the strong administration of Louis XIV's mercantilist finance minister Jean-Baptiste Colbert and his *grande ordonnance forestière* of 1669, France became the European leader in forest policy, a position that it held throughout the eighteenth century. However, in the long run, the French centralist system was not well adapted to forest problems, which often required immediate response to poachers, decisions about local species, and the adjudication of petty squabbles between different users. At the end of the eighteenth century, the German states took over leadership in forestry, a success conditioned by German political polycentrism.[7] In the hundreds of states and free cities that dotted the former Holy Roman Empire, a plurality of regional approaches to forestry arose; it was the only way to obtain real practical progress. While German nationalists complained about German particularism, or *Zersplitterung*, Wilhelm Pfeil (1783–1859), a leading Prussian teacher of forestry, emphasized that German scientific forestry was, in contrast to French forestry, "exclusively the product of the German partition into different states."[8] In this way, the rigid Prussian dogmatism of Hartig's forestry rules was counterbalanced by other forestry schools and approaches that came out of the mixed forests of central and southwest Germany.

By engaging in institutional environmental history, historians can thus move onto solid ground in evaluating the cultural and material factors that have shaped environmental change. By contrast, an approach that focuses merely on how historical actors have imagined Germany's Nature resembles something of a tragicomedy. No nature exists that is genuinely and solely German, that unifies Germany, and distinguishes it from other nations. Christian Cay Lorenz Hirschfeld (1742–1792), the best-known German garden theorist of the classical age (although this professor of philosophy never seems to have been a practical gardener), propagated a "German garden," a place of tranquility and subdued passions, and he grumbled at the "creeping vermin among the French," which stimulated sexual fantasies. Yet, he never was really able to define of what the "German" quality in gardening actually consisted.[9] And the most famous German landscape garden of Hirschfeld's time—the "Gartenreich" of Prince Leopold Friedrich Franz von Anhalt-Dessau, an attempt to transform a whole territory into a garden—was originally motivated by the desire to inspire the erotic longing of the princess, a desire that was not fulfilled by the garden but by the daughter of the gardener. This prince was a peaceful German patriot who despised the war heroes of his time, rather than a national chauvinist.[10]

The ideological search for Germany's Nature continued into the nineteenth and twentieth centuries, with equally questionable results. Wilhelm Heinrich Riehl (1823–1897), author of the *Natural History of the German People* and one of the founding fathers of German national nature romanticism, described the genuine Germans as the free people of the woods. Yet he was not so blind as not to realize that Germany was divided rather than united by nature, a mosaic of different forest types hardly distinguishable from those of bordering nations.[11] And while Riehl argued passionately that the truly German forest is the wild forest, during his lifetime Germany was the center of a great reforestation movement that created artificial, uniform forests. Even more so, Riehl's own influence was important in establishing the first professorships for forestry at the University of Munich. The coincidence of German forest romanticism and the German reforestation movement is striking, but the relationship between these two tendencies is complicated, and the concepts of German nature connected with them were contradictory.

The problem of basing German identity on a stable and uniform natural world continued into the twentieth century. During the first years of the Nazi dictatorship, a film entitled *The Eternal Forest* (*Der ewige Wald*) was produced that solemnly preached the message that the Germans are a sylvan people and therefore eternal like the (sustainably managed) forest. "It is the forest from where we come and by which we live," the movie preached. But the expensive film did not become a success. Even Adolf Hitler does not seem to have been content with the movie's message. Allegedly, he grumbled that the forest was a retreat for weak ethnic groups; and seen from a historical point of view, he was not entirely wrong.[12] For anyone who knew German forests, they were a mirror of Germany's diversity—historical as well as natural—not of German unity. In no way did German nature offer a template for a totalitarian state.

For the nature lover, it may be a relief to learn that the ideological manipulation of nature for nationalism was never really successful, at least not in Germany. Moreover, nature could serve a variety of ideological ends, not just those of nationalist conservatives and fascists. A typically German phenomenon was the alternative medicine movement (*Naturheilbewegung*), which by 1900 was nearly as popular as the environmental movement that blossomed seventy years later. Both were driven by the belief into the healing powers of nature to some degree. But the alternative medicine movement was, on the whole, remarkably devoid of a narrow German chauvinism.[13]

Let us turn from the imagination of Germany's Nature to the realities of its environmental transformation. Of primary importance in evaluating German environmental history is the concept of sustainability, which has a long tradition in European history, especially in German forestry. Donald Worster, one of the deans of American environmental history, professes that he is deeply skeptical of the term "sustainable development," which the 1992 United Nations Rio de Janeiro Conference (the first Earth Summit) elevated to the supreme environmentalist goal for the world economy. Worster dislikes this term in politics as well as in environmental history because he suspects that "sustainability" is a mere catchword to justify the unrestrained exploitation of nature. From his point of view, preservation of unadulterated nature is a better goal than "sustainable development."[14] Indeed, "sustainability" is originally an economic, not an ecological concept; it presupposes nature as an object of human use. The history of German silviculture shows that this concept is multivalent and that its interpretation has been dominated by economic interests. One main problem is that sustainability focuses on protecting particular aspects of the natural world, especially those that can be quantified, rather than entire ecosystems. Therefore, sustainability runs the risk of being one-sided. Moreover, "sustainable development," as the Earth Summit has defined it, links environmental policy in the developing world to foreign aid, yet such aid has in many cases been the culprit for environmental destruction in these areas.

It is easy to understand Worster's skepticism against the background of American experience, where sustainable development has never been a historical reality over a long period and on a large scale. However, for an analysis of the environmental history of the Old World, the criterion "sustainability" makes more sense and has more substance. In Europe, the continuity of villages, towns, and institutions for centuries makes a historical analysis of sustainability feasible. And the chances of a genuinely sustainable management of resources are greater when there is a high degree of local self-sufficiency and limited dependence on external forces, factors that pertain to the European case. Many parts of Europe have been characterized by a relatively high degree of environmental continuity and, in return, of local and regional autonomy since medieval times, if not since antiquity.[15] When reading the famous stories of the rise of European trade and the spread of colonialism, one should not forget that most European regions mainly depended on their domestic resources until the nineteenth or even the twentieth century. This self-sufficiency extended at least to the most important resources, such as grain and firewood, and held true especially for a country like Germany, which only had colonies for a few decades. The historian Helmut Jäger sees the German term *Nachhaltigkeit* merely as a fashionable Americanism, as a translation of "sustainability." But the term has a long German tradition, mainly in forestry, where it has been used for several hundred years—in a variety of contexts, to be sure, revealing a history of the ambiguity and the manipulative possibilities of this concept as well.[16] But there is no better alternative, even today.

The goal of sustainability has a long tradition in some Central European saline forests, for example. As early as 1661, the chancellor of Bad Reichenhall, an old Bavarian salt-works city, stated, "God created the woodlands for the salt-water spring so that the woodlands might continue eternally, like the spring. Men shall behave accordingly: do not cut down the old trees before the young ones have grown up."[17] In this preindustrial form, sustainability became the secular form of eternity. The precondition for this kind of sustainability was the autonomous saline town that needed huge masses of wood, which lived on its own forest resources and which was accustomed to having salt-works running constantly over many centuries. In many mining towns there was no such spirit of sustainability because of the dramatic ups and downs of the mines.

Certain peculiar traits of German and European environmental history become even clearer by comparison with other regions of the world. In the last decade, environmental historians have become increasingly well informed about many chapters of East and South Asian environmental history, offering the opportunity to compare these areas to Europe. Just take the two recently published anthologies on China, India, and Southeast Asia, *Sediments of Time* and *Nature and the Orient*, both products of cooperation between Western and Asian scholars, which contain the results of extensive regional research.[18] Because of this new abundance of information, a comprehensive comparison between "East" and "West" in environmental matters has become much more complicated than Lynn White's famous Christmas speech of 1966 on the "historical roots of our ecological crisis."[19] Whoever studies this new mass of literature may doubt whether a well-founded comparison will be feasible at all in the future. But, in the end, one point seems to be even clearer than before: with regard to the institutional treatment of forests, there has been a fundamental difference between China and India on the one hand and Western and Central Europe on the other hand, at least for the last five hundred years.

Since the medieval age the protection of forests had been a manifestation of power in Europe; in Asia, on the contrary, it was not. One would hardly expect that the environmental history of the world contains major features that are so distinctive. But as far as I can see, the evidence is overwhelming. To be sure, one may also find a love of trees in ancient Chinese literature, and certain traditions of forest protection existed in China as well. Nicholas K. Menzies has investigated these traditions: the imperial hunting reserves, the Buddhist temple and monastic forests, the Cunninghamia groves of some peasant communities. But on the whole it remains clear that these examples were exceptional. "The trend of government policy during the late Imperial period was to open land for settlement and to stimulate permanent agriculture, not to exclude the population," according to Menzies. Therefore, "administrative authority was rarely exercised to reserve forested land as government property," as was the case for many European governments.[20] Not the conservation of forests, but the clearing of forests was a manifestation of power in Asiatic cultures. "Traditional Chinese thought exhibited a definite bias against forests and the cultivation of trees," observes Eduard B. Vermeer.

"Forest areas were seen as hideouts for bandits and rebels, beyond the reach of government authority, where uncivilized people lived their wretched lives without observing the rules of property. In this view, the clearance of forests and agricultural reclamation brought safety and political and cultural progress."[21] Certainly this attitude was widespread in Europe as well, but for a long time it competed with an appreciation for the need to conserve forests.

In India, traditions of forest protection have been somewhat better developed than in China. Yet on the whole the situation seems to be similar. It is true that an anthology on Indian forest history starts with the forest protection edict of the Mahrat king Shivaji of about 1670 A.D.: "The mango and jack trees in our own kingdom are of value to the Navy. But these must never be touched. This is because these trees cannot be grown in a year or two. Our people have nurtured them like their own children over long periods. If they are cut, their sorrow would know no bounds."[22] But in the entire scholarly literature, this edict appears as a rather unique aberration. Apart from the Indian tradition of cultivating mango and other fruit trees, the edict did not refer to any institutional traditions of forest protection. Madhav Gadgil and Ramachandra Guha, the authors of an ecological history of India, praise the alleged traditional Indian harmony with nature, but they, too, do not present any sources on forest protection in precolonial India.[23] (Or do these sources exist in the archives of the Mughal period, but are written in old Persian which most Indian historians cannot read?) The contrast between the paucity of Indian sources and the immense mass of forest protection documents in Central and Western Europe from the sixteenth century onward could hardly be sharper.

The causes for this discrepancy are manifold. Often the royal hunting passion is said to have been the main motivation for European forest protection, but I believe that the high usability of the woodlands for pasture in extensive farming regimes, which lasted from prehistoric times to the nineteenth century, has been even more salient. Again, questions of livestock seem to have been at the core of the problem. For governments, shipbuilding was, as a rule, the primary interest that made forest protection a matter of utmost priority, whether in Venice, in Portugal, in Colbert's France or in John Evelyn's England. Even in early Ming China, the building of a fleet gave a uniquely strong, though transitory, impetus to a gigantic reforestation project.[24] This motivation, however, was lacking in most German regions, since Germany did not establish a colonial empire, and then only briefly, until the late nineteenth century. What seems to have been particular for Germany was a different force for the institutional control of woodcutting: the mining interests that used wood and charcoal as the energy base for mining and smelting. In addition, the forest economy of Central European saltworks, in contrast to the sea saltworks of Western and Southern Europe, worked by solar power and therefore did not need forest rights. Additionally, peasants have frequently been charged with being the enemies of the forest, but this accusation tends to be a one-sided evaluation from the perspective of governmental forestry. The peasants had their own woodlands for pasture, firewood, and building, which might have been of inferior quality seen from the forester's viewpoint, but were superior in regard to biodiversity.

One could challenge this analysis by asking whether the sharp institutional contrast between East and West in the treatment of forests corresponds to a contrast in actual practice. Was it really the forest laws and forest administrators that protected the woodlands in the past, rather than the unwritten customs and the interests of the people? And were authorities able to enforce their edicts? I admit that this is a difficult problem with many open questions. A number of forest historians have suggested that the true history of forest laws is the history of the violation of these laws. Forest history has also often been written as the history of forest destruction, at least before the great reforestation movement of the nineteenth century. In my earlier work, I repeatedly discussed this pessimistic view of premodern forest history; there are good reasons to be cautious with regards to many forest destruction stories in Europe.[25] Oliver Rackham has again and again ridiculed these stories, remarking that the deforestation storytellers forgot the simple fact "that trees grow again."[26] At least in most Western and Central European regions the forest easily regenerates itself even without artificial reforestation. Limiting the human use of the forest sufficed to ensure widespread forest cover. Under conditions of this kind, governmental forest protection could succeed with relative ease. The European tradition of institutionalized forest protection was surely favored by European ecology. But the decisive point was probably the fact that, in spite of innumerable forest conflicts between governments and peasants or other forest users, there was a mutual interest in the conservation of the forests up to a certain degree and—notwithstanding countless violations—a certain acceptance of the regulation of forest use.[27]

In comparing German environmental history to that of other European countries, other patterns of differentiation emerge. The relative German lack of colonies, for example, has been decisive in Germany's conservation of natural resources. Regarding the ecological aspects of colonialism, two important works gained worldwide attention, Alfred W. Crosby's *Ecological Imperialism* and Richard H. Grove's *Green Imperialism*.[28] Today, colonialism appears to be one of the most discussed subjects in international environmental history. Nevertheless, the significance of colonial expansion for environmental developments within Europe has remained a more or less neglected problem. Reading Crosby, one gets the impression that imperialism has been an ecological success story, at least from the European point of view. But Crosby does not tell the whole story. Europe not only colonized the New World with grain, cattle, and sheep, but was colonized itself with the potato, maize, and not least by the phylloxera that destroyed traditional European viticulture. The potato encouraged strong population growth and undermined European traditions of birth control. Maize increased soil erosion and did not fit into the traditional crop rotation systems. In many respects, European ecology was disturbed, not stabilized, by colonialism, as invasive species and pathogenic microbes moved in a two-way street between the metropole and the colonies.

Richard Grove's work pointed to the colonial origins of modern environmentalism. As urbanized Europeans encountered "pristine" environments abroad, it spurred demand for nature conservation and environmental restoration at home. Is colonial history after all, at least seen from the environmentalist standpoint, a story with a happy end? Considerable doubts remain. The history of ideas presented by Grove is not identical with a history of actions and real effects.[29] But even his history of ideas seems to be ambiguous. If one carefully scrutinizes several important points in Grove's argument, one repeatedly discovers that the true origin of colonial environmental awareness lay in Europe, not in the colonies. Poivre looked at Asian agriculture with the eyes of a French physiocrat, rather than having his views of nature shaped by the Asian experience. Alexander von Humboldt's fears of deforestation in South America originated in his German homeland where fears of that kind had become a real mass psychosis at the end of the eighteenth century.[30] In the spring of 1790, the young Humboldt undertook a journey on the Rhine together with Georg Forster, then a famous global traveler. Forster's report contains long reflections on the imminent danger of wood shortage, which might even cause northern peoples in the end to emigrate to the south.[31] In a recent publication, Grove himself pointed out that Hugh Cleghorn, one of the founding fathers of Indian forestry, and several other pioneers of colonial environmentalism were influenced by their Scottish background, by the experience of a country "already made barren by the evils of the English."[32] The criticism of the ruthless soil exhaustion caused by North American farmers was inspired by comparing such exploitation to the pattern of traditional European agriculture and the European agrarian reforms of the late eighteenth century.[33] Environmental change within Europe, not the colonial experience, shaped environmental perception in the modern age.

On the whole, the repercussions of colonialism on the European environment do not appear to have been fortunate for either the colonies or the metropole, for it delayed the introduction of necessary, sustainable resource management methods for those countries able to exploit their colonial possessions. The lack of a durable tradition of forest protection in leading colonial powers such as Spain, the Netherlands, and Great Britain was apparently conditioned by the ease with which these countries were able to import masses of timber from their colonies or other regions of the world. In contrast, the German states developed scientific forestry at a time when Germany had no colonies and was forced to live on its own wood resources. Furthermore, the omnipotence of the Mesta in Spain during the sixteenth and seventeenth centuries was connected with the rise of Spanish colonialism. As for England, the mass importation of Peruvian guano from the nineteenth century onward thwarted the efforts of agrarian reformers to improve the inherent sustainability of agriculture. The colonial world trade, moreover, threatened the traditional European balance between field and pasture by encouraging specialization and nurtured the illusion of unlimited resources. The full consequences of this development, however, belong to the postcolonial period, which probably experienced the deepest ecological change in history.

In contrast to the pessimistic view of European environmental attitudes that prevails in Lynn White's spiritual approach to environmental history and in other works on deforestation, the institutional approach reveals a European and especially German success story similar to the one told by Eric L. Jones in his *European Miracle*.[34] But I fear that this is not the end of history. In spite of all the achievements of environmental policies, there is no ground for optimism. Human institutions, even if they are effective, are never fully adapted to the complexity of environmental problems. One of the most impressive lessons of history has always been the insight that just the success itself may become the cause of decline in the long run. To be sure, it was the relatively stable ecological conditions of Western and Central European soil and the relatively effective institutions of these countries that made the rise of industrial civilization possible. Only a region with rich wood and water resources and, at least to a certain extent, sustainable methods of forest use was able to enter a path of unlimited growth in energy-intensive industries with high water consumption. Coal did not start the industrial revolution; the pit-coal only continued a development that had begun on the base of charcoal and wood. Moreover, only countries with effective urban and national institutions devoted to resource management and capable of overcoming at least the worst damages caused by industry were able to make industrial development a self-sustaining and popular process. But sustainability remains an illusion in an economy that every year consumes the fossil resources grown over millions of years.

In thinking about sustainability, perhaps we should learn the lessons of Chinese environmental history, which is far better documented than any other non-European history. Mark Elvin describes the history of the Chinese Empire as "three thousand years of unsustainable growth."[35] Perhaps he goes too far in this harsh evaluation; reading the anthology *Sediments of Time*, I tend to think that the environmental decline of China is fully documented only for the last three *hundred* years. Based on the present state of research, I prefer an interpretation of Chinese history that differs a little from Elvin's. It seems that Chinese agriculture for many centuries embodied a high degree of ecological stability. This stability was founded mainly on three factors: the wet rice cultivation which in its traditional form needed little or no manure, the highly elaborated system of terraces which stopped soil erosion, and the systematic use of "night soil," or human excrement, for fertilization.

It was mainly the last point for which Justus von Liebig, the great German chemist, praised the Chinese as being the wisest people on earth, since they returned to the soil all they had taken from it.[36] I presume that for a long time a high degree of inherent sustainability did indeed exist in Chinese agriculture and may partly explain Chinese cultural continuity over the millennia. But it was just this stability that encouraged continuous population growth and concealed the elements of unsustainability: population pressure, deforestation, erosion, even desertification in marginal regions and, above all, the growing loss of ecological reserves. The environmental crisis of China might anticipate the environmental crisis of Western civilization, a crisis aggravated precisely by its long-term success.

And there might be particular German lessons to be learned from environmental history. There seems to exist a particular German style of environmental politics deeply connected with Prussian-German political as well as bureaucratic traditions, a style characterized by a relatively complicated and systematic juridical and bureaucratic method that contrasts to the ad hoc approach we frequently find in the United States,[37] as well as in other European nations.[38] As Frank Uekötter has shown in his comparison of policies to combat air pollution in the United States and Germany, sluggish German bureaucracies reacted only slowly and cumbersomely to the growing problem of industrial smoke. In America, by contrast, a smoke inspector was installed in American cities, who—if he was a fighter and able to win over public opinion—attacked the smoke problem sometimes as heroically as a sheriff in the Old West. Not rarely, he succeeded, at least as long as the smoke nuisance was readily perceptible with ears and noses and not much science was needed. After the mid-twentieth century, however, the limits of this ad hoc approach became clear as the need for a systematic and scientific approach to air pollution required professional environmental bureaucracies. History seems to show that a blanket judgment as to whether bureaucratic or informal approaches to environmental regulation are more successful is not possible, since the advantage of yesterday may turn out to be a disadvantage today or in the future. Therefore, a critical reexamination of the bureaucratization of German environmentalism is an important task for environmental historians.

Notes

1. Ricarda Huch, *Erinnerungen an das eigene Leben* (Cologne: Kiepenheuer und Witsch, 1980), 93.
2. Wagner's interest in bird protection was the theme of a prize-winning essay by Daniel Jütte (Stuttgart) at a recent history student competition. Körber-Stiftung, ed., "Genutzt—geliebt—getötet. Tiere in unserer Geschichte," *Spuren suchen* 15 (2001): 18.
3. Reinhard Johler, "Vogelmord und Vogelliebe: Zur Ethnographie konträrer Leidenschaften," *Historische Anthropologie* 5 (1997): 20, 23, 27. For a very instructive Italian counter-voice against the Germanic bird-protection racism see Alfred Barthelmess, *Vögel—Lebendige Umwelt. Probleme von Vogelschutz und Humanökologie geschichtlich dargestellt und kommentiert* (Freiburg: Karl Alber, 1981), 146–151.
4. Joachim Radkau, "Die Nukleartechnologie als Spaltstoff zwischen Frankreich und der Bundesrepublik," in *Frankreich und Deutschland—Forschung, Technologie und industrielle Entwicklung im 19. und 20. Jahrhundert*, ed. Yves Cohen et al. (Munich: C. H. Beck, 1990), 302–318.
5. Heinrich Oberrauch, *Tirols Wald und Waidwerk: Ein Beitrag zur Forst- und Jagdgeschichte* (Innsbruck: Universitätsverlag Wagner, 1952), 21.
6. Conrad Totman, *The Green Archipelago: Forestry in Preindustrial Japan* (Berkeley: University of California Press, 1989).
7. Heinrich Rubner, *Forstgeschichte im Zeitalter der industriellen Revolution* (Berlin: Duncker und Humblot, 1967).
8. Karl Hasel, *Studien über Wilhelm Pfeil* (Hannover: Mitteilungen aus der Niedersächsischen Landesforstverwaltung 36, 1982), 137. For a general reevaluation see Joachim Radkau, "Das 'hölzerne Zeitalter' und der deutsche Sonderweg in der Forsttechnik," in *Nützliche Künste. Kultur- und Sozialgeschichte der Technik im 18. Jahrhundert*, ed. Ulrich Troitzsch (Münster: Waxmann 1999), 97–117.
9. Alexandra Mittmann, "'Gott schuf die Welt, und der Mensch verschönert sie'— Naturauffassung, Gartenideal, und Menschenbild in Christian Cay Lorenz Hirschfelds *Theorie der Gartenkunst*" (M.A. thesis, University of Bielefeld, 2001).
10. Michael Niedermeier, *Erotik in der Gartenkunst* (Leipzig: Edition Leipzig, 1995), 163–182; Friedrich Reil, *Leopold Friedrich Franz von Anhalt-Dessau nach seinem Wesen und Wirken* (Dessau: Karl Aue, 1845, rpt. Wörlitz: Kettmann, 1995), 4.
11. Joachim Radkau, *Natur und Macht*, 2nd ed. (Munich: C. H. Beck, 2002), 262–263.
12. Heinrich Rubner, "Naturschutz, Forstwirtschaft und Umwelt in ihren Wechselbeziehungen, besonders im NS-Staat," in *Wirtschaftsentwicklung und Umweltbeeinflussung (14.–20. Jahrhundert)*, ed. Hermann Kellenbenz (Wiesbaden: Steiner, 1982), 122.
13. Cornelia Regin, *Selbsthilfe und Gesundheitspolitik. Die Naturheilbewegung im Kaiserreich (1889–1914)* (Stuttgart: Steiner, 1995), 252–268.

14. Donald Worster, "Auf schwankendem Boden. Zum Begriffswirrwarr um nachhaltige Entwicklung," in *Der Planet als Patient. Über die Widersprüche globaler Umweltpolitik*, ed. Wolfgang Sachs (Berlin: Birkhäuser, 1994), 95–106.
15. At a time when Third World examples of autocentric development had been discredited, Dieter Senghaas presented several European countries as historic patterns of appropriate and autonomous development: Dieter Senghaas, *Von Europa lernen: Entwicklungsgeschichtliche Betrachtungen* (Frankfurt am Main: Suhrkamp, 1982). For a thorough explication of that approach, see Ulrich Menzel, *Auswege aus der Abhängigkeit* (Frankfurt am Main: Suhrkamp, 1988).
16. Helmut Jäger, *Einführung in die Umweltgeschichte* (Darmstadt: Wissenschaftliche Buchgesellschaft, 1994). About the many different meanings of *Nachhaltigkeit* in German forestry, see Wiebke Peters, "Die Nachhaltigkeit als Grundsatz der Forstwirtschaft, ihre Verankerung in der Gesetzgebung und ihre Bedeutung in der Praxis" (Ph.D. diss., University of Hamburg, 1984).
17. Götz von Bülow, "Die Sudwälder von Reichenhall," *Mitteilungen aus der Staatsforstverwaltung Bayerns* 33 (1962): 159–160. The question of whether sustainability was really achieved obtained in the saline forests of Reichenhall remains controversial even today, as is the whole question of early modern wood famine in Germany. See Alfred Kotter, "Holznot um 1600: Die Energieversorgung der Saline Reichenhall," in *Salz Macht Geschichte*, ed. Manfred Treml et al. (Augsburg: Haus der Bayerischen Geschichte, 1995), 186–192.
18. Mark Elvin and Liu Ts'ui-jung, eds., *Sediments of Time: Environment and Society in Chinese History* (Cambridge: Cambridge University Press, 1998), and Richard H. Grove et al., eds., *Nature and the Orient: The Environmental History of South and Southeast Asia* (New Delhi: Oxford University Press, 1998).
19. Lynn White, Jr., "The Historical Roots of Our Ecological Crisis," *Science* 155, no. 3767 (1967): 1203–1207.
20. Nicholas K. Menzies, *Forest and Land Management in Imperial China* (New York: St. Martin's Press, 1994), 44.
21. Eduard B. Vermeer, *Population and Ecology along the Frontier in Qing China*, in *Sediments of Time*, ed. Elvin and Ts'ui-jung, 247–248.
22. Madhav Gadgil, "Deforestation: Problems and Prospects," in *History of Forestry in India*, ed. Ajay S. Rawat (New Delhi: Indus Publishing Company, 1991), 13.
23. Madhav Gadgil and Ramachandra Guha, *This Fissured Land: An Ecological History of India* (New Delhi: Oxford University Press, 1992).
24. Jacques Gernet, *Die chinesische Welt* (Frankfurt am Main: Insel, 1979), 331.
25. Joachim Radkau and Ingrid Schäfer, *Holz—Ein Naturstoff in der Technikgeschichte* (Reinbek: Rowohlt, 1987), 59–65, 157–159; Ingrid Schäfer, *"Ein Gespenst geht um": Politik mit der Holznot in Lippe, 1750–1850* (Detmold: Naturwissenschaftlicher und Historischer Verein für das Land Lippe, 1992).
26. Radkau and Schäfer, *Holz*, 156.
27. Christoph Ernst, *Den Wald entwickeln. Ein Politik- und Konfliktfeld in Hunsrück und Eifel im 18. Jahrhundert* (Munich: Oldenbourg, 2000), 345.
28. Alfred W. Crosby, *Ecological Imperialism: The Biological Expansion of Europe, 900–1900* (Cambridge: Cambridge University Press, 1986); Richard H. Grove, *Green Imperialism: Colonial Expansion, Tropical Island Edens, and the Origins of Environmentalism, 1600–1860* (Cambridge: Cambridge University Press, 1995).
29. Isabelle Knap, "Die Entstehung von Umweltbewusstsein in Kolonialgebieten? Eine Untersuchung des Umgangs der europäischen Kolonialmächte mit Natur und Umwelt in der frühen Neuzeit am Beispiel von Mauritius, Java, Ambon und Südafrika" (M.A. thesis, University of Bielefeld, 2003), arrives at conclusions con-

trary to Grove. Dietrich Brandis, the German-born general inspector of the forests of British India and one of the heroes of the Grove thesis, resigned at the end of his career when he wrote the British government did not realize the advantages of good forest management—it only realized the forest revenue and the chance to arrange jobs for six to twelve young Englishmen every year. Indra Munshi, "Sir Dietrich Brandis: ein deutscher Generalforstinspektor in Indien," in *Der deutsche Tropenwald: Bilder, Mythen, Politik*, ed. Michael Flitner (Frankfurt am Main: Campus, 2000), 51.
30. Joachim Radkau, "Holzverknappung und Krisenbewußtsein im 18. Jahrhundert," *Geschichte und Gesellschaft* 9 (1983): 513–543.
31. Georg Forster, *Ansichten vom Niederrhein* (1791; rpt. Stuttgart: Reclam, 1965), 56–57.
32. Richard Grove, "Scotland in South Africa: John Crumbie Brown and the Roots of Settler Environmentalism," in *Ecology and Empire: Environmental History of Settler Societies*, ed. Tom Griffiths et al. (Seattle: University of Washington Press, 1997), 144.
33. Radkau, *Natur und Macht*, 210.
34. Eric L. Jones, *The European Miracle: Environments, Economies, and Geopolitics in the History of Europe and Asia*, 2nd ed. (Cambridge and New York: Cambridge University Press, 1987).
35. Mark Elvin, "Three Thousand Years of Unsustainable Growth: China's Environment from Archaic Times to the Present," *East Asian History* 6 (1993): 7–46.
36. Justus von Liebig, *Chemische Briefe* (Leipzig: C. F. Winter, 1865), 498–500.
37. Frank Uekötter, "Von der Rauchplage zur ökologischen Revolution. Eine politische Geschichte der Luftverschmutzung in Deutschland und den Vereinigten Staaten von Amerika 1880–1970" (Ph.D. diss., University of Bielefeld, 2001).
38. Radkau, *Natur und Macht*, 421–422.

Chapter 2

Conviction and Constraint

Hydraulic Engineers and Agricultural Amelioration Projects in Nineteenth-Century Prussia

RITA GUDERMANN

Extensively used grazing and barren lands—too wet or too dry for intensive agriculture—characterized large parts of the German lowland plains at the beginning of the nineteenth century. Middle-class inhabitants of nearby villages and towns regarded these lands as little more than infertile deserts or putrid swamps containing miasmas. But the rural population living on the edge of poverty depended on them. These "barren" lands were vital to the traditional rural mixed economy, serving as the commons where the poor lived and supported themselves through subsistence farming, wage earning, and domestic textile production.[1]

During the course of the nineteenth century, however, a new understanding of these marginal lowland areas landscape developed because of a demand for agricultural progress. From the perspective of many middle-class observers, swamplands where the poor lived no longer seemed to be the source of unhealthy miasmas and heathland no longer seemed to be infertile desert. Instead, both landscapes came to be regarded as potential goldmines that only had to be used correctly to reveal their real value. Driven by an optimistic belief in progress, an enlightened bourgeois public advocated privatization, intensification of agricultural use, irrigation, and drainage of wetlands to stimulate the cultivation of marginal land. Influenced by economic liberalism and rationalized farming, the Prussian state implemented agricultural reforms beginning in 1807 to meet these demands. The government established private ownership of land, and in doing so, fostered a new entrepreneurial approach to agriculture. Communal property was divided up and the grip on barren lands tightened. Peasants were no longer allowed to farm extensively, but were urged to maximize profits by intensively cultivating the land. Drainage and irrigation provided ways to solve economic and social problems in rural areas. According to one of the most prominent amelioration officers of nineteenth-century Prussia, such improvements in land use entailed nothing less than "snatching a currently depressed and barren region away from neglect, awakening the manifold powers and wealth slumbering within, and bringing them to life in an attempt to turn this area into fertile, blossoming cultivated land, especially for those inhabitants who are so dependent on it, by paying attention to the circumstances in any way favored by nature and not letting state funds lie around unused."[2]

No one stood in a better position to develop the "national resource," water, than hydraulic engineers, an emerging professional group that willingly placed its expertise in the service of the nation. This chapter will consider two aspects of these professionals' involvement in agricultural reclamation projects: first, the way in which these technicians understood themselves and their tasks and second, how they responded to conflicts and unexpected complications encountered in carrying out amelioration projects. Examples used in the study will focus on the Prussian provinces of Brandenburg and Westphalia, two areas characterized by a mixture of dry and sandy soil and unregulated rivers—features that posed a variety of challenges for engineers.

Swamps or Goldmines? The Nineteenth-Century Debate on Agricultural Intensification and Mechanization

From the end of the eighteenth century and increasingly in the nineteenth, technicians and engineers had asserted their optimistic views on the possibilities of agricultural improvement in the lowlands in discussions about water and the landscape. Memoranda, pamphlets, journal articles, textbooks, and handbooks reveal a good deal about engineers' professional self-image and their understanding of the tasks they intended to oversee. The plans of technicians and engineers to harness water for human use captured the public imagination in the nineteenth century because they promised a scientific and technological solution to the problem of unsatisfactory river control. Recurring floods, with their devastating effects on farmsteads, demonstrated with depressing clarity how little hydraulic engineering in the past had achieved. Technicians reasoned that in the future, however, more efficient farming, coupled with sustained control of waterways, held out the promise of almost endless increase in agricultural production. Engineers promised to dry out swampy areas through dykes and river and stream regulation while simultaneously irrigating dry, sandy soils, providing a constant supply of good water and reclaimed farmland for a growing agrarian population. Hydraulic engineers also underscored the advantages for the nation of a well-organized inland navigation system. They argued that water would be an ideal source of energy for a multitude of tasks because it did not require special maintenance.

With so many potential uses for water, engineers felt called upon "to transform the lack and the abundance of water in a country into water-wealth," applying engineering standards to nature rather than gazing passively upon it. "Should he [the engineer] achieve this," one technician wrote, "and should he use all of the forces of the water carefully according to his needs, he will thereby enrich his country, increase its creditability, and at the same time, open up inexhaustible resources for the arts and sciences. Economizing water in order to use it on demand, controlling the impact of flowing waters, preventing the standstill of water: these are the detailed tasks which will lead to an abundance of water and wealth in nations if applied properly."[3] Engineers thus believed that they could transfer the concepts of maximum efficiency and productivity from the mechanical to the organic world.

Among the more traditional and common tasks for hydraulic engineering was riverbank protection. After the establishment of private property in the early nineteenth century, the natural flow of the water was perceived to be a problem. As one contemporary explained, "It is part of the particularity of rivers that they continually push their beds forward . . . by gnawing soil from one bank and simultaneously adding it to the opposite one. . . . As cultivated land increases in value the more valuable those layers of soil become that mark the boundary of the rivers. And hence, the more the law of property will be fastened upon them, and the more a guarantee against all kinds of thieves will therefore be claimed."[4] Such reasoning convinced engineers that it was essential to restrict rivers' indiscriminate and "unlawful" claim to the land along its banks. Their task of building-up riverbanks thus included the protection of adjacent land from flooding.

Their efforts, however, did not stop there. Correction of existing drainage conditions was considered necessary, but involved greater interventions in natural hydraulic systems. River regulation provided the most dramatic example of the achievements of hydraulic engineering.[5] Even though interventions such as dykes, harbors, and drainage systems already had been constructed during the Roman Empire and the Middle Ages, technicians had not attempted large-scale river regulation projects before the nineteenth century. The rectification of the Upper Rhine River by hydraulic engineer Johann

Gottfried Tulla, which began in 1817, was thus considered a milestone in the art of engineering. The project aimed to prevent flooding in the rift valley at the foot of the Alps by narrowing the riverbed to a width between 200 and 250 meters, by cutting off the numerous braids, oxbows, and islands that punctuated the river's course, and by constructing embankments at points vulnerable to flooding. These activities opened up new agricultural areas and prepared the way for the industrialization of the Rhine, but the result was a river shorn of its geographical and biological diversity. Tulla's ideal river resembled a canal rather than a meandering, living river.[6]

In confronting the challenges of river regulation, engineers developed new, heroic ways to portray themselves and their tasks that relied on telling war metaphors to describe humankind's battle against flowing water. In the minds of engineers, a river's current appeared to be a natural force with a soul and with destructive anger that sought to destroy human culture. Victor von Domaszewski, an agricultural and hydraulic engineer, described the dangers for people living next to a river hindered by boulders and ice floes: "The current does not know itself anymore! Its water roars and raves, foams and scrambles, grows, races and grows so suddenly that the highest banks are too low for it, too narrow for it . . . now the angered stream has no more mercy! Not on the valleys it loves, [or] the houses and cities, which otherwise are reflected in a lovely way on its calm surface. And not on the faithful inhabitants of the valley, who loved it dearly and sang of it many times." This anthropomorphic view of nature gave a moral dimension to ordinary hydraulic patterns; rivers and other waterways were portrayed as the enemies rather than inert geographical entities. Consequently, Domaszewski understood his comment as "an indictment by a well-traveled and experienced . . . hydraulic technician against the criminal sand and gravel banks, which cause the most terrible devastations every year and whose high spirits are unfortunately continually growing."[7]

Water was particularly likely to be used as a symbol of nature's challenges to human civilization. After all, humans "were always either supported or disturbed by this element, and often it seemed to offer humans help with the execution of their highest possible calling, namely of regulating nature in all her unwillingness and wildness."[8] Technicians asserted that their interventions would minimize the effects of weather, which had always been outside the realm of human influence. In making such a claim they implied that their work was capable of supernatural feats, of liberating human beings from their age-old dependence on and fear of natural forces. Confronting nature's challenges also seemed to build societies' spiritual character and physical vigor, satisfying a kind of sporting ambition by training "human minds and strength, thereby keeping humans from sinking into bodily and intellectual laziness," as the engineer H. von Pechmann commented.[9]

But nature fought back fiercely when human intervention was flawed, such as when flats were dried out without providing proper drainage. "Thus it often happens that the power of the unsettled element breaks dams with its raging anger and, full of revenge, reclaims areas that humans had taken from it through hard work over many years," noted one contemporary description of floods.[10] In such a case an inundation appeared to be a punishing flood, one that was evidence of nature's revenge against amateurish human intervention.[11]

The highest aim of hydraulic engineering was to impose order on nature. In technicians' minds the peaceful and slow-flowing water at the lowest point of an area should not be hindered in any way by shallows and sand banks, boulders, trees, narrowing of the river, salient jagged banks, or mills' weirs. "If a river shows such an appearance, then it can be said that it is in its normal state," noted one engineering handbook, but such a baseline needed human correction to reach a more perfect level.[12] In order to reach this level, engineers considered it acceptable to intervene in every single water cycle through irrigation or drainage. Technicians disagreed, however, about how this ideal could be attained. Some insisted that harmful waters needed to be "disposed" of as quickly as possible by straightening and deepening the streambed, channeling them as quickly as possible to the sea. Other engineers criticized this plan; such measures, they argued, failed to use water's powers fully.

Despite theoretical disagreements, waterway builders remained optimistic about the possibilities of harnessing waterways for human uses. As their credo indicated, "Wherever great water building projects were carried out effectively, they have always been beneficial, despite the enormous costs." Success could be quantified, usually "expressed by a sum of money."[13] Despite such a materialist view of hydraulic engineering's advantages, however, technicians were not guided solely by economic concerns; their primary motives reflected social and cultural considerations that were merely couched in economic arguments.[14] For example, they argued that water regulation would improve the hygienic conditions of a region by getting rid of any harmful miasmas in swamplands. Such projects, they also asserted, would improve the fertility of farmland, and consequently, lessen poverty. By advocating the social benefits of riparian correction, hydraulic engineers hoped to demonstrate why every great nation needed their expertise for eliminating swamps and moors.

Engineers' disapproval of unproductive landscapes also reflected contemporary aesthetic notions, which viewed swamps, moors, bogs, and unregulated streams as disease-ridden eyesores in need of human improvement.[15] But by the end of the nineteenth century, engineers began to express a growing sensitivity toward nature's beauty in line with the neo-Romantic impulses of that era.[16] Technicians were fully aware that their large-scale drainage and irrigation projects destroyed unique landscapes. Their ambivalence about such natural destruction was expressed by Professor Max Popp, an ardent proponent of amelioration projects and the head of the Moor Research and Control Station at Oldenburg (Moorversuchs- und Kontrollstation Oldenburg). Because of economic difficulties and the political priority of maintaining the food supply, Popp wrote in 1924 in *Contributions to the Care of Natural Monuments* (*Beiträge zur Naturdenkmalpflege*) that the cultivation and colonization of "wastelands" should be accelerated. Reasoning that another thirty to one hundred years would pass before the remaining moors and heathlands would be completely cultivated, Popp asserted that later generations would concern themselves with the protection of nature. "By then," he concluded, "we should be able to care for the ideals of our people so that the spirit of materialism does not prevail, but rather, a sense of what is beautiful, good, and true can be reawakened in the spirit of the people [*Volksgeist*]." Popp suggested that "for the present, all areas worthy of being protected as natural monuments should be taken care of and only used if arable wasteland—which does not possess the same value—is cultivated."[17] At the turn of the century, therefore, when large-scale projects that would rectify all German rivers and build new canals were being put in place, hydraulic engineers worried little about possible aesthetic or ecological consequences of their interventions. They were fully confident that amelioration projects would bring advantages for their nation.[18]

In view of the many anticipated improvements that would result from river regulation and canal building, the state experienced an increase in support for large-scale hydraulic engineering projects during the nineteenth century. Technicians, in particular, expressed dissatisfaction with water laws that hindered progress on reclamation and agricultural improvement projects. After the division of the commons in the early nineteenth century, it had become necessary to negotiate with a large number of new owners to carry out rectification projects over large areas. Achieving agreement among this myriad of owners proved difficult; it seemed to some that "every amelioration project is bound to fail, even one which seems promising, if only one land owner refuses to agree on the project."[19] In light of the difficulties of negotiating with private landowners, technicians, like agricultural economists and social reformers, turned to the state to support hydraulic engineering measures by providing easier access to recently privatized lands.

In asking the state for assistance, however, technicians put themselves in the line of fire in political discussions. Conflicting individual interests eventually proved to be one of the main obstacles to engineering improvements that had been implemented since the middle of the nineteenth century. Because amelioration measures were initiated by the state, and often against the will of peasants affected by them, local people grew distrustful of the aims of engineers. Every person involved was afraid "to voice his satisfaction with a project and thus the whole undertaking. If one judges it only by the voiced complaints, it seems to have failed quite thoroughly. But eventually, years later, the general increase in wealth proves the opposite to be true." Though no peasants' voices commenting openly on the hydraulic projects can be found in the archival material, technicians' efforts constantly met with discontentment, making it difficult for them to continue pursuing what they deemed to be the common good.

An additional challenge stemmed from competing claims on water. It was difficult to decide if water "was to be used for irrigating meadows, for powering hydraulic machines, for supplying water for waterways," or for flooding ditches that would supply water to urban industries. Would water "be pumped up high for the advantage of hydraulic machines and thereby turn higher lying estates into swamplands," or would "every hydraulic machine have to give up part of its claims to the slope in order to improve the situation for the estates."[20] To address these multiple claims on water, technicians argued from a purely economic point of view. They insisted that a river fully under their control—one that no longer would flood adjacent lands but that could be used simultaneously by farmers, industries, and traffic—would harmonize conflicting needs, providing a technical solution to a political problem. Engineers promised to synthesize arguments and competing claims that were voiced in economic, social, and aesthetic debates about the landscape. Their promise to solve anticipated social conflicts by technical means made their contribution to public debates irresistible. Their struggle to win influence within the Prussian state, and to gain recognition as vital professionals, eventually developed a momentum of its own in the debate over the path that modernization should take.

Despite, or perhaps because of, its noble aims, hydraulic engineering presented itself as a profession battling against great difficulties from the start. Just as the construction of a "normal" state of waters inevitably involved conflict, so too was there tension between the theory of river rectification and its practical implementation. The theory asserted that mathematical calculations, straight lines, and even, bending curves were the desired outcome; in practice, winding, irregular rivers possessed an inherent dynamic that could not be conquered entirely.

Many engineers ignored or failed to recognize the negative side effects of waterway construction, some of which were so severe they led to the failure of systems entirely. Even well-known hydraulic engineers had to admit that, in the past, certain projects had produced devastating results to which the public was not blind. This led the hydraulic engineers to insist on carefully studying local conditions and taking them into account in construction. These challenging responsibilities, they insisted, should only be assigned to highly qualified experts.

In the 1870s, many of the engineering projects became the target of vehement criticism, particularly because hydraulic engineering had been carried out on a massive scale throughout Germany since the middle of the century. Debates emerged, for example, in the *Journal of Civil Engineering* (*Zeitschrift für Bauwesen*) between critics and supporters of contemporary hydraulic projects. One noteworthy example is the debate that developed around the theories of the former Prussian construction official, Dieck. As inspector of mechanical engineering in the coal and steel industry in Westphalia, Saarbrücken, and the Rhineland, Dieck had studied the effects of river regulations for thirty years. In an anonymous pamphlet published in 1875, Dieck challenged fundamental scientific assumptions of his profession when he referred to contemporary methods of river regulation as irrational. He postulated that as long as the engineers did not take into account the internal dynamics of water in their mathematical models, their efforts were bound to fail. Former swamps would turn into deserts, while drained areas would be flooded.

Such criticism from within their own ranks prompted hydraulic engineers to respond with equally vehement arguments in favor of more extensive regulation. A year after publication of the anonymous pamphlet, the engineer J. Schlichting dismissed the work as "the general reasoning of a person full of a pessimistic world view." The author, Schlichting asserted, failed to consider "that up to now no German river has been fully regulated. At best one can speak of only piecemeal regulation." Moreover, the guidelines that had been applied to hydraulic engineering projects thus far "are from a time when the goals and purposes of regulation were entirely different from now. Furthermore, river regulation has never had an adequate sum of money at its disposal and the authorities have failed so far to find a homogeneous approach to different projects and their implementation."[21] Dieck's pamphlet also criticized other aspects of modernization, such as railways, mining, forestry, and the stock market. Schlichting warned that the ill-conceived opinions expressed in the publication could cause great harm if uninformed people were to embrace them. According to Schlichting, waterways had declined in significance largely because of insufficient funding and increased competition from railroads. "Not the water regulation system," he asserted, "but the lack of funds and the competition to waterways posed by railway lines constructed over the last thirty years, have altered traffic and trade and reduced the importance of waterways."[22]

This debate on river rectification was thus polarized between those who proposed abolishing ambitious regulation projects entirely and those who demanded immediate and more consistent regulation. Dieck's passionate appeal, with numerous references to history, religion, and philosophy, did not stand a chance against the cool rationalism of his opponents. The debate about the failures of hydraulic engineering did not lead technicians to oppose regulation and amelioration measures. On the contrary, they demanded even more large-scale and drastic measures, such as combining irrigation and drainage. They would have to wait until far into the twentieth century, however, before realizing their utopian visions of environmental improvement.

The struggle to establish hydraulic engineering as a respected profession proved difficult in Prussia, a state shaped by tradition and particularism. Before the creation of the German Reich in 1870, the hopes of engineers often foundered. Their organizations, such as the Society of German Engineers (Verband Deutscher Ingenieure, or VDI), founded in the 1850s, were merely tolerated. Civil servants, trained as legal and administrative experts, continued to be the real elite within the state.[23] Nevertheless, in Prussia and more generally in continental Europe, hydraulic engineers gained acceptance and approval in tandem with large state-sponsored construction projects. They established their own organizations, journals, project standards, and codes of conduct, a process of professionalization described by Charles McClelland and others.[24]

After the establishment of the Reich, the demand for additional canals reached a new high, giving hydraulic engineering additional influence. Increased industrialization compelled some to challenge the supremacy of the railway system, which had been expanded at the expense of river and canal building. River transportation systems in France, England, Belgium, and especially the United States, also generated renewed interest in Prussia's waterways. Furthermore, initiatives to improve the sales potential of agricultural products (which had been overshadowed by trade and industry) relied on hydraulic engineering as did schemes to develop cheaper transport routes for mass goods, such as fertilizer for farming. At the time, the cost of transporting these products via railroads was considered to be too expensive. Such considerations helped to justify more and even larger improvement projects and supported the construction of drainage and irrigation systems. Engineers hoped that the enormous cost of constructing canals would be partially compensated for by the benefits associated with soil amelioration.

The most important argument in favor of hydraulic engineering in the latter half of the nineteenth century was the claim that it would improve the nation's well-being. As one engineer noted, "The rise and decline of each country always keeps pace with the conditions of water in that country! Where the waters, even if forced, live in peace with the population, where the waters oblige all human needs, there is affluence, there is even wealth!"[25] The utopias envisioned by hydraulic engineers imagined the possibility of not only connecting the fragmented regions of Germany, but also of unifying all nations. Hydraulic engineering seemed to advocate measures against war and offered an optimistic project for the future, one in which engineers obviously would play an important role. "The issue of the water economy touches on all of the primary relationships between the life of a nation and a people and can only be solved through the cooperation of national economists, statesmen, and engineers."[26] Thus, during the nineteenth century, the foundations of a mechanistic solution to social and economical problems were laid.

Digging Up the Treasure: Practical Conflicts and Constraints of Amelioration Projects

Analyzing the professionalization and technological optimism of Prussia's hydraulic engineers helps us to understand the enthusiasm and faith in progress that accompanied nineteenth-century amelioration projects. Their involvement developed into a two-pronged program of waterway correction and land reclamation. First, engineers devised ways to guard against periodic flooding of unregulated rivers. Second, they intensified the use of waterways and riverbanks and the development of heath, lowlands, and peat moors for agricultural, commercial, and eventually, industrial purposes. The programs to improve these areas were driven by the belief that privatization, intensified use, and amelioration would result in greater profits through far less effort. From the middle of the century, these projects were implemented in almost all of Prussia's glacially formed lowlands, changing the appearance and ecological function of former moor and heath landscapes forever.

The implementation of large-scale drainage and irrigation projects marked the end of the traditional rural cooperative where moors and heathland were treated as commons. Rural cooperatives were replaced by melioration cooperatives that were established on the private lands of new property owners and operated under the direction of state officials and technicians. In effect, a new class of competitive agricultural entrepreneurs replaced the village community, which traditionally had used a significant portion of its available resources collectively. Along with the other achievements of rationalized agriculture, drainage and irrigation successfully revolutionized traditional methods of farming by opening up marginal lands and providing a steady supply of water to agricultural fields. After the middle of the nineteenth century, Germany was no longer a "developing nation."

In developing these projects, the Prussian state relied on an army of well-trained and willing technicians and engineers who were keen to place their skills in the service of society. At the top of this hierarchy stood the "state amelioration inspector," a position created in 1857 in all of Prussia's provinces when it had become clear that additional authorities were needed to cope with the state's expanded responsibilities in agriculture. These inspectors were in charge of the implementation of the large-scale state amelioration projects. In addition, cultivation technicians were employed, either privately or by the state, to serve as certified meadow builders, drainers, and surveyors. Amelioration engineers, and the lower-ranking professionals who aided them, formed an occupational group with its own views on the treatment and use of the land. They often found themselves in conflict with other interested parties such as farmers, millers, fishermen, and later, environmentalists.[27] Despite being civil servants or state employees, they pursued their own interests and advocated a unique worldview that differed from other administrative officials.

Prussian authorities set high standards for this new occupational group of engineers. In addition to having classical training as engineers, they were expected to be readily mobile and constantly in training, for example by visiting model sites abroad.[28] Initially authorities struggled to find engineers willing and able to carry out the relatively new and demanding amelioration projects.[29] Even the prospect of favorable working conditions and pay did not alter this. According to some sources, railway construction attracted "all decent, well-qualified builders," making it difficult to hire competent individuals for amelioration projects. Better pay and fewer risks made railway building much more attractive for aspiring engineers.[30] Furthermore, the employment of subordinate cultural engineers was seasonal and subject to fluctuations in the economy. Little wonder that qualified men viewed this line of work as risky and lacking in security.[31]

One of the outstanding features of such pioneers in engineering was their awareness that they were a kind of "developmental aid worker," an elite participating in a unique way in bringing "progress" to rural areas. Technicians and surveyors considered the preparatory work for amelioration to be a real test of their creativity.³² Even though several amelioration projects failed on the first attempt, a consciousness remained of having created something significant. Such optimistic attitudes permanently altered the outlook on the future.³³ With little outside help, technicians had to cope with a range of difficulties on construction sites, and inconveniences and risks were a part of their daily work. They often labored on impassable terrain, standing in knee-deep water and sometimes in danger of sinking into low moors.³⁴ Additionally, they must have worried about the health risks of miasmas evaporating from the swampy soil in such areas.³⁵ Other professional groups were far less daring; clergymen, teachers, and customs officers considered it a death sentence to be transferred to the lowlands as a requirement in their career.³⁶

Amelioration engineers were far more familiar with the land and its people than administrative civil servants who oversaw the development projects from behind their desks. During construction for the drainage of the source of the Ems River, the state amelioration inspector Wurffbain reported, without any attempt to hide his feeling of belonging to a superior social class, that he met only "inhabitants living at the lowest level of culture, mostly ill, vegetating in utter apathy."³⁷ The rural population often had no understanding for the work being done and tried to hinder it as much as they could.³⁸ Surveyors, in particular, directly experienced the resentment of locals who sometimes refused to give directions or stole their poles and stilts.³⁹ This behavior utterly astonished technicians who remained unaware of why the rural poor resented their presence. But the reasons should have been readily apparent. Though Wurffbain was right in his observation that poverty, illness, and deprivation dominated some peasants' life, he failed to see the enormous social differences within rural society.

The state amelioration projects that Wurffbain participated in were a component of an economic program that strengthened rural entrepreneurs—who were mostly larger farmers in Brandenburg and Westphalia—and relied on compulsory governmental measures. The program completely overlooked the needs of small-scale farmers, even though smallholders formed the majority of the rural population. In the mixed economy of this social class, former moors and heathlands had played an important role as areas for grazing animals, hunting small game, acquiring extra firewood, or gathering berries and wild mushrooms. The majority of the poorer peasants, living in the midst of the supposedly "barren" swamps, moors, or heathland, had thus acquired an intimate knowledge of the natural conditions of their habitat. Such knowledge did not make them affluent—many of them had less than enough to get by on—but it did provide them with a livelihood and a place in the rural society.⁴⁰ The privatization of the commons and the implementation of draining and irrigation projects meant the expulsion of these rural peasants from the land by new landowners. Clearly, improvement of the land failed to remedy the poverty of those who formerly had benefited from the seemingly barren land.

Overlooking these social inequalities and strongly influenced by a belief in progress shared among the enlightened bourgeois public, technicians concluded that rational capitalistic farming provided the solution for the country's economic and social problems. From their perspective the rural population simply seemed to be stubborn and uncultivated rather than giving voice to real concerns about economic displacement and ecological degradation. In their eyes, the poor were simply unable to recognize the noble deeds that technicians performed in the service of their country.⁴¹ For this reason, engineers often saw themselves as missionaries carrying out important work for the benefit of savages.

Furthermore, technicians sensed that the affected population was "in every possible way intent on financial blackmail," as rural inhabitants tried to escape responsibility for producing and maintaining paths and water drainage ditches.⁴² In addition, private owners and communities sought to make amelioration associations pay for costs unrelated to engineering projects, such as the repair of rotting or damaged bridges, and "considered the funds for public construction to be goldmines that have to be exploited as much as possible."⁴³ Based on these observations, engineers increasingly applied their own strict standards of reasonable economic behavior when dealing with the rural population. But the obstacles and strain of their work did not shake their pride in being engineering pioneers.⁴⁴ Quite the opposite: the more difficult their tasks became, the more self-assured they became. Some of them even identified themselves with Frederick the Great, under whose "powerful scepter" future-oriented measures were carried out.⁴⁵

Engineers who had witnessed the poverty of the population firsthand were motivated by humanitarian concerns as well as more obvious professional politics to continue their work. They reasoned that if one wanted to solve all of the problems of flooding and drought and their potentially devastating effect on humans, then one had to seize humankind from nature's tyranny and force all of nature to succumb to human will.⁴⁶ They aimed at nothing short of complete cultivation of barren and infertile land.⁴⁷ In addition, they insisted that the population living in an improved area had to rigidly follow the instructions of engineers in the reclamation and maintenance of new agricultural fields.⁴⁸ In the process of establishing a new cultural administration for the state, it was often unclear who was responsible for which tasks. In some instances this meant that new inspectors received instructions from different authorities and had to have their plans reviewed by technicians unfamiliar with hydraulic engineering.[49]

Although administrative civil servants always had claimed economic rationality as their ultimate source of authority, technicians and engineers increasingly claimed technical intelligence solely for themselves. Especially in the early years of hydraulic engineering, this attitude of technical officials caused considerable conflict with traditional governmental elites, particularly when decisions had to be made about allocating funds for intended projects. If possible, administrative civil servants reduced technicians' cost estimates and questioned the necessity of large-scale sites and massive building techniques. They thereby forced technicians to be satisfied with less-than-perfect technical solutions and to accept temporary measures.[50] Lower-ranking technical personnel sometimes had to wait months before being paid.[51] Such conflicts continued even during construction.[52] For example, the amelioration inspector Wurffbain protested when the local government of Minden tried to reduce the rate of reimbursement for traveling expenses from 2 to 1.5 Thalers per mile. He explained, "My technical effectiveness consistently has been outstanding for the past six years and I have always been paid very well for it." He added that he had to disagree and "continually and effectively pursue the matter on principle."[53] In 1852, he complained to the same local government in Minden about the belated payment of his personnel. "It is a catastrophe for me," he wrote, "that I have to continually plea and beg for every single source of funding. Such conditions can ruin even the bravest and most effective man."[54] To push through his claims, Wurffbain emphasized that he had been entrusted with handling immense sums of tax money. In November 1855, for example, the Westphalian amelioration inspector supervised twelve projects in Westphalia and eight more in Saxony, projects that comprised an area of 288,000 acres and were funded in the amount of 1.5 million Thalers.[55] Approximately sixty-three miles of rivers had to be regulated, a number of canals had to be constructed, and almost three miles of road had to be built.[56]

Engineers maintained that solutions were adequate only when they were as large-scale and permanent as possible, and when they were carried out according to the latest developments in technology. "In such highly important drainage matters that deal not only with the betterment of state cultivation but also with the improvement of the atmosphere and the health of the people, one should not calculate the cost of the preparatory work too anxiously," Wurffbain advised.[57] At the same time, he argued, the hydraulic engineer could not accurately predict all of the particulars beforehand, especially in difficult cases where it was also important to appease all interested parties. The initial euphoria therefore soon changed to a form of down-to-earth pragmatism against the contested limitations imposed by the state bureaucracy. "We further the material interests of human beings and do not concern ourselves with any executive powers. Almighty God makes everything good and whoever has faith in him will never despair."[58]

Especially in the early years of their profession engineers defended themselves against not only administrative civil servants but also the bourgeois public, which questioned the purposes of their projects.[59] According to technicians, public concern was based on "utter ignorance" and hindered their success considerably. Frustrated by continued misgivings, one engineer wrote, "Sensible people do not plan projects with extraordinarily extreme or extraordinarily simple cases in mind. No one thinks of not building a house because it could be turned over by a tidal wave or ruined by an earthquake. By God, I think I have run this amelioration project in such a way that even nontechnicians can make sense of it."[60] Buoyed up by the optimism of a profession with growing influence, technicians resisted the skepticism of the traditional elite. Though not readily apparent, this situation offered engineers the opportunity to express their self-confidence and work ethic. As Wurffbain explained, "I was very aware of what I wanted and was asked to implement it and I therefore did not mind working very hard to reach the goal."[61]

Lower-ranking cultivation technicians initially had to struggle to gain recognition for their expertise and to be paid adequately for their work. Even the most advanced farmers "could not be convinced that this business also required special knowledge. They thought that if they were able to mark out a parallel line with three sticks and some string then the meadow was built and they therefore did not want to pay someone more than a wage laborer's earnings."[62] Arguing against this, the meadow builder A. Reinicke emphasized what was involved in interfering with waterways even on a limited scale. "No soil work needs to be calculated and carried out as exactly as this part of meadow building, because water will find even the smallest mistake."[63] Some hydraulic engineers, such as the most famous meadow builder Vincent, achieved an unchallenged position. But it was not until the last decades of the nineteenth century that cultivation engineers were acknowledged as experts.

Amelioration engineers derived most of their support from those with "intelligence and capital" in rural areas: large landowners, often from the bourgeoisie, who had purchased moorland and heathland with the expectation that property values would increase after amelioration. The Brandenburg amelioration inspector Koehler expressed approval that primarily large landowners were affected by the amelioration project on the Havel River.[64] There was, in effect, an alliance of "technical and agricultural intelligence" that made possible the pursuit of ambitious projects without consideration of traditional ties and financial obstacles.[65] To ensure that the majority of the rural population benefited from amelioration, technicians eventually emerged as ardent supporters of cooperative thinking. Cooperatives, which had the right to expropriate the land of reluctant peasants, offered the possibility of implementing large-scale projects and obtaining necessary loans to finance them. Thus technicians hoped "for an awakening of a spirit of association—of sharing costs—and of bringing sites to life with united power from which all will benefit, even if not in the exact same way."[66] At the same time, engineers pursued the educational goal of making inhabitants give up "their anxious assessing and dividing of advantage into large and small." From their point of view, only the envy of neighbors convinced committed farmers to sacrifice the implementation of beneficial amelioration measures.[67]

After several years of practical experience technicians realized self-critically that they had to develop a degree of social competence in order to successfully finish an amelioration project. "A technician has to acquire a certain practical overview . . . to overcome certain difficulties that exist on paper and thereby make implementation be seen favorably by large landowners."[68] Thus, Wurffbain believed that one of his most important tasks involved being "a relentless informer."[69] Instead of confronting state authorities with self-confidence, he was now convinced that complex plans could be carried out only "through the combined and clear efforts of landowners in large areas under the protection of the higher intelligence of the state authorities."[70] In effect, increased conflict with people affected by engineering measures had made even administrative civil servants aware of the vast gap between the theory and actual implementation of amelioration projects. Intervening in the water system of an area with a traditional agricultural structure contributed to a host of complex problems, as drainage or irrigation measures resulted in numerous, often unintentional consequences for the ecology and economy of an area. Therefore, traditional authorities gratefully accepted the promises of technicians to solve recurring problems with technical means.

In the end, the goals of administrative civil servants and technicians were compatible. This is evident in the fact that the administration was lenient in handling the numerous problems and "growing pains" associated with amelioration projects. The few reports of failed drainage and irrigation measures did not automatically condemn hydraulic engineers. For example, the following assessment was written about Wurffbain's projects in Westphalia: "It makes sense that technicians are not at fault for the initial failures of drainages; at most it can be said that the projects were carried out twenty-five years too early, and that one should have waited for a population with more economic understanding and better supplied with help."[71] According to another observer, one could only accuse technicians of a little too much enthusiasm when faced with such complex projects.[72]

In the long run, innovative hydraulic engineering posed as a kind of qualification machine for upwardly mobile engineers and technicians who were willing to take on challenging improvement projects. Such projects trained them in their profession, but also paved the way for hydraulic engineering to emerge as a respected field in society at large.

How does the history of hydraulic engineers in Prussia compare to Wittfogel's theory of the interdependence of power on water and power on people, as recently discussed by Joachim Radkau?[73] German hydraulic engineers, as shown by examples from the Prussian provinces Brandenburg and Westphalia, indeed tried to recommend themselves as a new functional elite by trying to gain control over water. They never gained control of the element, however, nor did they gain total power over their people. In the face of technical failures and local opposition, their rebukes only became shriller. To explain the unsuccessful implementation of amelioration measures, engineers tended to blame the complex difficulties in implementing drainage projects, the lack of cooperation from unthankful peasants, and the lack of government money and compulsory implementation measures. Indeed, hydraulic engineers were often more successful at developing successful rhetorical claims about technical efficacy and professional expertise than in carrying out concrete projects.

To the same degree that the expertise of these new professionals was recognized, the knowledge of laymen was devalued. The history of amelioration measures initially was characterized by very daring improvement proposals made by laymen with knowledge of hydraulics and an intimate understanding of the local area. Their practical skills came into play during early reclamation efforts, especially when authorities failed or inadequacies of hydraulic measures in the past became clear. As one critic of hydraulic engineering noted, "Nature is showing us that the Havel is a very sick river that has also suffered a great deal due to the many hydraulic doctors who exacerbated the river's poor condition."[74] It was very easy, however, for engineers to dismiss the proposals of the laymen as "sanguine" and to overwhelm them with technical jargon.[75] The Potsdam High Official Wendler, who was involved in such disputes between local elites and outside experts, concluded that "this is a conflict between empiricists who obtain their knowledge from experience, and technical-minded people who derive their knowledge from the sciences. The latter have left the battlefield and hidden behind a large shield and won."[76]

Wendler's summary is a good description of how events developed. In the small world of the heaths and moors of Westphalia and Brandenburg, one can detect the great confrontation between an innovative and aspiring social group and the inertia of circumstances that resulted from centuries of balancing economic and ecological conditions. Often, the proponents of faith in progress forgot to consider the high social costs of their plans for pushing through modernity. In contrast to America, there was no "Wild West" to conquer in the densely populated Prussian regions.[77] Even extensively used areas played an important role in the economy of rural societies.[78] But as soon as land amelioration engineers became entrenched administratively and in the public consciousness, it became almost impossible to shake the monopoly of this profession over rural land use development. Social problems related to water resources and decisions about access to such resources were indeed solved in the following decades by intensified efforts to manage the water supply. As the critics feared, almost every hydrological scheme resulted in unintended consequences, which could not always be alleviated through technological fixes. In particular, the number of wetlands shrank to the degree that interventions in river systems increased—a trend that is currently being partially reversed through further technological interventions.

Hydraulic engineering has thus imposed constraints that have continued into the present. It was not until end of the nineteenth century that another well-educated and committed social group dared to defy the engineers' visions of complete water control—bourgeois environmentalists. For instance, the Berlin art professor Ernst Rudorff complained in a seminal 1880 paper that land improvement involved "turning the checkered, graceful land into a barren, closely cropped, regularly quartered pattern of maps as much as possible.... Creeks, which have the bad habit of meandering around, have to deign to run straight in ditches."[79] By this time, the national debate on water resources and their use had been heavily influenced by the progressive vision of a would-be elite of technicians. It therefore took the environmentalists tremendous effort to plant their vision of a "natural" balance between men and water into the public mind—but that is another story.

Notes

1. Rita Gudermann, "'Mitbesitz an Gottes Erde—Die ökologischen Folgen der Gemeinheitsteilungen," *Jahrbuch für Wirtschaftsgeschichte* 2 (2000): 17–42; and by the same author, *Morastwelt und Paradies. Ökonomie und Ökologie in der Landwirtschaft am Beispiel der Meliorationen in Westfalen und Brandenburg, 1830–1880* (Paderborn: Schöningh, 2000).
2. Rasch Wurffbain, *An die Herren Grundeigenthümer des Bühlenbrinks, der Bocker-Heide*

und des Lipper Bruchs (Wiedenbrück, 1849), 9ff., in Staatsarchiv Münster (STAMS), Oberpräsidium 1662, vol. 2, 99ff.
3. V. v. Domaszewski, *Das Wasser als Quelle der Verwüstung und des Reichtums. Nach der Natur geschildert* (Vienna: Waldheim, 1879), 5.
4. K. Arnd, *Die Gewässer und der Wasserbau der Binnenlande in naturwissenschaftlicher, technischer und staatswirthschaftlicher Beziehung* (Hanau: König, 1831), 3–4.
5. Klaus Kern, *Grundlagen naturnaher Gewässergestaltung. Geomorphologische Entwicklung von Fließgewässern* (Berlin: Springer, 1994), 111, 116–118.
6. On Rhine regulation, see Christoph Bernhardt, "Flussbau und Überschwemmungen am Rhein im 19. Jahrhundert" (unpublished manuscript, Berlin, 1995); Mark Cioc, *The Rhine: An Eco-Biography, 1815–2000* (Seattle: University of Washington Press, 2000).
7. V. v. Domaszewski, *Das Wasser als Quelle*, 24–26.
8. Arnd, *Die Gewässer und der Wasserbau*, 1.
9. H. von Pechmann, *Praktische Anleitung zum Flussbaue*, vol. 1 (Munich: Lindauer, 1825–1826), 1.
10. "Überschwemmung," in *Meyers Großes Conversations-Lexicon für die gebildeten Stände*, vol. 12 (Hildburghausen, 1853): 1038–1041, here 1039.
11. On the century-old topoi of the threat of the four elements see Gernot Böhme and Hartmut Böhme, *Feuer, Wasser, Erde, Luft. Eine Kulturgeschichte der Elemente* (Munich: C. H. Beck, 1996), 269–275.
12. F. C. Schubert, *Landwirthschaftlicher Wasserbau. Handbuch für Land- und Forstwirthe, Cultur- und Bautechniker* (Berlin: Wiegand, Hempel, Parey, 1879), 62.
13. V. v. Domaszewski, *Das Wasser als Quelle*, 5–6.
14. H.-L. Dienel, *Herrschaft über die Natur? Naturvorstellungen deutscher Ingenieure 1871–1914* (Stuttgart: Verlag für Geschichte der Naturwissenschaften und der Technik, 1992), 181.
15. For a more detailed discussion of these aesthetic views, see Gudermann, *Morastwelt*, 155–162.
16. Dienel, *Herrschaft*, 184–185.
17. Max Popp, "Welche Bedeutung besitzt die landwirtschaftliche Ödlandkultur in Deutschland, und was hat die Naturdenkmalpflege von ihr zu erwarten?," *Beiträge zur Naturdenkmalpflege* 10, no. 1 (1924): 16–34, here 30–33.
18. Raymond Dominick, *The Environmental Movement in Germany. Prophets and Pioneers, 1871–1971* (Bloomington: Indiana University Press, 1992), 4.
19. Arnd, *Gewässer*, 227.
20. G. Hagen, *Handbuch der Wasserbaukunst*, vol. 2 (Königsberg/Pr.: Bornträger, 1841–1865), 339–340.
21. J. Schlichting, *Zur Schiffbarmachung der Flüsse. Kritische Beleuchtung der Schrift "Regulirung oder Canalisirung der deutschen Flüsse"* (Berlin: Ernst und Korn, 1876), 4.
22. Ibid., 5.
23. Charles E. McClelland, *The German Experience of Professionalization. Modern Learned Professions and Their Organizations from the Early Nineteenth Century to the Hitler Era* (New York: Cambridge University Press, 1991), 65–70, 91–94.
24. McClelland, *Experience*, 65–70, and Kees Gispen, *New Profession, Old Order Engineers and German Society, 1815–1914* (Cambridge: Cambridge University Press, 1989). See also R. A. Buchanan, "Gentlemen Engineers: The Making of a Profession," *Victorian Studies* 26 (1983): 407–429; Eckhart Bolenz, *Vom Baubeamten zum freiberuflichen Architekten. Technische Berufe im Bauwesen (Preußen/Deutschland) 1799–1931* (Frankfurt am Main: Peter Lang, 1991), 45–64 and 107–131.
25. Domaszewski, *Wasser als Quelle*, 6.

26. A. Dieck, *Die naturwidrige Wasserwirthschaft der Neuzeit. Ein Mahnruf* (Wiesbaden, 1879), iii.
27. Susan C. Anderson and Bruce H. Tabb, eds., *Water, Leisure and Culture: European Historical Perspectives* (Oxford: Berg, 2002), and Susan C. Anderson and Bruce Tabb, eds., *Water, Culture, and Politics in Germany and the American West* (New York: Peter Lang, 2000).
28. STAMS, Oberpräsidium 1662, vol. 1, *Schreiben* of 22 June 1834, Bl. 2ff.
29. Letter to the ministry of the interior and the ministry of finances, 18 December 1847, GSTA, I. HA, Rep. 87 F, MLw 4507 (M). About the reclamation engineer shortage see "Bericht über die in den Jahren 1855 und 1856 im Preußischen Staate ausgeführten Drainirungen, nebst Verzeichniss der Draintechniker," *Annalen der Landwirthschaft in den Königlich Preußischen Staaten* 16, no. 32 (1858): 249–261, here 252–253.
30. Letter from the Oberpräsident Vincke to Arnim v. Bodelschwingh, 24 April 1844, GSTA, I. HA, Rep. 87 F, MLw 4507 (M). Letter from Vincke to the ministry of the interior and the ministry of finance, 18 December 1847, GSTA, I. HA, Rep. 87 F, MLw 4507 (M).
31. E. Kloepfer, *Fest-Schrift zur Feier des 100jährigen Bestehens der früher Großherzoglich Hessischen später Königlich Preußischen Landes-Kultur-Gesellschaft für den Regierungsbezirk Arnsberg* (Hagen, 1909), 68.
32. STAMS, Oberpräsidium 1662, vol. 1, Schreiben from 22 June 1834, Bl. 2ff. See also Wurffbain, *Nachrichten über Landes-Meliorationen, insbesondere über die Melioration der Boker-Heide in der Provinz Westfalen durch Ent- und Bewässerung* (Berlin: Ernst und Korn, 1856).
33. *Rechenschafts-Bericht über die Fortschritte und den Zustand der Melioration der Bockerheide, vermittelst Anlage eines 4 1/4 Meilen langen Kanals von der Lippe bei Neuhaus bis wieder in die Lippe unterhalb Lippstadt, Entwässerung und Bewässerung der in dem Verbande der Meliorations-Sozietät der Bockerheide liegenden Grundstücke. Vom 1. November 1850 bis Ende Februar 1853* (Paderborn, 1853), 7.
34. Wurffbain, report, 5 June 1852, GSTA, I. HA, Rep. 87 F, MLw 4509 (M), Bl. 147ff.; Wurffbain, Report in the name of the Baukommission, 12 January 1851, GSTA, I. HA, Rep. 87 F, MLw 4508 (M), Bl. 245f.; Wurffbain report, 1 March 1852, GSTA, I. HA, Rep. 87 F, MLw 4509 (M), Bl. 101ff.
35. Letter from Wurffbain to the Minden District Government, 5 June 1852, GSTA, I. HA, Rep. 87 F, MLw 4509 (M), Bl. 147ff., here Bl. 149.
36. Schmidt, *Therapie*, 177.
37. Letter of Wurffbain to the Minden District Government, 5 June 1852, 149.
38. Wurffbain in his report on the amelioration of the Bokel-Mastholter-Niederung, 1 August 1852, quoted in R. Michaelis, "Die Melioration der Bokeler- und Mastholter-Niederung," *Annalen der Landwirthschaft in den Königlich Preußischen Staaten* 21, no. 42 (1863): 153–169, here 161.
39. Wurffbain, letter, 20 June 1856, STAMS, LA Warendorf 348; Letter of the Potsdam District Government to Hagen, 24 July 1832, BLHA, Pr.Br. Rep. 6 B, LA WH 833; Wurffbain in his report on the amelioration of the Bokel-Mastholter Niederung, 1 August 1852, quoted in *Denkschrift aus Anlaß des 75jährigen Bestehens der Sozietät zur Regulierung der Gewässer in der Bokeler und Mastholter Niederung (Bokel-Mastholter-Sozietät) am 11. Juni 1930* (Rietberg: Vahle, 1930), 4. For the Brandenburg example, see O. Roeder, *Die Meliorationen im Havellande. Bericht, erstattet an Sr. Excellenz. dem Königl. Preussischen Minister für die landwirthschaftlichen Angelegenheiten, Herrn Dr. Friedenthal* (Berlin: Wiegandt, Hempel, Parey, 1878), 35.

40. On the multiple uses the peasants made of the barren land see Gudermann, "Mitbesitz."
41. GSTA, I. HA. Rep. 87 F, MLw 4509, Bl. 149. Wurffbain in his report on the amelioration of the Bokel-Mastholter Niederung, 1 August 1852, quoted in Michaelis, "Melioration," 161. See also A. Reinecke, *Der Standpunkt des Wiesenbaues und Vorschläge über die bei Wiesen-Meliorationen zu befolgenden Grundsätze* (Lippstadt: Staats, 1870), 5–6.
42. Wurffbain, report, 1 March 1852, GSTA, I. HA, Rep. 87 F, MLw 4509 (M), Bl. 108–109.
43. Ibid., Bl. 110.
44. Wurffbain, report, 29 October 1852, GSTA, I. HA, Rep. 87 F, MLw 4509 (M), Bl. 232.
45. G. M. Kletke, *Die Rechtsverhältnisse der Landes-Kultur-Genossenschaften in Preußen. Nach den Entscheidungen und Verordnungen der höchsten Spruch- und Verwaltungs-Behörden bearbeitet* (Berlin, 1870), 1. See also Roeder, *Meliorationen*, 60.
46. Wurffbain, "Die Melioration des Münsterlandes. Mit einer hydrographisch-geognostischen Uebersichts-Karte," *Archiv für Landeskunde der Preußischen Monarchie* 2 (1858): 305–367, 339.
47. Wurffbain, "Melioration," 342.
48. Wurffbain, "Melioration," 349.
49. Report of the Oberpräsident, 6 May 1852, GSTA, I. HA, Rep. 87 F, MLw 4509 (M), Bl. 124ff.
50. Letter of the Minister of Agriculture to the Oberpräsident, 6 March 1849, GSTA, I. HA, Rep. 87 F, MLw 4507 (M).
51. Letter of Wurffbain to Minden Municipal Government, 23 August 1852, in Staatsarchiv Detmold (hereafter STAD), Reg. Minden, M 1 III E 46, Bl. 242f.; Petition of the bookkeeper Arnold, 26 September 1858, STAD, Reg. Minden, M 1 III E 48, Bl. 259f.; Letter from Wurffbain to the Oberpräsident, 29 August 1834, STAMS, OP 1662, vol. 1, Bl. 16f.
52. STAD, Reg. Minden, M 1 III E 46, Wurffbain, Letter to Minden Municipal Government, 15 July 1850, B. 74f. See also GSTA, I. HA, Rep. 87 F, MLw 4507 (M); STAD, Reg. Minden, M 1 III E 46, STAD, Reg. Minden, M 1 III E 47 passim, and STAD, Reg. Minden, M 1 III E 46, proof of payments to Wurffbain from Minden Municipal Government, 29 April 1851, fol. 102.
53. Wurffbain to Minden Municipal Government, 15 July 1850, in STAD, Reg. Minden, M 1 III E 46, 74f.
54. Letter from 23 August 1852, STAD, Reg. Minden, M 1 III E 46, 243.
55. Letter to Oberpräsident, 11 December 1851, GSTA, I. HA, Rep. 87 F, MLw 4509 (M), 97f.
56. Wurffbain's list from November 1855, STAMS, Reg. Münster 1582, 257ff.
57. Wurffbain an MLDF, 21 August 1850, GSTA, I. HA, Rep. 87 F, MLw 4508 (M), 232.
58. Wurffbain to Wehrmann, 26 January 1851, GSTA, I. HA, Rep. 87 F, MLw 4508 (M), 235.
59. Wurffbain, report, 1 March 1852, GSTA, I. HA, Rep. 87 F, MLw 4509 (M), 196f.
60. Wurffbain to Wehrmann, 6 February 1850, GSTA, I. HA, Rep. 87 F, MLw 4508 (M).
61. Wurffbain, report, 1 March 1852, 196f.
62. Quoted in Kloepfer, *Fest-Schrift*, 68.
63. Reinecke, *Standpunkt*, 10f.
64. Letter to the Ministry of Agriculture, 25 May 1883, GSTA, I. HA, Rep. 87 F, MLw 3774 (M).

65. Wurffbain, *Nachrichten*, Sp. 46.
66. Wurffbain in his report on the amelioration of the Bokel-Mastholter-Niederung, 1 August 1852, quoted in Michaelis, "Melioration," 162.
67. Wurffbain, "Melioration," 364 and 235–236.
68. Ibid., 357.
69. Wurffbain, report, 29 October 1852, GSTA, I. HA, Rep. 87 F, MLw 4509 (M), Bl. 232.
70. Wurffbain, "Melioration," 367.
71. L. Brenken, "Die Boker Haide," in *Die Entwickelung der Landeskultur in der Provinz Westfalen im 19. Jahrhundert*, ed. E. Haselhoff and H. Breme (Münster: Der Westfale, 1900), 83–94. See also Keller, *Weser*, 130.
72. Brenken, "Boker Haide," 80f.
73. Joachim Radkau, *Natur und Macht. Eine Weltgeschichte der Umwelt* (Munich: C. H. Beck, 2000), 107–114.
74. C. F. Wendler, *Die Havel-Überschwemmungen und deren Verhütung. Denkschrift* (Potsdam: author's publication, 1855), 5.
75. Berring, report, 22 February 1837, BLHA, Pr.Br. Rep. 2 A, Reg. Potsdam I LW 1364, 2ff.
76. Wendler, *Havel-Überschwemmungen*, 3.
77. Anne Vileisis, *Discovering the Unknown Landscape: A History of America's Wetlands* (Washington, D.C.: Island Press, 1997).
78. Martina de Moor, Leigh Shaw-Taylor, and Paul Warde, eds., *The Management of Common Land in North West Europe, c. 1500–1850*, Comparative Rural History of the North Sea Area Series, no. 8 (Turnhout: Brepols Publishers, 2002).
79. Ernst Rudorff, "Ueber das Verhältniß des modernen Lebens zur Natur," *Preussische Jahrbücher* 45 (1880): 262–276, here 262.

Chapter 3

A Sylvan People

Wilhelmine Forestry and the Forest as a Symbol of Germandom

MICHAEL IMORT

ONE OF THE MORE flattering stereotypes about Germans is that they have a special relationship with "their" forest. This affinity for the forest, often called *Waldgesinnung, Waldbewußtsein*, or by similar terms best translated as "forest-mindedness," supposedly makes Germans more likely to visit the forest and be particularly concerned about its health and aesthetics. Today, this alleged affinity is mainly interpreted as a cultural construction nourished by nostalgic idealizations of the forest in literature, art, and music that continue to form an important part of the German educational canon.[1] During the first half of the twentieth century, however, German public discourses were replete with ethnic or *völkisch* interpretations that presented forest-mindedness not as a learned cultural pattern, but as a national characteristic of Germans that was supposedly the result of two thousand years of coevolution between forest and people. In other words, forest-mindedness was presented as a defining racial element of Germandom. Völkisch authors in particular seized upon this idea and developed a normative analogy in which the ecological organization of the forest served as a blueprint for the political organization of a "New Germany." Among German foresters, the most popular vehicle for this völkisch argument was the "sustainable forestry" *(Dauerwald)* idea, a silvicultural concept originally conceived as the first "ecological" challenge to the tenets of scientific forestry. Attracted by its völkisch connotations, the Nazis soon seized upon the idea of sustainable forestry and mandated it as forest policy for the entire Reich. After the end of the Nazi regime, however, the association of ecologically aware sustainable forestry with völkisch ideology and Nazi policy had the ironic effect of delaying the successful implementation of ecologically aware forestry methods in Germany until the 1990s. In the first part of this essay, I examine the cultural construction of the forest into the national landscape of Germandom and the role foresters played in the subsequent völkisch interpretation of the forest ecosystem as a political analogy for the German nation. In the second part, I show how the shadow of this völkisch interpretation of the forest has affected the ecological transformation of the actual forest landscape of Germany over the course of the twentieth century.

The Construction of the Forest as a Symbol of Germandom

Foreign visitors are frequently surprised to find that the famous Black Forest is not the sublime sylvan wilderness they imagined it to be. How can it be, they ask, that the supposedly forest-minded Germans do not see the glaring contradiction in celebrating the Black Forest as their quintessential Nature when in fact it is an enormous, rigidly patterned spruce plantation dissected by well-maintained gravel roads and trails on which throngs of hikers stroll from the car park to the nearest inn and back? The answer is that Germans resolve this cognitive dissonance by blending two conceptualizations of the forest, one emphasizing its vestigial wildness, the other its evident orderliness.

On the one hand, Germans visit and celebrate the Black Forest and the forest in general because it is the closest approximation to pristine nature that their country has to offer after centuries of intensive landscape modification. For Germans, then, the forest more than any other domestic landscape manifests nature. Moreover, the forest is a landscape fairly devoid of indicators of industrialization and technology, offering Germans a spatial and temporal escape hatch to an idyllic and nostalgic Germany of yore. On the other hand, Germans celebrate the evident order, straightness, and tidiness of the managed forest landscape as an embodiment of "typically German" characteristics. In this view, nature has not only been improved but also made recognizably German by the imposition of "German order": the symmetric rows of trees of equal dimensions and flawless appearance are seen as visible expressions of industriousness, neatness, and other supposed national characteristics of the German people.

While these conceptualizations may appear contradictory in that they celebrate the evidently dissimilar categories of primeval and "silvi-cultured" forest, both ultimately converge to signify the forest as German. Thus, when asked about their "relationship" with the forest, individual Germans are often eclectic in their blending of aspects from both conceptualizations, yet most will concur that the forest—wild or ordered—represents the quintessential German landscape.[2] In contrast to foreign observers, Germans are thus not necessarily perturbed by the apparent tension between conceptualizing one and the same forest as German wilderness and German order. The reason for this lies in the long-established parallel tradition that both ideas of the forest have in German culture, one being rooted in Romantic aesthetics, the other in scientific forestry. Both traditions take us back to the nineteenth century.

Writing Roots for the German Nation: Early Nineteenth-Century Intellectuals and the Germanization of the Forest

The construction of the forest as a symbol of Germandom began late in the eighteenth century when German writers of the Storm and Stress (*Sturm und Drang*) movement and the early Romantic school aestheticized the forest into a positive topos. Since Classical times, the forest had been associated with the forces of evil: the realm of darkness, sorcery, and beasts, it was feared as a moral vacuum that defied the progress made by human civilization.[3] The Romantics celebrated the forest precisely because it was such an asymmetric, disharmonious, and non-civilized landscape, a place where humans could commune with God by experiencing His sublime nature. In their works, early Romantic writers such as Friedrich Hölderlin, Ludwig Tieck, Novalis, Wilhelm Heinrich Wackenroder, and the Schlegel brothers August Wilhelm and Friedrich thus re-scripted wilderness into sublime landscape, turning the forest from the antithesis of Enlightenment into the epitome of Romanticism.

Yet this reevaluation of the forest was more than just another reorientation of aesthetics; it was also a deliberate political project to unify the German-speaking people against the occupation of their Heimat by Napoleon. Seizing upon the poetic Romanticization of the forest, intellectuals such as the publicists Ernst Moritz Arndt and Friedrich Ludwig Jahn, the playwright Heinrich von Kleist, and the painters Caspar David Friedrich and Georg Friedrich Kersting, politicized the forest into the core of a nascent German national identity. These intellectuals effectively Germanized the forest by reworking two descriptions of the Germanic forest and its inhabitants by the first-century Roman historian Tacitus.

In his *Germania*, Tacitus idealized the "barbarian" Germanic tribes as fierce and freedom-loving so as to highlight what he saw as the servility and degeneracy of his fellow Romans. The connection between forest life and vigor that Tacitus suggested led the Romantics to celebrate the forest as the ancestral home of the German people and as the source of strength for their future resurgence.[4] The second motif was the Germanic chieftain Arminius (18 BC–AD 19) as described by Tacitus in his *Annals*. Hermann, as he is known to Germans, defeated the Roman army in a battle in the Teutoburg Forest of northwestern Germany. The *Hermannsschlacht* (Hermann's Battle) effectively ended Roman imperial ambitions in northern Germany. In the long term, it also prevented the Latinization of most of Germany and established the Rhine as the linguistic frontier between Romance and Germanic cultures—a border that had been violated by Napoleon's invasion of Germany. For the Romantic nationalist project of inspiring Germans to unite and rise against Napoleon, Hermann thus represented a prime example of Germanic heroism against foreign domination, just as the primeval Germanic forest that had brought forth such a hero was a timeless preserve of "Germanness."[5] In their scholarly, artistic, and literary works, the Romantics popularized this nationalist project by representing the forest as the spiritual conduit between a heroic Germanic past and future united Germany.

For instance, philologists such as Herder and the Brothers Grimm argued that the similarity of forest folklore across the German lands was proof of a common Germanic past that united all German speakers. At the same time, painters such as Friedrich and Kersting used trees and forests as symbols of Germany and so gave visual currency to the idea that the forest represented Germanness. Meanwhile, writers such as Friedrich Gottlieb Klopstock and Heinrich von Kleist celebrated Hermann's victory as an inspiration for Germans to think of themselves as a nation and to repeat his feat, this time against the Napoleonic armies. Finally, looking beyond liberation, publicists such as Arndt and Jahn argued that only an extensive "Germanic" forest cover could protect the emerging unified German nation from future French incursions. No one, however, was more skillful and effective at expressing the idyll of the German forest than Joseph von Eichendorff, the "poet of the German forest," whose works are part of the school curriculum to this day.[6] Writing forest-laden poetry and novellas through the entire first half of the nineteenth century, Eichendorff created the definitive map of the emotive topography of the German forest. He also inspired composers such as Franz Schubert and Robert Schumann (as well as Ernst Rudorff, the founder of the Heimat protection movement) to turn his poems into songs and piano pieces. Composers such as Albert Lortzing, Carl Maria von Weber, Johannes Brahms, and Richard Wagner further animated the forest as setting and actor in their operas, while their songs and piano pieces carried "sough of the forest," or *Waldweben*, into the salons of the emerging bourgeoisie.[7] By the middle of the nineteenth century, the German forest had come to stand for fervent love of the fatherland, for mythical ideas of a primeval Germanic freedom rooted in natural law, for liberation from foreign dominance and struggle for national unity, for heroism, vigor, and manliness. It was portrayed as soulful, melancholic, and brooding, but also as vigorous, primeval, and indomitable—and those qualities were also ascribed to Germans as national characteristics: after all, were trees and humans not rooted in the same soil?

Naturally, not everyone agreed with these generalizations. Karl Marx, for example, scoffed that "goodhearted enthusiasts . . . of German blood and liberal thinking are seeking the history of our freedom . . . in the primeval Teutonic forests. But what distinguishes the history of our freedom from that of the wild boar if it is to be found only in the forest?"[8] Yet even such ridicule only underlines how prevalent a theme the Germanized forest already was. After the founding of the Kaiserreich in 1871, the state itself recognized the symbolic potential of the Germanized forest and put it to work toward the creation of an official myth of origin for the German nation. Just as the Brothers Grimm had hoped, the forest proved suitable for providing a unifying history that transcended regional differences and emphasized cultural commonalities between the provinces of the German empire. And once again, Hermann and his forest battle emerged as a bracketing element in this official German forest history. In 1875, Kaiser Wilhelm I unveiled the first national monument of the Kaiserreich: a gigantic statue of Hermann on a wooded upland ridge in the Teutoburg Forest. The bronze colossus of twenty-seven meters faced west, whence the Roman legions had come and where Germany's "arch enemy" France was looming beyond the Rhine.

Meanwhile, interest in the forest as the presumed site of Germanic history was stirring in the scholarly community as well, with numerous scholars attempting to reconstruct the forest life of the Germanic tribes through the study of Germanic language, custom, and history.[9] Building on the cultural Germanization of the forest established by poets, painters, and composers, academics now tried to do their part by seeking evidence for the claim that the forest had determined German national characteristics and shaped German history. One of the most influential figures in establishing this new nexus between the forest, nationalism, and social science was the sociologist, folklorist, and novelist Wilhelm Heinrich Riehl (1823–1897). In his multivolume *The Natural History of the German People as a Foundation for German Social Policy, 1851–1869* (*Naturgeschichte des Volkes als Grundlage einer deutschen Sozialpolitik*), Riehl argued that the character of the German people had been determined by the environment they lived in: the primeval German forest had brought forth an equally primeval stock of hardy and resourceful Germans. In his own day, he continued, the "sociopolitical value" of the forest was greater than ever as it was the only remaining environment where the national character of the German people was not constantly eroded by agricultural satiety and urban comfort, making the forest a "truly magnificent preserve of our most unique national custom." From those "remains of Germanic freedom in the forest [*Waldfreiheit*] which seem so out of place amidst our modern conditions, a profound influence emanates on custom and character of all classes of the nation." In his history of nature, as in his novellas and novels, Riehl frequently juxtaposed field and forest as "tame" and "wild" forms of land use, which were needed to complement one another: intensive agriculture, industrialization, and progress were necessary to keep abreast with other modern nations, but it was from the forest and its primeval population that Germany again and again received the strength and character that made and sustained it as a nation: "A people must die without the hardy backwoodsmen who constantly replenish the natural, raw power of its national character."[10] In a similar vein, the geographer Friedrich Ratzel wrote in his *Anthropogeographie* of 1891 about "forest peoples," or *Waldvölker,* as living in such close interconnection with the forest that "the nature of the forest interlaced with their entire being." Among them Ratzel counted the Germanic tribes who had "emerged directly from the forest onto the stage of history."[11] Both Riehl and Ratzel thus looked at the forest from an environmental determinist perspective and stressed the social-hygienic benefits of the forest for the health of the body politic of the German nation. In this view, the German nation had a sound racial foundation in its forest past, while its continued exposure to the bracing forest environment kept it from succumbing to the softening influences of civilization.

By contrast, other authors emphasized the military importance of the forest for the defense of the nation against its external enemies. Harking back to the defeat of the Roman legions in the dense Germanic forest, these authors depicted the forest in military diction. During the Wars of Liberation, Jahn and Arndt had already demanded a "forest shield," or *Bannwald,* be planted on the densely populated French border. Arndt had even warned that, "without its forest, Germany will be no more."[12] In 1844, the Prussian forester Friedrich Wilhelm Frömbling took this idea further by suggesting Germany be guarded from within through a network of self-sufficient militia villages dispersed across an impenetrable "defensive forest," or *Wehrwald.* Quoting the example of Hermann's battle, Frömbling recommended this "forest fortification" as a "territorial defense suited to the German national character" as it built on the hardiness of German forest-dwellers and their familiarity with their surroundings.[13] This, Frömbling argued, would give the defenders a great advantage over any foreign attackers who, like the Romans two thousand years earlier, were unaccustomed to the forest environment. This argument reemerged during the First World War and again in the 1930s, when foresters posited that the forest landscape was a vital battlefield ally for the German armies.[14] During the nineteenth century, however, Frömbling was one of the few foresters who ventured to comment on the military, political, or cultural importance of the forest for the German nation. Indeed, some foresters insisted that "the customary effusiveness about the forest be left to poets and aesthetes" altogether.[15] Foresters thus kept aloof from the debate while patriotic intellectuals used the forest to rally anti-French sentiments, invent a heroic Germanic past, construct a unified German cultural history, and symbolize political unity. They presented the Germanized forest as the first and last line of defense of a "rooted" Germanic culture and identity against foreign domination and the softening influence of "civilization." As such, the forest was seen not only as a symbol of Germany, but increasingly as code for the alleged superiority of Germandom. The stage was set for the racialization of the German forest into the völkisch forest. This time, however, foresters would be among the vanguard.

The Forest as Educator: Wilhelmine Foresters and the Construction of the Völkisch *Forest*

For the most part, the early nineteenth-century aestheticization and mythification of the forest as the quintessential German nature took place in the notional rather than the material forest. That is to say, the imaginary forest of Romantic poetry and music often bore little resemblance to the real forest landscape: for the construction of the Romantic German forest it was neither necessary nor desirable to point to a corresponding material landscape. By contrast, the early twentieth-century construction of the völkisch German forest relied heavily on imagery, both positive and negative, taken from a new forest landscape that existed in actuality: the *Schlagwald* (age-class forest), created and maintained by professional foresters trained in scientific forestry. Beginning in the late eighteenth century, German territorial rulers had fostered the formulation of scientific forestry at hastily established academies and university departments. Simultaneously, they founded state forest services that soon implemented the new forestry methods in the state-owned forests. In the long run, though, a new regime of strict state supervision of private forest owners and standardized training for all foresters at state-run institutions would ensure that public and private forests alike would be managed according to the tenets of state-sanctioned scientific forestry.

Traditionally, German forest historians have seen this development as the rational reaction of the authorities to a timber shortage caused by overexploitation, which in turn resulted from population growth and industrialization. More recent interpretations argue that the territorial rulers manufactured the specter of a wood shortage in an attempt to expand state power and increase their share of revenue from the increasingly valuable forest.[16] Whether real, feared, or outright invented, the effect of the discourse about the much discussed timber shortage was that it gave the princes leverage to turn the forest from a communally owned resource into a private and unencumbered property.[17] More important, the property reform was the necessary first step toward the imposition of a new comprehensive forest management scheme that was based on scientific forestry.

Beginning around the year 1800, then, the appearance and composition of the German forest began to change according to the prescriptions of scientific forestry.[18] The first step was to clear and replant the understocked forests with frugal yet fast-growing species such as Norway spruce (*Picea abies*) and Scotch pine (*Pinus sylvestris*). The new plantations were then managed according to a rotation plan that used staggered clear cuts to ensure a continuous and predictable supply of wood in perpetuity: in effect, scientific forestry pioneered the concept of sustainability in resource management.[19] The rapid success of German foresters in ordering the German forest landscape found worldwide recognition, making Germany the undisputed leader in the new science of forestry.

Yet the new silviculture pioneered in Germany was very much rooted in the mechanistic worldview of the Enlightenment: symmetry and uniformity underpinned the new forestry, leading to even-aged, monocultural plantations where uneven-aged, mixed forests had once grown.[20] In the second half of the nineteenth century, the ecological consequences of the extensive monocultures became painfully evident: the ratio of one-third coniferous to two-thirds deciduous trees had been reversed; the concomitant reduction in species diversity had impoverished the forest flora and fauna; repeated planting of conifers had acidified the forest soils; and the resulting uniform stands possessed little resilience to insects, storms, and fire. The consequence was a wave of large-scale stand failures that not only threatened the ecological balance in the affected regions, but also flooded the market with low-grade timber and so depressed prices. Simultaneously, many among both foresters and the general population began to resent the uninspiring aesthetics of the uniform timber rows of scientific forestry.[21]

By the mid-nineteenth century, some German foresters began to suggest alternative "back-to-nature" models of forestry that called for a return to uneven-aged, mixed stands, as well as a balance between the exigencies of timber production and ecological and aesthetic considerations.[22] By the 1880s, the lines were drawn for a vigorous debate between "scientific" and "back-to-nature" forces in German forestry. Superficially, the subject matter of the debate was the ideal structure of the forest and how it should be managed, yet some foresters soon extended the discussion to the question of what sort of larger worldview was informing "scientific" forestry on the one hand and "back-to-nature" forestry on the other. Scientific foresters described the forest mainly in mathematical terms of volume and value, with some even insisting that "the purpose of the forest is primarily to provide income to its owner. . . . Everything else is incidental."[23] Appalled by such "materialistic" diction, back-to-nature foresters championed what they termed "idealistic" and "German" goals such as organicism, aesthetics, harmony, and permanence. Their idea of a harmonious and organic *Mischwald* or mixed, permanent, and uneven-aged forest without clearcuts became a clarion call for those disaffected with the "non-nature" (*Unnatur*) of the modern German forest monocultures.[24]

Before long, some back-to-nature foresters began to think about possible links between the structure of the German forest and that of German society. Emil Adolf Roßmäßler, one of the founding fathers of the back-to-nature forest movement, wrote,

> The forest in particular is a manifestation of social organization in the plant kingdom that is more pronounced in Germany than in the warmer countries of Europe. Not only do the proud trees withdraw from the company of lower classes of plants by gathering and huddling in the forest. They also segregate from one other as conifers remain separate from deciduous trees. Why, spruce even stays away from pine, beech from oak.

Detecting in this a "distinct simile" of human society, Roßmäßler continued:

> It would provide us with surprising entertainment if we were to compare the mutually exclusive social associations of Germans with those of German plant life. I will leave it to my readers, however, to find among the various human congregations the respective counterparts of the self-sufficient beech forest, the noble oak forest shielding its inferiors, or the plebeian willow thicket by the banks of the river.[25]

Moreover, in the eyes of other back-to-nature foresters, the forest not only passively reflected the stratified structure of German society, but also exerted a "furthering influence on consciousness and character of the *Volk*."[26] Because of this "hygienically beneficial influence of the natural forest on body and soul," they argued, "the preservation of the German forest has arisen as a question of national importance"[27] —a view evident in Riehl's writings. And similar to Riehl, back-to-nature foresters argued that it was only the "authentic"

German forest that could act as a counterforce to the loss of identity, or *Vermassung*, in German society. This meant that the German forest needed to be preserved not as a carpet of spruce or pine, but in all its former variety. The task of foresters, then, was to restore the forest to its authentic, natural state not merely for the forest's sake, but for the good of the country: a "truly German" forest was necessary to safeguard the health of the German body politic as the country went through the convulsions of industrialization.

Evidently, many of the ideas promoted by back-to-nature foresters overlapped with those prevalent among the various life reform or *Lebensreform* movements of the German educated middle classes (*Bildungsbürgertum*) of the late nineteenth and early twentieth centuries, such as anthroposophy, organic farming, vegetarianism, and Heimat protection. It would appear that back-to-nature foresters shared with these movements what Sieferle has called the "anti-industrial affect of Romanticizing politics," condemning the "mechanistic" phenomena of urbanization, technological progress, liberal economics, and the secularization of society as causing the cultural disintegration of the German nation.[28] To counter this perceived decline, the back-to-nature movements wanted German life in all of its aspects to return to its former supposedly "organic" ways, just as back-to-nature foresters advocated a return to an organic forest that was authentic, aesthetic, and ecologically healthy. Moreover, both movements were fighting a losing battle in that they resisted changes in the human and natural environment that were accepted or even welcomed as signs of modernization by a majority of the population. As a dissenting minority, if a very vocal one, back-to-nature activists could publish their outrage and so stoke the intellectual debate, but do little to prevent those changes in practice beyond their personal sphere of influence. For example, only a handful of forest estate owners switched to back-to-nature forestry and no state forest service implemented it. This left back-to-nature forestry to be practiced at the local district level, where individual proponents were able to devise back-to-nature silvicultural management plans, at times even clandestinely and in violation of declared policies. Similar to the human-centered life reform (*Lebensreform*) movement, back-to-nature forestry thus led a fringe existence that was publicized well beyond its practical importance.

In view of these similarities, can back-to-nature forestry be seen as another expression of the cultural pessimism and antimodernism that is said to have pervaded the educated German middle class and particularly its back-to-nature movements at the turn of the twentieth century? Or should we follow the recent argument that, rather than being a one-dimensional expression of regressive antimodernism, the back-to-nature movements actually represented a forward-looking attempt to co-opt or at least influence the inevitable advent of modernization and so temper its effects on the people and landscape of Germany? Thomas Rohkrämer suggests that some groups attempted to work toward an "other modernity" where technology would either be controlled by a strict code of civic ethics or harnessed for the purposes of achieving a more humane world. Similarly, William Rollins argues that the primarily aestheticizing argumentation of the Heimat protection movement did not arise out of a myopic and irrational Romanticism, but rather out of the (very rational) calculation that the German public was best mobilized for goals of Heimat protection by alerting them to the ubiquitous decline in their aesthetic and emotional experience of nature.[29]

Within the contours of this late nineteenth-century discourse about nature and society, most of the publishing back-to-nature foresters were inspired by a mission to save not only the German forest, but also Germandom. While there were some who saw the "failure" of scientific forestry as corroborating the virtues of socialism and internationalism, most took a conservative, even völkisch stance and depicted age-class forestry as the flawed outcome of an equally flawed process of modernization. In their eyes, back-to-nature forestry should re-create an "authentic" German forest that could be a natural counterweight or even antidote to the landscape modification wrought by industrialization and urbanization. As we will see shortly, some went even further and suggested this authentic forest serve as a blueprint for an organicist and völkisch Germany. These normative analogies of forest and society foreshadowed the eventual appropriation of the forest for Nazi propaganda.[30] Before we examine some examples of writings by back-to-nature foresters and how they were fashioned into "commonsense" arguments for a völkisch state that proved so useful to the Nazis, we should briefly consider some of the reasons why Wilhelmine foresters tended to lean to the political right and cultural pessimism.

The socialization of most Wilhelmine foresters, both scientific and back-to-nature, resembled that of other academically trained civil servants in Germany, but it also involved some stages that were unique and so helped form a distinctive and firmly conservative esprit de corps among foresters.[31] Frequently born into a family of "forestry tradition," the aspiring forester would attend a humanistic *Gymnasium* and then move on to the forestry academy, where he would become a member of an exclusive academic fraternity of foresters. Before and during his studies, the young forester had to fulfill twelve years of mandatory army service in one of the elite ranger units created specifically for foresters by Frederick the Great in 1740. Upon graduation and discharge from active service, he would become a civil servant in one of the state forest services (which offered the great majority of positions) as part of a military-style hierarchy complete with uniforms and rank insignia. Passing from one institution to the next, the aspiring forester would learn not only technical knowledge but also the professional, social, and political codes of conduct that produced, in the words of the longtime president of the National Association of German Foresters, a distinct *"homo foresticus"*: conservative, obedient, state-minded (*staatstreu*), and elitist.³² Hence, even if Wilhelmine forestry was not quite "the entrenched command-economy-paramilitary complex"³³ of one recent critic, the integration of forestry, military, and conservative social institutions was certainly intense and more often than not encouraged, if not outright required, a conservative ethos.

In fact, the conservative ethos of foresters can be shown to have continued through much of the twentieth century. For example, in his history of the forestry academy in Hannover-Münden during the Weimar period, Peter-Christoph Schleifenbaum has identified a strong "monarchistic, nationalistic conviction" among forestry students that expressed itself in a level of resistance against reform initiatives that was "unsurpassed at any other Prussian university."³⁴ Another illustration of the traditionally conservative attitude of professional foresters is the fact that all ten foresters ever elected to Prussian or German parliaments were members of conservative or right-wing parties.³⁵ Finally, with reference to foresters in the Federal Republic, Werner Pleschberger has discerned a "forestry ideology" or a way of thinking that lends itself to an "ideological exploitation of the forest for regressive bourgeois thinking determined by a passion for order."³⁶ It is only in the last twenty years that the mainly conservative ethos of foresters has been somewhat tempered by the increasing numbers of students who are leaning toward the Green movement and whose main motive for studying forestry is "the environment."

In comparison to many members of nature conservation and life reform groups, then, foresters' professional ethos and political affiliations during the Wilhelmine period led many of the publishing back-to-nature foresters to take a conservative stance in their writings that frequently revealed a distinct tone of cultural pessimism. In fact, many of the most widely read writers even took an outright völkisch perspective, building their argument on such authors such as the "founder" of the völkisch movement, Paul de Lagarde (1827–1891), and his "prophet" Julius Langbehn (1851–1907).³⁷ Both Lagarde and Langbehn shared with Riehl an emphasis on the role of the forest landscape in the formation of the national character. For example, in his *Deutsche Schriften* (*German Writings*) of 1878, Lagarde exhorted Germans to heed the voices of their collective past that emanated from the forest: to hear "the beeches and oaks speak ... of the German faith" once again, Germans had to find their way back to the forest and the land, because it was there that "the manliness of the nation grows quietly like the beech in the forest."[38] Similarly, Langbehn, in his hugely successful *Rembrandt as Educator* (*Rembrandt als Erzieher*) of 1890 blamed the perceived social and cultural disintegration of Germany on the "soullessness" of contemporary education against which he held up Rembrandt as an exemplar of "healthy German strength of soul" that was "anchored in a strong sense of rootedness and soulfulness."[39]

Langbehn's views in particular were taken up by völkisch back-to-nature foresters and reworked into the vision of a future Germany whose social structure was to be patterned after the structure of what these foresters saw as the authentic, natural German forest. In their eyes, modern scientific forestry was altering the German forest for the worse, just like the process of modernization was weakening the body politic of Germany. By returning the German forest to its "natural" state, these foresters hoped to make an important contribution to returning the German body politic to its former integrity, for to them the forest was both the source and the result of unique völkisch strength. If politicians understood and managed German society like a "natural" forest, the argument concluded, a Germany would reemerge that was just like that forest: strong because of its functional unity, resilient because of its structural diversity, and productive because of its cooperative hierarchy. And just like the "natural" forest was more ecologically stable and productive in the long term compared to its "scientific" counterpart, so a Germany that was organized according to this model would eventually outlast its more industrialized neighbors.

One of the first foresters to develop this völkisch analogy between German forest and German society in detail was Rudolf Düesberg (1856–1926), an avid proponent of back-to-nature forestry who had published several articles on the subject in the 1890s. In 1910, he published *The Forest as Educator* (*Der Wald als Erzieher*), a book that took its title and tone directly from Langbehn's *Rembrandt als Erzieher*. Like Langbehn, Düesberg called for the re-education of a German nation he saw in cultural and social decline. Instead of Rembrandt's art, however, Düesberg proposed the German forest as the "educator," suggesting that "the laws governing the structure of the forest apply equally to a rationally-ordered human society."[40] Düesberg argued that the German forest and the German people were a product of the same Heimat: "deeply rooted, sedentary, risen to greatness in the struggle with rough climate and through hard work on poor soil. Hence, the social order of the forest . . . forms a model for those institutions necessary for the strengthening of Germanness. In this manner, the forest can become the educator of the German *Volk*."[41]

Düesberg cautioned that the age-class forest of scientific forestry was not suitable for instructing Germans as to their proper social order: age-class forestry violated "the nature of the forest" because it no longer allowed trees of different species and ages to grow together, but separated them into pure stands of distinct age-classes. The resulting stands could not make complete and efficient use of the incoming sunlight and thus were less productive. Nor did they enjoy the formative phenological influence the different members of a mixed collective had on one another: instead of becoming a differentiated community, they grew into a homogenized mass. Düesberg detected a "desperate similarity" between the formative conditions obtaining in even-aged forest monocultures and those in a "utopian social-democratic state," where "louts always flourish at the expense of nobler characters" because they were not checked and formed properly by the community.[42]

Why were German foresters practicing scientific forestry then? Because, Düesberg held, modern foresters had allowed themselves to be misguided by the way of thinking of "homeless nomads"—a Jewish conspiracy was infiltrating German forest administration, causing righteous German foresters to stray from the path. To Düesberg, Germany and its forests were losing their cooperative Germanic character and becoming dominated by the "foreign worldview of the nomad."[43] By contrast, he saw back-to-nature forestry methods as based on the "rootedness of the German race in the soil," which "ran in their blood" and conferred upon Germans a "special appreciation for land and soil, for Heimat and Fatherland."[44] Consequently, so Düesberg, the German forest, economy, and people should all be managed according to the same management principles taken from back-to-nature forestry: German tree species, companies, and citizens alike should be fostered, their foreign counterparts excised. Next, the thus rarefied collectives should be reorganized after the model of the traditional German cooperative system: a forest that was uneven-aged, mixed, and selectively cut based on biocoenotic principles (*Gruppen-Plenterwald*);[45] a state that was corporatist in structure and governed through occupational estates (*berufsständischer Staat*); and a nation that was organized as an ethnically pure "national community" (*Volksgemeinschaft*). Eventually, this cooperative restructuring would help Germany achieve not only political sovereignty and economic self-sufficiency but also völkisch domination. For, once so organized, the forest would not only be healthier and more aesthetic, but would also yield an inexhaustible supply of timber and other resources. Likewise, the corporatist German state not only could keep Social-Democratic anarchy at bay, but its economy also would become strong enough to have Germany's neighbors do its bidding. Finally, the collective will of the national community would push Germany's borders to the East, forcing the expulsion of the Polish people to a reservation in "one of the bastard states of South America."[46]

Düesberg thus used the forest to exemplify the idea that the community, not the individual, was the basic unit of German society. At the individual level, trees served as placeholders in which Germans were to recognize themselves. At the collective level, the forest community became a simile of the national community. At the political-economical level, the dichotomy of "Germanic" forestry versus "Semitic" husbandry was the base line for drawing racial boundaries between "us" and "them": idealistic Germans who were rooted in the soil and lived cooperatively so as to further the common good as opposed to materialist Semitic nomads who roamed the land in their capitalist pursuit of personal profit. Finally, Düesberg regarded the Slavic peoples as so inconsequential that he reduced them to an object: without further justification, they could simply be removed to make room for the Germans.

In hindsight, Düesberg's 1910 analogy reads like an ominous vision of the Third Reich—and it was partly such hindsight that led both the völkisch movement and the Nazis to use his interpretation of the forest to buttress their ideologies. Thus, Düesberg's *Der Wald als Erzieher* was probably less of a forward-looking inspiration for the völkisch movement than a retrograde justification for the "New Germany": building on the affinity of the German people with the forest, the Nazis pointed to the established "commonsense" analogy between forest and nation to "prove" that National Socialism was the "natural" form of government for Germany. This process had its beginnings during the Weimar period, when a variety of authors, both foresters and nonforesters, reworked the völkisch analogy between forest and nation into a poignant critique of the democratic state. Some authors stated matter-of-factly that "it is the way to the national community that the forest shows to those who are willing to look."[47] Others would simply "remind" their readers of the "obvious" parallels between canopy layers and social stratification, between care of the forest edge and care of the ethnic German borderlands, between natural reseeding and eugenics, and between the forest organism and the organism of the Volk.[48] Still others used the forest for rather elaborate anti-Semitic parables:

> Once there were strangers that immigrated to a well-ordered state. They were of a foreign race, shrewd, and given to a life without work. They were neither enough of a warrior-people, nor numerous enough to be able to conquer; but they were gifted and always on the lookout for an opportunity to appropriate for themselves what others had produced. With qualities like that, they had no choice but to become parasites.
>
> And so the strangers became parasites in the greater states. While the others worked the soil, created assets and practiced the ancient

> trades in customary harmony, the newcomers effortlessly feathered their own nests. They secured a foothold, began to sap, and grew ample and rich. Everything about them was foreign: their livelihood, their morals, their garb, their customs. So foreign, indeed, that they did not seem to be from this world. God Himself had chosen them, people said, and reserved a special fate for them.
>
> But the old established inhabitants who were rooted firmly in the soil defended themselves against the strangers. An unprecedented time of suffering began with a thousand persecutions and just as many counter-ruses, and it continues to this day.
>
> One spoiled life for the other. But why continue this story—by now everyone has guessed what is meant here. Who would not be thinking of mistletoe and forest trees?[49]

Once again, the "Jewish conspiracy" was portrayed as threatening harmonious German social life just as the harmonious forest was strangled by parasites. The tongue-in-cheek suggestion that "everyone has guessed what is meant here" indicates how established the analogy of forest and Volk was by the 1920s already. In 1929, the professor of forestry Franz von Mammen wrote a National Socialist sequel to *Der Wald als Erzieher* that appeared in an enlarged version in 1934. Mammen wanted "to enable everyone who lives and acts within the National Uprising to find out all that the forest can teach us, particularly at this juncture . . . I sincerely hope that my simple contribution will do its part towards making tree and forest the educator of the German Volk in the Third Reich."[50] With such a clear premise, it is not surprising that Mammen's book was but one continuous sermon about the völkisch importance of the German forest. As one reviewer noted, Mammen had assembled "everything" written about "the ethical and aesthetic importance of the forest to Germandom."[51] In chapter 1, "World Views Fighting in the Forest from Cradle to Grave," we find subheadings such as "Identity and Diversity," "Socialism and Individualism," and "Struggle and Harmony." Chapter 3 deals with "Parties in the Forest," chapter 4 with "Foreigners in Forest and Volk," and chapter 5 with "Harmony and Unity in Forest and Volk." In these chapters, the incantations of the völkisch forest as educator spanned all aspects of life in the New Germany: from "racial vigor" to "genetic improvement"; from "duty and sacrifice" to "*Führer*-principle" and "will to power"; from "rootedness of the strong" to "weeding out of the weak and sick"; from "education on the basis of biology" to the "struggle for existence" between the "degenerate" (*Entartete*) and "the strong"; from "common good before personal profit" to "service to the community"; and from "father's house" to "mother's breast."[52]

Time and again Mammen pointed to the forest and individual tree species to share "their lessons for man and Volk": there were spruce, fir, and pine as the "peasants and labourers" of the forest. "Thinly dispersed" among those were ash, linden, elm, and maple as the aesthetes who were allowed to contribute beauty and grace to the working life of the "comrades" (*Volksgenossen*). Towering above all were the "hard and noble oak trees" as the "leaders of the Volk," victoriously fighting off the "burrowing wasps and worms" of the obsolete parliament and its parties.[53] In these parallels, the boundaries between forest and nation were dissolved: the forest and the Volk were finally one, ostensibly functioning according to the same eternal laws.

Thus, the beginning of this conflation of forest and nation was the nineteenth-century professional debate among foresters about the relative virtues and shortcomings of scientific forestry and back-to-nature forestry. Soon the debate transcended the discipline and entered the sociopolitical and socioeconomic domains, eventually producing interpretations of the forest as an analogy of the "New Germany." While the ideas of back-to-nature foresters found considerable resonance on paper, scientific forestry remained unchallenged as the practical paradigm of German forestry. This situation changed radically in 1933, however, when the Nazis mandated a comprehensive back-to-nature forestry concept as forest policy for the entire Reich. For the first time, back-to-nature forestry, along with its by now substantial ideological connotations, was poised to make the leap from the notional to the material forest and so change the actual forest landscape of Germany.

The Dauerwald Concept and the Ecological Transformation of the German Forest Landscape

In 1920, Professor Alfred Möller of the Prussian forestry academy at Eberswalde consolidated the various critiques of scientific forestry into a complex new silvicultural school of thought called *Dauerwald* or sustainable forestry.[54] The primary idea behind sustainable forest silviculture was to manage the whole "forest organism" rather than individual trees, which meant never cutting the entire stand, but removing selected trees on a continual basis.[55] The intention was to create and maintain stands that were continuous-cover, mixed, uneven-aged, selectively cut, naturally regenerated, multilayered—in short, a diverse forest structure that was more stable in ecological and economic terms and produced the maximum sustainable yield in terms of quality and value rather than mere volume of wood.

Soon after its publication, Möller's idea of the sustainable forest was embraced enthusiastically by those German foresters who saw Germany's defeat in the First World War and the economic depression of the early 1920s as an opportunity to reform the declining forestry sector according to the tenets of back-to-nature forestry. However, as the economy and hence the timber prices improved with the implementation of the Dawes Plan, the temptation to cut entire stands proved hard to resist for cash-starved forest owners. Support for the sustainable forestry idea seemed to dissipate almost as quickly as it had formed—until the incoming Nazi government surprisingly decided to found its forest policy on the concept of sustainable forestry.

This decision was partly the result of the eminently völkisch affinity that Hermann Göring, the second man in the Nazi state, felt for the forest. A passionate hunter and outdoors enthusiast, Göring had been introduced to the idea of sustainable forestry in 1931 during a hunting visit to the forest estate of Walter von Keudell, a retired Prussian Minister of the Interior and prominent practitioner of sustainable forestry. When Göring took personal control of the entire German forestry sector in 1934, Keudell became his deputy minister in forestry affairs. Within months, several laws and decrees required German forest owners to manage their forests according to the principles of back-to-nature forestry in its sustainable forestry incarnation.[56] Among other things, forest owners could no longer cut conifer stands under fifty years of age or clear-cut more than a fraction of the stand. Moreover, they had to cut in short intervals of three years, and always select the lesser trees so as to allow only the best specimen to reseed. Finally, forest owners were obliged to introduce more broadleaf species and promote a mixed species composition and uneven age-structure. The new rules contravened several tenets of scientific forestry, while promoting an approach to forestry that answered almost all of the demands of back-to-nature foresters. This approach is also rather similar to "close-to-nature forestry," or *naturnahe Waldwirtschaft*, practiced since the 1990s by all German state forest services. Ironically, then, we might conclude that it was the Nazis who pioneered the application of ecologically aware forestry in Germany. But is this assessment correct? Did the regime truly implement ecological concepts in the forest, or did they merely use the notion of sustainable forestry as a propagandistic smokescreen for war preparations? The answer, as is so often the case, lies somewhere in between.

The regime's preference for the sustainable forestry approach was not based primarily on ecological reasoning. Rather, the Nazis found the völkisch overtone of the sustainable forestry idea very promising for their own propaganda as it allowed them to point to the forest as a "natural" model for their vision of the German national community: like the sustainable forest, the national community was supposedly an eternal collective in which the individual worked toward the greater good but was ultimately dispensable. In a programmatic speech to the national convention of German foresters in 1936 that was cited and reprinted ad nauseam, Reichforstmeister Göring expounded at length on the "obvious" analogy between sustainable forestry and the National Socialist idea of the indigenous German community:

> Forest and people are much akin in the doctrines of National Socialism. The people is [sic] also a living community, a great, organic, eternal body whose members are the individual citizens. Only by the complete subjection of the individual to the service of the whole can the perpetuity of the community be assured. *Eternal forest and eternal nation are ideas that are indissolubly linked.*[57]

Evidently, the concept of sustainable forestry held great propagandistic potential for the Nazi regime and it was for this reason rather than their ecological benefits that the policies of sustainable forestry were never officially revoked, even though it soon became obvious that the booming German economy and the beginning war preparations demanded more wood than could ever be harvested under the limitations on clear cutting imposed by the Dauerwald model. The regime solved this dilemma by keeping the regulations on the books while issuing a string of decrees that weakened them in practice. For example, as early as September 1935, Göring increased the mandatory cutting quota in all publicly owned forests to 150 percent of the sustainable yield; only one year later, all private forests were subjected to the same quota. When Keudell refused to oversee this piecemeal dismantling of the sustainable forestry policy, he was swiftly replaced with the solicitor Friedrich Alpers, a party stalwart and SS officer. On 1 December 1937, the new Generalforstmeister issued a comprehensive decree that relaxed the limitations on clear cutting (to satisfy the lumber demand) while retaining the main ecological principles of the idea of sustainable forestry: foresters were still obligated to establish a mixed, site-adapted, and uneven-aged forest wherever possible. To differentiate the new policy from the stricter concept of sustainable forestry, it was relabeled "close-to-nature" forestry (*naturgemäße Waldwirtschaft*). For the remaining years of the Third Reich, the new policy committed German foresters to supply lumber and other raw materials for the war effort with little regard for the integrity of the forest.

In view of this obvious tension between pretense and reality, what was the ecological legacy of the era of sustainable forestry? Forest historian Heinrich Rubner for one speaks of a "crisis of sustained-yield forestry" during the Third Reich, charging that the regime used the *Dauerwald* label to veil dangerous levels of overcutting.[58] There can be no doubt that the German forest was being overexploited even as the regime was trumpeting its commitment to sustainable forestry to the world, yet I would argue that this policy left a long-term legacy for the German forest that was ecologically beneficial. Even though they must be called mainly symbolic, the regulations for sustainable forestry influenced German forestry in general and the future forest laws of the Federal Republic in particular in that they literally forced German foresters at all levels to finally consider the forest ecosystem rather than the trees as the basic management unit of forestry. This opened a chance for the tenets of back-to-nature forestry to enter the mainstream of forestry and thus challenge the historical dominance of age-class forestry.[59] In fact, some German forest historians have called the era of sustainable forestry under Keudell the "halcyon years of German forestry" during which "progressive" laws were passed.[60] After 1945, the back-to-nature ideas enshrined in these regulations were either retained (in the Federal Republic, some laws remained in effect until they were superseded by the Federal Forest Law of 1975) or incorporated into new legislation. In part, this continuity is due to the fact that the same senior administrators were involved in forestry legislation before and after 1945; on the other hand, after almost a century of literary discussion, many German foresters were finally ready to accept some of the ecological and economic merits of back-to-nature forestry and see its basic principles inform mainstream German forestry—so long as the term *Dauerwald* itself was not used.[61]

Such trepidation was the result of the odious association of the concept of sustainable forestry with völkisch forest-mindedness that the "eternal" propaganda of the Nazi regime had created. While foresters now embraced various back-to-nature ideas such as mixed, site-adapted, and uneven-aged stands, they were intransigent in their opposition against a mandatory policy of sustainable forestry. Undaunted, twenty-one sustainable forestry proponents formed a "Working Group for Close-to-Nature Forestry" in 1951 that tried to relaunch ideas of sustainable forestry under the name of close-to-nature forestry (*naturgemäße Waldwirtschaft*), but the label of "Nazi-forestry" stuck and the movement remained marginal.[62] In the forestry literature, back-to-nature foresters sighed that they were saddled with "the close-to-nature legacy of the Thousand Year Reich," that colleagues insisted on "pinning the ideological mark of Cain" on them, and that the *Dauerwald* term represented a "weighty burden."[63] This burden was certainly not lightened when East Germany promulgated a variation of sustainable forestry as its forest dogma in 1951. This furthered the suspicion of the mainstream that sustainable forestry was a tool of authoritarian regimes that should be shunned in democratic West Germany—an opinion promptly surfacing in West German propaganda against the "Sovietized Forest" issued by the Ministry for All-German Affairs in Bonn.[64] In short, Dauerwald had become "a word no-one dared to touch anymore."⁶⁵ Meanwhile, the Dauerwald concept continued to thrive in Switzerland, for example, where the movement was not sunk by the albatross of Nazi association.⁶⁶

Within West German forestry, we thus find the peculiar situation that the ideas of the Dauerwald lived on in various disguises but were not allowed to be called by their proper name. Moreover, while the forest laws made programmatic statements about the ecological transformation of the German forest, the forest administrations did not necessarily implement this vision. It took the catastrophic effects of acid rain in the 1980s and windstorms in the early 1990s before back-to-nature forestry was truly integrated into the forest policies of the German state forest services. The devastation suffered by the age-class monocultures stood in stark contrast to the relatively light damage observed in the mixed, uneven-aged forests of close-to-nature forestry. Beginning with the Saarland in 1987, German state forest administrations one by one changed their official silvicultural policy to close-to-nature forestry. After almost 150 years of debate, back-to-nature forestry had officially arrived even in the land of its birth.

Ironically, no one seemed more surprised about this sudden success than back-to-nature foresters themselves.⁶⁷ In their flagship publication *Der Dauerwald*, one forester voiced the movement's

> utter surprise how quickly close-to-nature forestry has found such broad acceptance. Although the Working Group has certainly fostered this development, it is ultimately not the merit of individual persons, institutions, or events. Instead, it is the result of the emergence of a broad-based ecological awareness, which, aided by the media, has risen to unprecedented heights.⁶⁸

The traditional affinity with the forest, it appears, is alive and well in Germany after all.

Notes

1. For critical appreciations of the phenomenon of German affinity with the forest, see Bernd Weyergraf, ed., *Waldungen—Die Deutschen und ihr Wald* (Berlin: Nicolaische Verlagsbuchhandlung, 1987); Albrecht Lehmann, *Von Menschen und Bäumen: Die Deutschen und ihr Wald* (Reinbek: Rowohlt, 1999); Annette Braun, *Wahrnehmung von Wald und Natur* (Opladen: Leske und Budrich, 2000); Albrecht Lehmann and Klaus Schriewer, eds., *Der Wald—ein deutscher Mythos? Perspektiven eines Kulturthemas* (Berlin: Dietrich Reimer, 2000); Albrecht Lehmann, "Waldbewußtsein. Zur Analyse eines Kulturthemas in der Gegenwart," *Forstwissenschaftliches Centralblatt* 120 (2001): 38–49.
2. See the qualitative empirical study in Lehmann, *Von Menschen und Bäumen*, as well as Braun, *Wahrnehmung von Wald*.

3. For reviews of the forest in the public imagination since Classical times, see Clarence J. Glacken, *Traces on the Rhodian Shore* (Berkeley: University of California Press, 1967); Roland Bechmann, *Trees and Man—The Forest in the Middle Ages* (New York: Paragon House, 1990); Robert Pogue Harrison, *Forests—The Shadow of Civilization* (Chicago: University of Chicago Press, 1992).
4. They did not, however, pay equal attention to the assessment by another first-century Roman writer, the satirist Juvenal (between AD 58 and 67–127), who had ridiculed the Germanic tribes as "acorn-burping"; see Hubertus Fischer, "Dichter-Wald—Zeitsprünge durch Sylvanien," in *Waldungen*, ed. Weyergraf, 13–25. The attraction of the *Germania* as a literary "proof" for the alleged forest heritage of Germans persisted well into the twentieth century. As Simon Schama relates in vivid terms, in 1943 the Nazis committed an SS-unit in Italy to forcefully "repatriate" the oldest surviving manuscript copy of the *Germania* because of the promise held by its beginning words: "*de origine et situ germanorum.*" See Simon Schama, *Landscape and Memory* (Toronto: Random House of Canada, 1995), 75–81.
5. According to Richard Kühnemund, Hermann's alleged exclamations "Rather death than slavery!" and "In unity there lies strength!" appeared over 130 times in German literature between 1500 and 1945. See Richard Kühnemund, *Arminius, or the Rise of a National Symbol in Literature* (Chapel Hill: University of North Carolina Press, 1953), xiii.
6. Heinrich Rubner, *Idealism and Forestry in Germany in the Time of Goethe (1756–1832)*. Proceedings of the XVII IUFRO Congress (Kyoto, 1981), 306.
7. For treatments of the forest theme in German art, literature, and music, see H. Schmidt, "Das Waldthema in der Musik," *Forstwissenschaftliches Centralblatt* 74 (1955): 219–235; Hermann Gusovius, "Wald und Kunst. Der Beitrag des Waldes zur Motivgeschichte in Malerei, Musik und Dichtung," *Forstarchiv* 28 (1957): 53–57; H. Galli, "Der Wald in der Musik," *Schweizerische Zeitschrift für Forstwesen* 115 (1964): 771–772; Hans Leibundgut, "Musik und Wald," *Schweizerische Zeitschrift für Forstwesen* 115 (1964): 731–759; Gottfried Schmiedel, "Der Wald in der Musik," in *Wald. Landeskultur und Gesellschaft*, ed. H. Thomasius (Jena: S. Fischer, 1978), 393–402; Elmar Budde, "Der Wald in der Musik des 19. Jahrhunderts—eine historische Skizze," in Weyergraf, *Waldungen*, 47–61; Helmut Schmidt-Vogt, *Musik und Wald* (Freiburg: Rombach, 1996). Regrettably, the most recent English-language volume on the nexus between music and German national identity does not cover this phenomenon. See Celia Applegate and Pamela Potter, eds., *Music and German National Identity* (Chicago: University of Chicago Press, 2002).
8. *A German Ideology* (1856) quoted in Fischer, *Dichter-Wald*, 23.
9. For example by the Grimm-pupil Wilhelm Mannhardt, *Wald- und Feldkulte der Germanen und ihrer Nachbarstämme* (Darmstadt: Wissenschaftliche Verlagsanstalt, 1875–1877).
10. Quoted in Friedrich Bülow, "Wilhelm Heinrich Riehl und der deutsche Wald," *Raumforschung und Raumordnung* 2 (1938): 547. For a discussion of Riehl's interpretation of landscape, see George L. Mosse, *The Crisis of German Ideology: Intellectual Origins of the Third Reich* (New York: Grosset and Dunlap, 1964), 19–24. For an abridged English version of Riehl's work, see David J. Diephouse, ed., *The Natural History of the German People* (Lewiston, N.Y.: Edwin Mellen Press, 1990).
11. Friedrich Ratzel, *Anthropogeographie* (Stuttgart: J. Engelhorns Nachfolger, 1882/1891), 313–314.
12. Ernst Moritz Arndt, *Ein Wort über die Pflegung und Erhaltung der Forsten und der Bauern im Sinne einer höheren d. h. menschlichen Gesetzgebung* (Schleswig: Königliches Taubstummen Institut, 1820), 71.

13. Friedrich Wilhelm Frömbling, *Deutschland's künftige Vertheidigung mit der Nationalwaffe. Die Feld- und Waldfortifikation für Deutschland* (Königsberg: Theodor Theile, 1844), unpaginated preface.
14. See for example Franz von Mammen, *Die Bedeutung des Waldes insbesondere im Kriege* (Dresden: Globus Wissenschaftliche Verlagsgesellschaft, 1916); Franz von Mammen and H. Riedel, *Die Kriegsnutzung des Waldes. Eine Anleitung zur Mobilmachung des Waldes* (Dresden: Globus Wissenschaftliche Verlagsanstalt, 1917); Alfred Willy Boback, *Wald und Landesverteidigung: Eine forstlich-wehrwissenschaftliche Betrachtung* (Berlin: Neudamm-Neumann, 1935); Victor Dieterich, "Beispiele der wehrpolitischen Bedeutung von Forst- und Holzwirtschaft," *Forstwissenschaftliches Centralblatt* 62 (1940): 121–135.
15. Philipp Geyer, *Der Wald im nationalen Wirthschaftsleben* (Leipzig: Duncker und Humblot, 1879), v.
16. See Joachim Radkau and Ingrid Schäfer, *Holz. Ein Rohstoff in der Technikgeschichte* (Reinbek: Rowohlt, 1987); Joachim Radkau, "Wood and Forestry in German History: In Quest of an Environmental Approach," *Environment and History* 2 (1996): 63–76. See also the paper by Joachim Radkau in this volume. James Scott concurs that the emergence of scientific forestry "cannot be understood outside the larger context of the centralized state-making initiatives of the period." See James C. Scott, *Seeing Like a State: How Certain Schemes to Improve the Human Condition Have Failed* (New Haven: Yale University Press, 1998), 12. See also Henry Lowood, "The Calculating Forester: Quantification, Cameral Science, and the Emergence of Scientific Forestry Management in Germany," in *The Quantifying Spirit in the Eighteenth Century*, ed. T. Frangsmyr, J. J. Heilbron and R. E. Rider (Berkeley: University of California Press, 1990), 315–342. For the development of timber prices, particularly the steep increase after 1780, see Heinrich Rubner, *Forstgeschichte im Zeitalter der industriellen Revolution* (Berlin: Duncker und Humblot, 1967), 106.
17. Until the late eighteenth century, medieval usufruct rights entitled peasants and other beneficiaries to quotas of fuel wood, offal wood, and other "by-products" from most publicly and privately owned German forests. Beyond diminishing the ecological ability of the forest to produce the desired large dimensions of saw wood, these rights also represented a legal encumbrance that the territorial rulers were keen to abolish. They did so by buying out each beneficiary's traditional rights in exchange for a small forest allotment. The remaining forest became the absolute and unencumbered property of the state or the ruler.
18. For a review of the literature on the historical geography of the German forest, see W. Schenk, "Preindustrial Forests in Central Europe as Objects of Historical-Geographical Research," in *Methods and Approaches in Forest History*, ed. M. Agnoletti and S. Anderson (Wallingford, UK: CAB International, 2000), 129–138.
19. For an English-language treatment of the principles of the age-class forest and its political and social consequences, see Christoph Ernst, "An Ecological Revolution? The 'Schlagwaldwirtschaft' in Western Germany in the Eighteenth and Nineteenth Centuries," in *European Woods and Forests: Studies in Cultural History*, ed. C. Watkins (Wallingford, UK: CAB International, 1998), 83–92.
20. See Lowood, *The Calculating Forester*, and Scott, *Seeing Like a State*.
21. One example of public resentment that combines political, economic, ecological, and aesthetic considerations comes from *Rheinpreussen*, the territories on both banks of the lower Rhine and in Westphalia that had been awarded to Prussia at the Congress of Vienna in 1815. The local farmers resented the conversion of their traditional beech and oak coppicewoods (which provided fuel wood, stall litter,

and pig feed) into spruce forests (which yielded only industrial timber) as the imposition of a foreign forest culture. In protest, the locals often uprooted the newly planted *Preussenbaum* or Prussian tree, forcing the authorities to call in the army to guard the young trees. See Karl Hasel, *Forstgeschichte* (Hamburg: Paul Parey, 1985), 209.

22. Gottlob König, *Die Waldpflege aus der Natur und Erfahrung neu aufgefasst* (Gotha: Becker'sche Verlags-Buchhandlung, 1849); Emil Adolf Roßmäßler, *Der Wald. Den Freunden und Pflegern des Waldes geschildert* (Leipzig: C. F. Winter, 1860); Karl Gayer, *Der Waldbau* (Berlin: Parey, 1880), and *Der gemischte Wald* (Berlin: Parey, 1886). "Back-to-nature" did not abandon the principles of sustainability and planned management based on nineteenth-century forest science but merely rejected the notion of even-aged monocultures. It promoted a more natural, i.e., mixed and uneven-aged, forest that was managed selectively rather than by clear cutting. In today's diction, "back-to-nature" forestry would be labeled *Waldbau auf natürlicher Grundlage* or *naturnaher Waldbau*, called "natural forest management," "close-to-nature forestry," or "ecoforestry" in English.

23. Professor of Forest Economics Max Endres at the Seventh Annual Meeting of the German Society of Foresters in Danzig in 1906, quoted in Georg Sperber, "Der Umgang mit Wald—eine ethische Disziplin," in *Ökologische Waldwirtschaft. Grundlagen—Aspekte—Beispiele*, ed. H. G. Hatzfeldt (Heidelberg: C. F. Müller, 1994), 49.

24. Rudolf Düesberg, "Wie erwachsen astreine Kiefern?," *Zeitschrift für Forst- und Jagdwesen* 25 (1893): 605.

25. Roßmäßler, *Der Wald*, 101–102.

26. A. Werneburg, "Über den geregelten Plänterbetrieb," *Zeitschrift für Forst- und Jagdwesen* 7 (1875): 441.

27. Franz von Mammen, "Heimatschutz im Walde," in *Heimatschutz in Sachsen*, ed. R. Beck et al. (Leipzig, 1909), 49–50.

28. Rolf Peter Sieferle, *Fortschrittsfeinde? Opposition gegen Technik und Industrie von der Romantik bis zur Gegenwart* (Munich: C. H. Beck, 1984), 61.

29. Thomas Rohkrämer, *Eine andere Moderne? Zivilisationskritik, Natur und Technik in Deutschland 1880–1933* (Paderborn: Schöningh, 1999); William Rollins, *A Greener Vision of Home: Cultural Politics and Environmental Reform in the German Heimatschutz Movement, 1904–1918* (Ann Arbor: University of Michigan Press, 1997).

30. This view is shared by Lehmann, cf. *Von Menschen und Bäumen*, 33–36.

31. See Irene Seling, "Die Leitideen der Dauerwaldbewegung aus sozialhistorischer Sicht," *Forst und Holz* 53 (1998): 728–732 and the ensuing debate over whether Wilhelmine foresters indeed represented a distinct sociological subgroup among politically conservative and nationalistic *Bildungsbürger*: Ernst Ulrich Köpf, "'Bildungsbürger' als Dauerwaldbewegte—ein Ärgernis?," *Allgemeine Forst- und Jagdzeitung* 170 (1999): 1–4; Gerhard Oesten, "Über die geschichtliche Dimension forstökonomischer Forschungsgegenstände," *Allgemeine Forst- und Jagdzeitung* 170 (1999): 5–11.

32. Lorenz Wappes, "Die Stellung des Forstverwaltungsbeamten im Rahmen der Gesamtverwaltung," *Deutscher Forstverein* (1926): 480.

33. Sperber, *Der Umgang mit Wald*, 52.

34. Peter-Christoph Schleifenbaum, "Die Forstliche Hochschule zu Hannoversch Münden, 1922–1939" (Ph.D. diss., Georgia-Augusta-Universität, Göttingen, 1987), 33.

35. Two were from the antidemocratic and antirepublican DNVP, six from the NSDAP, one from the conservative CDU and one from the center-right FDP. See Friedrich

Borkenhagen, *Deutsche Försterchronik* (Strassenhaus: Wirtschafts- und Forstverlag Euting, 1977), 175.
36. Werner Pleschberger, "Forstliche Ideologie: zur Kritik eines unzeitgemässen Weltbildes," *Centralblatt für das gesamte Forstwesen* 98 (1981): 44. See also Peter Glück and Werner Pleschberger, "Das Harmoniedenken in der Forstpolitik," *Allgemeine Forst Zeitung* 37 (1982): 650–655.
37. The terms are Mosse's. For a discussion of Lagarde and Langbehn in the context of the völkisch movement, see chapter 2 in Mosse, *The Crisis of German Ideology*. A more in-depth treatment of their cultural criticism and its extension by Arthur Moeller van den Bruck in his *The Third Reich* (*Das Dritte Reich*, 1922) can be found in Fritz Stern, *The Politics of Cultural Despair: A Study in the Rise of the Germanic Ideology* (Berkeley: University of California Press, 1961).
38. Paul de Lagarde, *Deutsche Schriften* (Göttingen: Dieterich'sche Universitätsbuchhandlung, 1878), 239, 321–322, and 391.
39. Julius Langbehn, *Rembrandt als Erzieher* (Leipzig: C. L. Hirschfeld, 1890; 50th ed., 1922), 37.
40. Rudolf Düesberg, *Der Wald als Erzieher* (Berlin: Parey, 1910), 139.
41. Ibid., iv–v.
42. Ibid., 57.
43. Ibid., 73 and 74.
44. Ibid., 182.
45. A biocoenosis is a community of organisms occupying a uniform habitat. A *Gruppen-Plenterwald* is a mixed, uneven-aged forest in which trees are managed in clusters rather than individually.
46. Düesberg, *Der Wald als Erzieher*, 192.
47. Walther Schoenichen, *Vom grünen Dom. Ein deutsches Wald-Buch* (Munich: Callwey, 1926), 205.
48. Theodor Künkele, "Die außerfachlichen Aufgaben des Forstbeamten," *Silva* 14 (1926): 413. *Grenzlanddeutschtum* refers to ethnic Germans living in regions bordering on Germany.
49. Raoul Heinrich Francé, *Ewiger Wald* (Leipzig: Richard Eckstein Nachf., 1922), 61. By 1930, Francé's book had reached ten editions with approximately 100,000 copies printed.
50. Franz von Mammen, *Der Wald als Erzieher. Eine volkswirtschaftlich-ethische Parallele zwischen Baum und Mensch und zwischen Wald und Volk* (Dresden: Globus Wissenschaftliche Verlagsanstalt, 1934), 3.
51. Lorenz Wappes, Review of "Der Wald als Erzieher," *Der Deutsche Forstwirt* 113 (1935): 750.
52. Mammen, *Der Wald als Erzieher*, 124–125.
53. See Mammen, *Der Wald als Erzieher*, 73, as well as 88–89.
54. Alfred Möller, *Dauerwaldwirtschaft* (Berlin: Julius Springer, 1921) and *Der Dauerwaldgedanke. Sein Sinn und seine Bedeutung* (Berlin: Julius Springer, 1922).
55. Möller's use of the term "forest organism" was criticized immediately by other foresters who suggested he use "community of life" (*Lebensgemeinschaft*) instead. In today's terminology, forest organism is best rendered as ecosystem.
56. Michael Imort, "'Eternal Forest—Eternal Volk': Rhetoric and Reality of National Socialist Forest Policy," in *How Green were the Nazis? Nature, Environment, and Nation in the Third Reich*, ed. F. J. Brüggemeier, Mark Cioc, and Thomas Zeller (Athens: Ohio University Press, 2005).
57. Hermann Göring, "Deutsches Volk—Deutscher Wald," *Zeitschrift für Weltforstwirtschaft* 3 (1935/1936): 656, grammar and emphasis in the original.
58. Heinrich Rubner, "Sustained-Yield Forestry in Europe and its Crisis during the Era

of Nazi Dictatorship," in *History of Sustained-Yield Forestry: A Symposium*, ed. H. K. Steen (n.p.: Forest History Society, 1983), 170–175.
59. Imort, "Eternal Forest—Eternal Volk."
60. Quoted in Heinrich Rubner, *Deutsche Forstgeschichte 1933–1945: Forstwirtschaft, Jagd und Umwelt im NS-Staat*, 2nd ed. (St. Katharinen: Scripta Mercaturae Verlag, 1997), 68. Peter Michael Steinsiek and Zoltán Rozsnyay, "Grundzüge der deutschen Forstgeschichte 1933–1950 unter besonderer Berücksichtigung Niedersachsens," *Aus dem Walde. Mitteilungen aus der Niedersächsischen Landesforstverwaltung* 46 (1994): 18.
61. Zoltán Rozsnyay and Uta Schulte, *Der Reichsforstgesetzentwurf von 1942 und seine Auswirkungen auf die neuere Forstgesetzgebung* (Frankfurt am Main: Sauerländer, 1978).
62. See Wolf Hockenjos, "Naturgemäße Waldwirtschaft als Ideologie. Ursprung und Hintergründe einer Unterstellung," *Der Dauerwald* 10 (1994): 24–33.
63. Wolf Hockenjos, "Forstideologisches aus Baden," *Allgemeine Forst- und Jagdzeitschrift* 166 (1995): 35; Sebastian Freiherr von Rotenhan, "Naturgemäße Waldwirtschaft im Blickpunkt der Forstpolitik," *Allgemeine Forst Zeitung* 46 (1991): 920; Hermann Wobst, "Geschichtliche Entwicklung und gedankliche Grundlagen naturgemäßer Waldwirtschaft," *Forstarchiv* 50 (1979): 24.
64. Anonymous, *Sowjetischer Wald. Raubbau und Unfähigkeit* (Bonn: Bundesministerium für gesamtdeutsche Fragen, 1951). The East German close-to-nature policy was repealed in 1961.
65. E. Röhrig, "Ein Beitrag zur geschichtlichen Entwicklung der naturgemäßen Waldwirtschaft," *Forstarchiv* 50 (1979): 122.
66. Wobst, *Geschichtliche Entwicklung*, 24.
67. This is not to say that the Working Group had not been growing steadily or even rapidly at times, especially during periods of depressed forest revenues—essentially repeating the situation that had favored the initial acceptance of the idea of sustainable forestry in the early 1920s. For example, membership rose from 250 in 1979 to 500 only four years later. See Wobst, *Geschichtliche Entwicklung*, 23, and Hilmar Schoepffer, "Die naturgemäße Waldwirtschaft und ihre Grundsätze—Darstellung der Entwicklung und Erläuterung des Begriffes," *Forstarchiv* 54 (1983): 47.
68. Paul Lang, "Naturgemäße Waldwirtschaft in der Forstpolitik der Bundesländer," *Dauerwald* 4 (1991): 32.

Chapter 4

Forestry and the German Imperial Imagination

Conflicts over Forest Use in German East Africa

THADDEUS SUNSERI

BETWEEN 1904 AND 1914 German officials demarcated thirty forest reserves in the Rufiji-Kilwa region of southeastern Tanzania encompassing some 74,000 hectares, one-tenth the extent of all forest reserves in German East Africa by 1914.[1] These reserves enclosed a one-hundred-mile stretch of coastal mangroves, including the entire delta of the Rufiji River, eastern Africa's biggest drainage system. In addition, some 445,000 hectares of game reserves had been demarcated in Kilwa and Rufiji districts that would eventually grow into the Selous reserve, today the largest wildlife reserve in Africa.[2] In areas where the colonial state did not establish forest and wildlife reserves it encumbered Africans with ordinances that limited access to forests and forest products and that curtailed hunting. These ordinances also regulated African agricultural practices in proximity to forests, and created artificial boundaries between agricultural land and forests that Africans themselves would not have recognized.

Germans entered into East Africa with a history of regulating forests in Germany and in the colonial territories of other European powers, notably those of the British and Dutch in South and Southeast Asia. In these colonial contexts they came to view the production practices of peasants and pastoralists as destructive of forests, and believed that the task of scientific forestry was to protect forests from people. Bound up with this view was a Eurocentric preconception of what constituted a proper forest. While arguably well over half of German East Africa was forested if savanna woodlands were considered, German scientific forestry privileged the closed canopy montane forests that made up less than 1 percent of the landscape, a figure that compared unfavorably to the approximately 26 percent of Germany that was forested.[3] German forest policy concentrated on protecting this 1 percent of forested landscape from Africans (and eventually from European settlers) during the twenty-five years of formal rule over East Africa.

If German policy had focused merely on protecting these parcels an argument might be made that environmentalism propelled German colonial forestry. This was Hans Schabel's conclusion after surveying German forest policy in Tanganyika: "This emphasis on environmental management (*Landeskultur*) in German East Africa was truly visionary and undoubtedly deserves to be considered the single most important legacy of German forestry involvement in East Africa."[4] Along similar lines Gregory Barton recently pronounced colonial environmentalism—Empire Forestry—to be an unequivocal gift of Western civilization to colonized peoples, an "environmental revolution still in the process of saving humans from themselves."[5] Both of these conclusions result from taking the rhetoric of imperial environmentalism too seriously without focusing on local case studies of how colonial forests were managed and how their management intersected with development policies. What imperialists did and what they said they were doing were often widely divergent, and could be obfuscated by the broad tapestry of colonial forest policy. My approach here is to focus on specific forests in one region of German East Africa, the Rufiji River basin, tying them to the social history and political economy of German colonialism. This approach shows that German policy on the ground ran up against the concrete material realities of East Africa.

This on-the-ground reality demonstrates the gulf that existed between metropolitan visions of environmentalism (encapsulated in the Heimat protection movement) and policy making in German East Africa. In William Rollins's critical view of imperial environmentalism, an early German environmental movement was "devoted to the goal of conservation in the German colonies" yet served in a contradictory way to "legitimate Germany's conquest and control of foreign territories."[6] While Rollins is correct that colonial officials sometimes invoked a pastoral vision of the German Heimat as the end-goal of colonial forestry, this discourse was negated by the fiscal and development needs of the colony. Likewise, Roderick Neumann's overview of the genesis of imperial nature protection sees a romantic Western vision of the African landscape as the starting point for the demarcation of national parks that provided "one important vision of what Africa 'should' look like."[7] While this vision was a powerful force in the creation of wildlife reserves by the 1930s in British-ruled Tanganyika, when an international conservation movement emerged as a strong voice in colonial policy, concrete African realities were more important in shaping German forest policy over the long term than were metropolitan ideals. Put in simple terms, forest policy began with a completely un-German and unromantic landscape, the mangrove forests of the East African coastal river estuaries.

Long exploited by East Africans in conjunction with Indian Ocean traders for their timber and fuel, German foresters were quick to bring these forests under imperial control and tinker with notions of scientific forestry, commercial exploitation, and social control. Policy with respect to the coastal mangroves offers a microcosm of German colonial forestry as a whole, one that had no relation to visions of the Heimat or an "Edenic vision of landscape."[8] While the case study here of the Rufiji basin includes the single largest extent of mangrove forests in East Africa, it also offers examples of more German-like montane forests as well as a savanna woodlands landscape called *miombo* that was not even included in the German rhetoric of environmentalism, but that played a significant role in the material limits of colonial rule. We learn as much about imperial environmentalism by looking at the landscapes that Germans did not consider worth preserving as we do examining the forests that were actually reserved for scientific forestry.

The Origins of German Imperial Forestry

The evolution of European states was closely bound to the control of forests. The scientific forestry that emerged in Germany in the eighteenth century was based on the premise that forests were scarce resources that were essential for state revenue.[9] The state should thus regulate forest use and sharply circumscribe rural people's access to forests for fuel, pasture, construction materials, fodder, and game. Officials sought to quantify how much wood was available in a given forest for immediate and future needs since timber was an economic resource important for industrial development. The science of forestry developed as a means of calculating wood quantity so that timber harvesting could mesh with fiscal and economic needs. Because there was a widespread belief that the consumption of wood was outstripping supply, German foresters spearheaded forest regeneration by dividing forests into plots that could be harvested and replanted in long-term rotations. This *Schlagwaldwirtschaft* (age-class forestry), which consisted of rotational planting and harvesting, viewed forests solely as an economic resource, creating uniform, non-diverse forests that could be methodically and easily harvested and monitored. By the mid-eighteenth century scientific opinion pointed to the additional importance of forests for the natural environment, especially

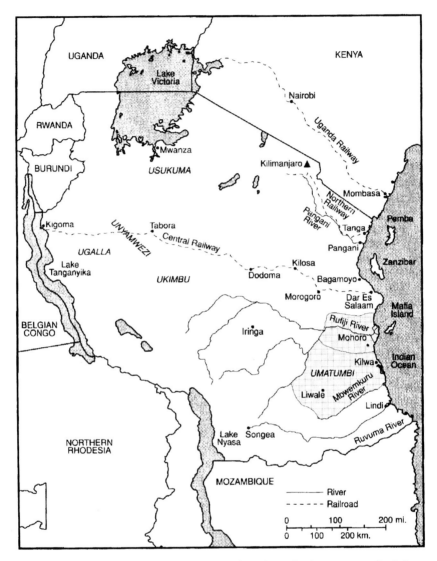

Figure 4-1. Map of German East Africa with Rufiji and Kilwa Districts Shaded

for preserving watersheds and regulating climate. Although this nascent environmentalism took a backseat to economic concerns throughout the nineteenth century, it bolstered the perception that the state needed to control forests for the common good. The assertion of state rights to the forests over the rights of villagers led to ongoing social conflict in Europe that sometimes dovetailed with revolutionary movements.[10] Transgressions of forest laws such as the theft of wood or the use of the forests as a commons were widespread crimes in nineteenth-century Germany.[11]

Long before the advent of Germany's empire in Africa, German foresters studied forest conditions in South and Southeast Asia and brought the knowledge they acquired back to German forestry schools. Germans oversaw Indian forests in the nineteenth and early twentieth centuries, and helped to create the forest departments in British Burma and Dutch-ruled Indonesia.[12] Arriving in Asian forest environments, Germans adapted their traditions of scientific forestry to colonial conditions. While foresters in Asia were confronted with far more diverse forests and landscapes than in Europe, their major goal remained the quantifying of timber for fiscal and commercial exploitation and the curtailing of peasant access to forests. By the turn of the century colonial forest regulations as developed in Asia were applied to German East Africa. These regulations included: a legal structure circumscribing wood cutting through licensing, co-optation of indigenous elites, and use of police powers; the creation of forest reserves that were off-limits to indigenous peoples' subsistence use; and the creation of plantation forests directed at sustained commercial forest management. The Asian experience convinced German officials that peasant agriculture and pastoralism, especially the use of fire to clear bush or woodlands for cultivation or grazing every few years, threatened forests.[13]

Although German foresters arrived in East Africa with an established template for regulating forests, they encountered unique environments and conditions that shaped new policies and the colonial state itself. One key difference between Asian and African colonialism was the population dynamic of German East Africa. Excluding the territories of Ruanda-Urundi in the Great Lakes region, which were virtually closed to labor extraction during German rule, the colony had only four million people.[14] The shortage of labor proved to be an obstacle to scientific forestry and other colonial development projects, including European commercial plantations and settlement, railway construction, and peasant cash crop production. German East Africa furthermore lacked animal power in much of the colony owing to the presence of sleeping sickness and other animal diseases that killed oxen and horses. In contrast, South and Southeast Asia had relatively high population levels and were regions where elephants, oxen, horses, and water buffalo could be used in production.[15] Since the colonial endeavor was premised on making German East Africa into a source of cash crops and industrial raw materials, especially coffee, sisal, rubber, and cotton, policy makers sought to channel African labor away from subsistence-oriented household production. Because forests were an essential part of the peasant subsistence and market economy, as well as havens from warfare and colonial development projects, a central premise of German forest policy in East Africa was to empty forests of people and concentrate them in towns and villages that could be overseen, taxed, and tapped for labor. As in Asia and Europe, state regulation of forests engendered social conflict.

Apart from the dictates of social control and development, colonial forest policy was shaped first by the fiscal needs of the state and secondarily by German views of conservation. While German foresters had a poor understanding of the long-term history of topography in East Africa, they understood the connection between forest cover and soil and climate change.[16] Believing that East African forest cover was at risk from human use, they sought to separate people from forests. German foresters also collected tree specimens and experimented with forest regeneration on trial forest plantations, and introduced Asian tree species into the East African environment. At the same time, the colony's poor balance of payments and dependence on the metropole for subsidies made the economic use of forests an overarching concern. This fiscal concern intersected with changing development paradigms, especially the shift from viewing the colony as suitable for white settlement in the 1890s to a focus on African cash crop production after 1906. Furthermore, as the colony was developed, it became a major consumer of forest products. The capital of Dar es Salaam generated an ongoing demand for timber and fuel that could not be met by the surrounding forests. Railway construction, the imperial flotilla, and such modest industries as commercial beer brewing created an ongoing demand for wood. While much of this demand came from overseas supplies, forest policy sought to substitute Tanganyikan wood for imported wood, and also aimed to compete for export markets.

German Perceptions of African Forest Use and Agriculture

Germans viewed African forests through the lens of their European experience, identifying high montane or coastal forests with at least 80 percent canopy cover as true forests worth conserving, while woodlands with less than 80 percent canopy cover were not considered to be forests. The exceptions to this observation were the coastal mangroves located in river estuaries that had no European counterpart, but which were well known from the German experience in Asia. The roughly 50 percent of the territory classified as miombo woodland was (with just a few exceptions) not demarcated as forest reserves. This meant that in much of German East Africa peasants and pastoralists were not forced off the land by forest policy, though these were marginal lands with sparse populations. However, miombo woodlands were widely targeted as game reserves that were far bigger than forest reserves. Legally Africans were prohibited from residing in game reserves, although the colonial state was too weak to enforce this provision in most cases.

Miombo woodlands in many ways encapsulated the fears that premised German scientific forestry. Germans believed miombo to be remnants of once great East African forests that had been destroyed by primitive African agriculture. Because they harbored tsetse flies that carried sleeping sickness, which in turn obviated a cattle economy (and plow agriculture) in half of the colony, Germans viewed miombo as a threat to their vision of civilization, and would have gladly cut them down completely if they had had the resources to do so. Instead, Germans practiced a policy of retrenchment that attempted to move scattered populations from the woodlands into concentrated villages where they could master the environment in close proximity to dwellings. Hopefully, concentrated settlement would over time create islands free of miombo throughout the colony where development could take place. One observer wrote, "The miombo forests are completely inhabited by countless tsetse flies that only disappear when extensive contiguous cultivated acres are created through the radical clearing of the forests."[17] Practically speaking, the labor power was not sufficient to master miombo, and ironically its extent increased under German rule as peasants sought to escape colonial taxation, mandatory cash cropping, labor compulsion, and myriad other colonial policies by moving away from administrative centers to live in scattered and isolated settlements in the woodlands. Indeed, miombo woodlands are the product of human interaction with the environment that long predated agriculture, when hunters used fire to burn the bush and woodlands in order to promote a grass cover that attracted game.[18] Miombo woodlands developed as fire-resistant species that benefited from fire-induced cracking of seeds that facilitated germination. Peasants who used fire to clear fields for farming therefore had a symbiotic relationship with the miombo environment. In contrast, pastoral economies directly threatened miombo woodlands. The devastation of East African pastoralism in the 1890s through a rinderpest pandemic therefore increased miombo cover and with it expanded the zones of sleeping sickness.[19]

Colonial forest policy and its assumptions about African resource use mirrored the German view of what it meant to be a civilized state. The German agricultural landscape was considered to be "cultivated land" (*Kulturland*) and Germans to be "cultured peoples" (*Kulturmenschen*), both terms implying a direct relationship between civilization and European forms of intensive agriculture.[20] Likewise, German use of forests was sometimes referred to as the "cultivated" or "plantation" forest.[21] In contrast, Africans were frequently referred to as "children of nature," practicing "aboriginal agriculture," and many of their forests were *Urwälder*, "jungles" or aboriginal forests. Yet the supposedly original state of Tanzanian forests did not prevent Germans from believing that African agrarian practices, especially the shifting agriculture that relied on bush burning to open up new fields every few years, had depleted the land of much of its forest cover and required the direction of the colonial state to set things right. While the residual use of fire to clear land for agriculture in parts of Germany was called "cultivated burning," Germans condemned African bush fallowing as "wild burning."[22] An early German forester in East Africa wrote "we have taken over a land in which the forests were devastated and therefore it is all the more necessary that we retain those that are left and protect them as best as we can."[23] Shortly after the turn of the century Governor Götzen wrote to the Foreign Office that African agrarian practices were "hostile to civilization" because they created a scarcity of wood and water and an arid environment antithetical to commercial agriculture. "In light of this situation," he wrote, "I hardly need raise the issue that one of the government's most pressing duties, let alone noblest, is to preserve the forests as well as found new forest stands in denuded parts of the landscape."[24] The first step in this direction was to bring African shifting agriculture under control.

The German condemnation of African shifting agriculture shaped the earliest legislation aimed at forest regulation.[25] Field burning was deemed responsible for a host of ills, including drying out of the land and mountain streams, hindering humus buildup and wood growth, and promoting a thorn bush that facilitated flooding in times of heavy rainfall. By reversing these trends scientific forestry would help replicate a German environment in the colony and thereby facilitate German settlement, which was central to the colonial agenda before 1907.[26] German officials worried about the scarcity of wood for construction near almost all administrative centers, which necessitated that European timber be imported at great expense. Field burning was also considered to be responsible for the destruction of natural products, such as gum copal and wild rubber, which were important export commodities in the late nineteenth century.

The first forest ordinance of 12 December 1893—three years after formal colonial rule had begun—addressed these concerns by prohibiting field burning in a one-to-two-mile radius of government administrative centers and by requiring rural dwellers to cut and rake grass and brush it into piles for burning rather than set fire to large parcels.[27] The ordinance prohibited villagers from burning around the banks of rivers and made rural communities collectively responsible for preventing fires in proscribed areas. Woodcutting for household and commercial use in the vicinity of government stations was also regulated in the 1893 ordinance by requiring a permit and by curtailing wood markets. In 1894 the forest assessor Krüger warned that in light of the growing building activity around the colony's towns, forests were disappearing.[28] The colonial administration therefore co-opted African *majumbe* (village headmen) to act as salaried forest wardens and police in the absence of sufficient state personnel. This gave headmen extraordinary authority over other villagers.[29] In Rufiji district a five-rupee fee was required each year for a woodcutting permit, a price almost double the hut taxes that were first implemented in 1898.[30] This earliest of forest ordinances recognized the weakness of the two-year-old colonial state in most of the colony by targeting only areas around administrative centers where wood for construction was necessary and oversight was feasible. As German East Africa was extended in the 1890s through conquest of the interior, forest laws radiated outward to less accessible regions. By 1898 majumbe were made responsible both for the collection of hut taxes and for maintaining designated forest complexes where field burning and woodcutting were prohibited.[31]

Colonial Forest Management in the Rufiji Region

The Rufiji region was subjected to forest regulations at an early date because of its fiscal importance as the biggest source of mangrove trees in the colony and its well-established commercial links to Zanzibar and other Indian Ocean destinations. The Rufiji is the largest river in East Africa, originating in the southwest highlands and draining an area of some 68,500 square miles.[32] The river separates into eight major arms as it empties into the Indian Ocean, creating a seventy-five-mile-wide delta that is ideal for the growth of mangroves. For one hundred fifty miles inland the Rufiji creates a flood plain that made the region a granary for coastal trade networks by the nineteenth century. Rufiji people, who had diverse origins as migrants and refugees from as far west as the Congo, favored rice production on flood lands, but had adopted maize and cassava as secondary staples alongside millet by the end of the nineteenth century. Stretches of coastal and high forests of the Kichi hills and Matumbi mountains south of the Rufiji and riverine forests along the river checkered the Rufiji landscape. Miombo woodlands blanketed the lowlands surrounding the hills. Rufiji people used *mitumbwi* dugout canoes to cross the river from villages to fields that dotted the riverbanks. Besides grain, a variety of forest products that included ivory, copal, wild rubber, and beeswax were exported down the river into Indian Ocean trade networks centered on Zanzibar island. By the late nineteenth century all the arms of the Rufiji delta were dotted with settlements where people specialized in cutting mangroves for sale to Indian and Zanzibari merchants while growing rice and cassava on delta islands.[33] Perhaps several hundred dhows arrived at delta ports annually to take on mangrove cargoes.

Germans took an immediate fiscal interest in the Rufiji delta. From 1885 to 1890, when the German East Africa Corporation (DOAG) held concessions to administer much of the mainland and coast on behalf of the German empire and the Sultan of Zanzibar, toll stations north and south of the delta were set up to tax exports of mangroves and other forest products.[34] When the German state took over the territory in 1891 it immediately brought the Rufiji delta under colonial control. Late in 1894 Governor von Schele enacted an ordinance that banned the cutting of mangrove poles for firewood in Rufiji region, so valuable were they as an export commodity.[35] In 1897 the Kilwa district officer (who oversaw the delta before 1900) recommended that as soon as forest oversight was firmly established villagers should be prohibited from expanding their agrarian lands in the delta.[36] In 1898 the German administration established a forest office to oversee thousands of hectares of delta mangroves that were divided into three forest districts.[37] In that year the administration contracted the Rufiji Industrial Corporation to erect a sawmill in the delta to process and market 4800 cubic meters of mangrove timber annually.[38] A secondary industry was to market mangrove bark to German chemical firms for processing into leather dye.[39] The German administration also attempted to market East African mangroves in South Africa as mine shoring and railway sleepers, hoping to compete with timber from the Dutch East Indies.[40] The Rufiji's importance as a source of mangroves led officials to create a separate administrative district centered on the river that extended from the coast to the upper Rufiji as far west as Kungulio, which would become the gateway to the Selous game reserve. Rufiji was the only district in German East Africa that originated as a regulated forest, with a trained forester, Karl Grass, as district officer for a decade. In asserting control over the Rufiji forests, the German government came to monopolize an export industry that had fed Indian Ocean trade networks for well over a millennium.

Forest officials planned to create a managed forest economy along the Rufiji, cutting and replanting mangroves and exotic tree species according to established rotations. Toward this end they began a trial tree plantation at Mohoro where Asian teak and local hardwoods were grown. However, the Rufiji forest economy was quickly overwhelmed by the demands of colonial wood consumption coupled with a limited labor supply. Rufiji was expected to supply wood to fuel the Imperial flotilla based in Dar es Salaam, as well as for the government hospital, a beer brewery, and government plantations. Seventy-two percent of Rufiji wood provided to government sectors was used for fuel rather than timber.[41] Plans to afforest the delta were frustrated by lack of labor. Even after the government implemented a hut tax in 1898 intended in part to get rural people to work for a wage, the forest administration suffered from a lack of labor for a type of work that was universally despised. Migrants recruited from the interior quickly deserted when confronted with work in delta mud. Rufiji delta dwellers themselves, whose mastery of mangrove harvesting predated German rule, were the mainstay of the workforce.[42] Sixty percent of the Rufiji forest department's costs were for its workforce, and the forest department was never able to overcome the bottleneck created by the labor shortage.[43] This would be a limiting factor on scientific forestry in German East Africa and meant that far more trees would be cut down than the administration was able to replant.

The Era of Forest Reserves

In 1903 forestry in German East Africa was perceived to be in a state of crisis. In that year missionaries reported to the Foreign Office in Germany that a thoroughly un-German regulation of the East African forests existed whereby cutting of tree stands was not followed by re-afforestation.[44] Admitting that to date replanting had yet to be undertaken while Africans continued to destroy the forests unabated, Governor Götzen and forest officials laid out a plan for a thoroughgoing state control of the colony's forests, which was "a matter of life and death for the colony."[45] The new policy aimed "to occupy as state property with all due haste as much reserved [forest] land as possible."[46] The proposed system of forest reserves would demarcate islands of mostly highland forest throughout the colony where hopefully the state could muster the oversight necessary to evacuate Africans and prevent fresh incursions into the forests, while targeting the available labor at specific forest locales. The forest reserve policy had several goals. It would allow unique tree species to be preserved, including forest products of economic importance like copal and wild rubber. It would provide the basis for the rotational planting and cutting of trees, including trials of economically important indigenous and exotic tree species. Reserves would enable a sustained forest use to meet the colony's timber and fuel needs, and hopefully facilitate an export trade in timber. Finally, forest reserves often encompassed sources of rivers and streams, the protection of which would help create a landscape suitable to intensive agriculture. With these goals in mind, in 1904 Governor Götzen promulgated a forest protection ordinance that empowered the administration to declare forest reserves on Crown Land, and district officials set out to assume control over as many forest parcels as rapidly as possible throughout German East Africa.

The administration lost no time in declaring all coastal mangroves, including the entire Rufiji Delta, as forest reserves in 1904.[47] At about the same time a government land commission demarcated a one-thousand-hectare parcel of coastal forest ten kilometers inland at Naminangu as a forest reserve, departing from a decade of controlling coastal mangroves, but leaving inland forests of the Rufiji region relatively untouched by colonial rule.[48] The many local residents who were present at the declaration were informed that Naminangu forest was henceforth under the protection of the district office and that it was prohibited to obtain forest products of any sort in the reserve. Similar coastal forests and highland forests in other parts of the colony were demarcated as forest reserves from 1904 to 1905.

While concerns about the environment, particularly the effects that deforestation had on climate conditions and water availability, were always part of German forest policy, the use of forests for development and profit clearly predominated in selecting forest parcels as reserves. This goal was clear in a 1905 report by the Chief Forester of German East Africa, Otto Eckert. Asserting that some 250,000 hectares of closed high forest existed in the colony encompassing about five million cubic meters of exportable timber with a value of 500 million marks, Eckert wrote "from the standpoint of the forest administration all of these aboriginal forests are ripe for harvest. Delaying their exploitation until a later date thus means a loss of state property."[49] Once these forests were cut down, Eckert argued, they could be replanted with new forests that could produce five times as much wood mass in the value of 2.25 billion marks, enabling the colony to gross some five million marks annually if the forests were regulated in a one hundred-year cycle. This was the classic age-class forestry (*Schlagwaldwirtschaft*) as described by Scott: "The forest as habitat disappears and is replaced by the forest as an economic resource to be managed efficiently and profitably."[50] The truth is that neither the colonial forest administration nor the government itself could muster the labor necessary to assume such dramatic control over East Africa's forests. However, lack of labor did not prevent the administration from moving forward with forest policy. Hopes were placed on a regulated forest economy to help reverse the negative balance of payments that plagued the colony and to emancipate it from the need to seek subsidies from the often anticolonial Reichstag in Germany.[51] If Eckert's projected forest revenue could actually be achieved, it would almost make up for the annual subsidy of roughly six million marks that the imperial government allocated to the colony at the turn of the century.[52]

German scientific forestry, then, was propelled by the fiscal needs of colonial development. Germans publicized the merits of East African hardwoods at the 1904 St. Louis Exposition, hoping to compete on the world market with Scandinavia and Asia in supplying wood for furniture and tools. At the same time the colonial government sought to substitute local woods for imports of timber from India, Germany, and Scandinavia.[53] After 1904 timber and wood fuel were needed for the construction of the Central Railway and the expansion of Dar es Salaam harbor facilities. German fiscal concerns were also directed at nonwood forest products. At the turn of the century wild rubber had surpassed ivory as the colony's principal export, reflecting the *Raubwirtschaft* (plunder economies) of early colonial rule in Africa.[54] The desire to control sources of wild rubber led to the establishment of the Liwale Forest Reserve in 1905 in a region known for rubber production.[55] This parcel of some 7500 hectares was an example of miombo woodland that German foresters did not even consider to be proper forest. The Ngindo people who occupied the area believed that the reserve was only created as an extension of a government rubber plantation founded a few years earlier. Exploiting wild rubber was also a motive in the establishment of the Tamburu Forest Reserve in June 1911 on a six-thousand-hectare parcel of coastal high forest fifteen kilometers south of the Rufiji. The forest protocol noted that valuable timbers were located in the forest as well as "a not inconsequential amount of rubber, whose protection warrants fiscal and economic interest."[56] While the Rufiji district officer informed local people that henceforth field burning and all use of forest products in the reserve was forbidden, it charged local majumbe headmen with guarding the forest from use, while awarding several headmen contracts to tap the wild rubber trees in the reserve. In asserting state control over rubber tapping, the colonial government displaced local people who had earned money to pay taxes in this way, and disrupted African rural and trade economies.

Forests and Rebellion in Rufiji Region

German forest conservation severed African peasants from a resource that had both cultural and economic importance. The German missionary Martin Klamroth, a member of the Governor's Council, advised the colonial administration on matters of African religion and culture and was well aware of connections between forests and the spirit world among the Zaramo and other Bantu-speaking peoples of southeastern Tanzania.[57] The Zaramo believed that spirits of the deceased resided in forests as *mwenembago*, which Klamroth translated as *Waldherr*, "master of the forest."[58] Connections between forest spirits and villagers were ongoing, mediated through men and women who were possessed in the forests and who then remedied a variety of social and physical ills such as sickness, infertility of people or the land, lack of rainfall, or the threat of locusts or other crop predators. Spirit possession took place in the forests around seven specific tree species. Forests were also places to be feared as the home of malevolent spirits called *shetani* or *kinyamkela*, who resided in the hollows of large trees.[59] It was strongly prohibited to fell such trees, otherwise spirits might come and reside in one's house. Villagers built small "spirit huts" near their fields and used other talismans or medicines to ensure agricultural fertility. Forests were furthermore sites of initiation and circumcision rituals that connected youth with their ancestors.[60] The spiritual importance of specific trees and the forests in general militated against overuse of the forests, and created a conservation logic among rural dwellers. To destroy the forests would mean inviting a host of ills that would endanger agriculture and social well-being. Crosse-Upcott's study of the Ngindo people of Kilwa district also points to this conservation logic.[61] The Ngindo situated their fields close to forests that were used for natural products that could be sold, like rubber, copal, honey, and wax. Individual clans guarded specific forest parcels that streamed behind homesteads "like the tail of a meteor," which substantiates recent assertions of the historical symbiosis between African agriculture and forest regeneration.[62] Crosse-Upcott believed that the Ngindo had a conservation ethos when it came to forests, writing "the Ngindo, despite their quiet village existence, are essentially forest-oriented."[63] In slightly modified form, this relationship to forests was shared by all peoples of southeastern Tanzania.

Forests were also fundamental to African rural economies, as exemplified by the Rufiji district.[64] People obtained wood for charcoal that was used for daily cooking, for iron forging, and for coastal industries that included sugar, salt, and coconut oil production. Household tools such as mortars and pestles, *mitumbwi* dugout canoes, and wood for hut construction came from the forests. In the flood plains north and south of the Rufiji river people built field huts on stilts where they resided during the annual flood season, and which they used to guard fields from birds, pigs, and other crop predators as crops ripened. Forest products like wax, copal, rubber, timber, ivory, game meat, and animal skins were bartered to coastal traders for cotton textiles or sold for cash that could be used to purchase food in times of famine. Forests offered a variety of foods that could be resorted to in times of famine.[65]

The use of forests as havens from conflict was also essential at the turn of the century. In the nineteenth century slave raiders from the coast and martial societies from the interior plundered the entire Rufiji basin for grain and people.[66] During this period inhabitants of the Rufiji region used the forests for refuge, carving out fields for crops and building isolated homesteads and villages in inaccessible regions that were hidden from view.[67] Forests offered havens that the exposed alluvial plain of the Rufiji did not, even if the agricultural productivity of the mountain forests was not as good. The pattern of using the forests for refuge continued during German rule. While colonial rule ended the predations of slave raiders and inland invaders, it created new burdens. Germans demanded that rural people provide labor for corvee projects such as road construction and forest labor, and did all they could to induce men to work as porters or as wage laborers on European plantations in the northeast highlands.[68] Workers were needed for the construction of two railway lines after 1905. Tax defaulters and penal laborers were often sent to do forest work.[69] By the turn of the century the colonial government also had inaugurated a program of peasant cotton production that intended to make the German textile industry self-sufficient. Many Africans reacted to these demands as they had to nineteenth-century raiding by moving into the forests where they could maintain food production. Already frustrated by a labor shortage, the German administration viewed these forest movements as a threat to economic development.

The Maji Maji rebellion that broke out in Rufiji and Kilwa districts in 1905 and spread to other parts of German East Africa in the next year was motivated in part by German forest policy.[70] Villagers who once could sell forest products to pay taxes or purchase commodities were denied that possibility by German regulations. Bans on field burning prevented peasants from opening up new lands for agriculture. The creation of forest reserves coupled with hunting laws also threatened the rural economy. Forests harbored a variety of crop predators that had begun to plague southeastern Tanzania after the turn of the century. Hunting laws introduced in 1898 sharply curtailed African use of nets or fire to hunt in collective parties, an important method of eradicating pigs and antelope before crops ripened.[71] The requirement that Africans purchase hunting licenses to shoot game hindered other forms of animal eradication. The ivory economy that had once enriched local big men and had been a mainstay of patronage networks was effectively closed by forest and hunting restrictions. Bush burning that had once enabled villagers to destroy the habitat and breeding grounds of insects, snakes, and crop predators near their fields was severely encumbered by forest and hunting regulations. When questioned as to why they participated in the Maji Maji rebellion, some people pointed to forest and hunting regulations as a major grievance.[72]

The rebellion was so named because rebels used a water medicine (*maji*) to protect themselves from German bullets, a spiritual recourse that was directly related to other uses of medicine ministered by spirit mediums with forest connections. The missionary Klamroth believed there was a direct connection between the rebellion and African forest cosmology. Following the rebellion Germans attacked spirit mediums and their shrines, and in so doing forest policy dovetailed with social control. Many of the forest reserves that were declared following the rebellion encompassed sites where medicines with a supposed warlike purpose could be obtained. For example, the Mpanga forest reserve south of the upper Rufiji River was believed to be an important distribution center for regional medicines.[73] Germans believed that the leadership of the Maji Maji rebellion had been based there. In creating Mpanga forest reserve and others at Naminangu, Utete, and Mtondo, Germans knowingly severed villagers from spiritual centers that they believed had been used to encourage rebellion. Before the rebellion some forest reserve declarations had stipulated that people were allowed to tend ancestral graves in the forests.[74] No forest declaration included this provision after the war.

Forest reserves facilitated social engineering by channeling people into open areas where they could be brought under colonial control. During Maji Maji, villagers had moved deeper into the forests in order to escape conflict and the forced requisitioning of food and labor by German forces.[75] The pattern of forest movement continued after the war. Matumbi villagers, for example, often women whose husbands had been killed in battle or who fled to escape the labor indemnities that followed the war, moved from open agricultural lands to occupy riverine strips called *matimbe* in the forested mountain highlands where they could concentrate on subsistence food production.[76] In so doing they disappeared from the colonial economy at a time when the German administration had targeted the Rufiji-Kilwa region for cotton production. With the spate of forest declarations between 1907 and 1914 precisely in these mountain regions, Germans sought to evict people from the forests. There was thus a very strong correlation between the declaration of forest reserves and social control, and forest reserves in some cases were not declared because of their specific ecosystems or fiscal value, but as a means of channeling people toward colonial development.

Forest Conservation and the Colonial Development Agenda

Forest policy in German East Africa shifted alongside the vision of colonial development. In the 1890s policy makers assumed that the colony was destined to be one of white settlement, particularly in the northeast highlands. Allowances were thus made for European planters to cut down montane forests to make way for coffee plantations, with a modest attempt to limit deforestation in the highlands. Indeed, before 1906 policy makers saw forest policy as facilitating the creation of a landscape of intensively cultivated land punctuated by managed forest parcels in imitation of Germany.[77] After 1906 policy makers changed their attitude about European settlement. The worsening labor shortage coupled with the Maji Maji rebellion persuaded officials that German East Africa must rely on peasant household agriculture as the mainstay of the economy. Planters proved themselves to be unable to meet Germany's demand for cotton and other industrial cash crops and therefore were regarded as more of a burden on the economy than an asset. Although settlers were discouraged from immigration to German East Africa, several hundred of them arrived anyway.

This shifting view of European settlement was reflected in forest policy. Forest laws promulgated in 1895 demanded that planters attend to forest conservation; however, they generally skirted forest ordinances despite the threat of fines and imprisonment. The revised 1898 forest ordinance aimed to "protect the interests of the plantations" while protecting watersheds by requiring coffee planters to maintain forests on mountain tops and ridges and in the vicinities of springs and to reforest those locations "with good indigenous woods and Eucalyptus." Governor Liebert wrote, "It should not be difficult to retain the larger contiguous forest complexes necessary for climate conditions while being attentive to the interests of individual plantations."[78] While the 1898 ordinance required that one-fourth of plantation lands be reserved for forests, it clearly deferred to the greater necessity of European settlement and coffee cultivation.

As the number of German plantations expanded rapidly in the colony from about fifty in 1898 to several hundred by 1910, moving from the highlands of the northeast to central and southeastern Tanzania, and as the central and northern railways were extended steadily, officials amended forest regulations in 1908 to curtail the unencumbered exploitation of forests on private and leased lands.[79] This departure mirrored the shift away from a white settlement policy. The 1908 ordinance empowered the administration to forbid deforestation on private and leased lands if it was deemed destructive to the common good, and prohibited cutting of trees with a diameter less than twenty-five centimeters, and one quarter of trees larger than twenty-five centimeters in diameter. On forest parcels over one hundred hectares, one-fourth of the trees were to be preserved. Landowners could substitute these provisions with a regulated plan of cutting and afforestation.

Planters and the government quarreled over control of forest parcels, as they did over many issues related to colonial development. When the German industrialist Heinrich Otto founded a cotton plantation of several thousand hectares near Kilosa in Morogoro district after 1907, he demanded that the plantation include the Khutu-Khutu forest reserve that adjoined the land.[80] Otto needed the forest to fuel two massive steam tractors and cotton gins that he imported from Germany. The Morogoro district officer gave way on these demands and granted Otto the forest parcel, a move that the government in Dar es Salaam quickly disavowed.[81] In ongoing negotiations over control of the forest, Otto sought to avoid provisions of the 1908 ordinance that demanded rational forest use by arguing that the reserve was not really a forest at all, but merely bush that was only made into a forest reserve to keep African farms from encroaching on his property.[82] The government's view, as articulated by Colonial Secretary Dernburg, was that the destruction of the forest parcel for plantation needs would destroy the water course and hinder irrigation.[83] At the very least, the treasury needed compensation for the loss of the wood to the Otto plantation. In the long run Otto succeeded in annexing a large part of the Khutu-Khutu forest, and the remainder gave way to the arrival of the Central Railway in 1910. The Khutu Khutu parcel is among a few created under German rule that no longer exists as a forest reserve, and it is notable that its disappearance was the result of European plantation agriculture and railway development rather than African shifting agriculture.

The 1908 ordinance was followed quickly by another in 1909 directed at African wood use that undertook a thorough state control of forest resources on Crown Land. The 1909 ordinance required government permission to use forest products on declared Crown Lands (which included forest reserves), "especially wood, bark, fibers, resin, rubber, leaves, blossoms, and fruits."[84] This ordinance prohibited peasants from settling, farming, and pasturing livestock in forest reserves. Outside of forest reserves fees were exacted for use of forest products, although exemptions could be made for hut construction and cooking fuel. The 1909 ordinance expedited the declaration of forest reserves throughout the colony, from roughly 222,000 hectares in 1908 to 750,000 in 1914.[85]

Although forest reserves encompassed less than 1 percent of the area of German East Africa in 1914, they were concentrated in regions of African agriculture and settlement, and hemmed in rural societies significantly. Even small forest reserves caused hardship to peasant farming, as the Tongamba forest reserve in Kilwa district exemplified. Encompassing a mountain forest of only 430 hectares, the 1912 Tongomba forest protocol specified that no land be set aside for adjacent villages since they "already have about twenty times their needs."[86] Twelve years later, however, after German rule had ended, the new British administration of Tanganyika Territory concluded that the villages near Tongamba had inadequate land for cultivation.[87] In order to prevent the villagers from moving away and disrupting British agrarian schemes, the forest borders were adjusted to allow villagers room for expansion.

The German administration occasionally made allowances for peasant access to forests. In 1912 coastal villagers in Kilwa district petitioned the district office for access to the mangrove reserves because they needed wood for salt manufacture, fuel, and construction of fish traps and huts.[88] A few miles north there was unrest among villagers near the coast who believed that they were no longer able to use mangroves. A delegation of forty villagers went to the Kibata district station to complain. In this area of the Maji Maji rebellion's outbreak, the district administration relaxed rules for mangrove use, but still required fees to obtain wood for construction or for sale to Indian merchants.

Although the 1909 forest ordinance empowered officials to protect reserves from peasant encroachment, the administration backed away from the vision of eliminating field burning outside of forests that had been the mantra of forest policy in the first decade of German rule. German officials came to realize that African production—subsistence or commercial—was dependent on shifting agriculture for the foreseeable future. Most German plantations, in fact, resorted to shifting agriculture as a cheaper means of dealing with the ongoing labor shortage and the expense of importing fertilizer. Forest policy concentrated on demarcating and protecting existing forests by requiring that villagers cut and maintain fire strips around reserves to protect against field burning. However, this regulation was by no means foolproof. In July 1914 the government forest plantation near Mohoro in Rufiji district was ignited when a local peasant named Mwalimu bin Kombo burned weeds around his mango trees.[89] The seasonal southeast monsoons blew the flames to the forest reserve 1.5 kilometers to the west, where it jumped a ten-meter-wide fire strip and burned section 23 of the forest plantation, destroying some thirty-two hectares of young teak stands. This fire on an experimental forest plantation was a severe setback to German efforts at scientific forestry. Mwalimu bin Kombo was detained while the case was investigated. Elders confirmed that for a decade the practice as overseen by German foresters had been that in the dry season protective strips would first be created around the forest reserve by cutting down grass and burning it several times before allowing local people to burn their own fields. However, because Kombo had burned his fields during a heavy wind, local officials believed that he was culpable of negligent arson. The central government's response to the incident was to point out that there was no general prohibition on wild burning outside forest reserves, therefore no criminal procedures could be brought against Kombo, and because he had modest property, a demand for compensation would be meaningless.

Despite prohibitions on the practice, there were many examples of peasants opening up fields in forests by using fire or otherwise encroaching on forestlands. The best surviving evidence comes from the Bunduki forest reserve that was demarcated in Morogoro district in 1906. In 1911 thirty-five people were fined two rupees each, the equivalent of two-thirds the annual hut tax, for laying out their fields in the forest reserve.[90] Other transgressions of the forest were treated more severely. In 1910 six men, three women, and one child were found guilty of damaging the Bunduki forest. Most of these cases involved farmers burning their fields adjacent to the forest, and the typical sentence was a one-rupee fine and one day labor in chains. A twelve-year-old child named Zinga received a two-rupee fine and one day in chains for grazing goats in the protective strip that bordered the forest. Understanding the distinction that was made between burning in forests and on their own fields, many of those accused of arson claimed that the forest fires had originated outside the reserve.

Although negligence resulted in many forest fires, there is evidence that arson was a means that peasants used to attack forest policy, a type of hidden resistance. The Kilwa penal records for 1908–1909 include seven cases of arson and many more for disobeying state authority.[91] Forest officials suspected that many forest fires were set intentionally, and complained of the lack of resources to effectively police forest use. In 1908 a land commission demarcated Kipo forest on the north bank of the Rufiji on some 2500 hectares of steppe forest, setting aside only 80 hectares for local villagers.[92] A battleground and haven for refugees of the Maji Maji war just a few years previously, in 1909 the Kipo forest was hit by a fire that authorities suspected was "premeditated arson."[93] In 1911 Rufiji officials reported that "in the last year no forest reserve has been completely protected from fire. In part these fires have originated outside of the reserves and jumped into the forest, but in part they are alleged to have been premeditated or set negligently inside the forests, yet we have been unable to capture a malefactor."[94]

The intrusion of forest reserves into the agrarian economy led some planters to voice common grievances with African peasants. In 1913 a handful of Rufiji planters complained that forest and hunting reserves harmed them and neighboring peasants by harboring elephants, vermin, and insects that damaged crops.[95] One planter proposed that the hunting preserve on the upper Rufiji be abolished. Forest reserves that dotted the Rufiji river furthermore hindered plantation development by making it difficult to procure wood for fuel and construction, and prevented plantations from expanding their cotton lands. Peasants complained that they lacked building and fuel wood, and that they were unable to obtain the wood of the *mkongo* tree to make the dugout canoes that they needed to traverse the river. The strict ban on using fire in forests thwarted peasants from collecting honey or wax. Finally, the forest reserves prevented villagers from collecting roots, mushrooms, fruits, and insects that they used as supplementary foods in times of scarcity, and medicines that they obtained from specific trees.[96] While colonial officials acknowledged that these complaints were valid, they concluded that "the discomforts caused by the forest reserves must be borne for the common good."[97]

AT THE END of German rule in East Africa forest policy had been modified significantly since the early years of colonialism, when a vision of scientific forestry had dominated the discourse. Indeed, the last years of colonial rule saw some divergence in thinking between colonial administrators and foresters. While greeting the 1909 forest ordinance as empowering the state to attack the predatory use of forest products on Crown Land, Chief Forester Wilhelm Holtz decried the threat to the forests that peasant agriculture and pastoralism posed outside of declared reserves, especially the continued use of fire to open up new fields.[98] Government officials concentrated on protecting forest reserves from destruction while giving peasants a fairly free hand outside of reserves in their methods of agriculture and subsistence use of forest products. The need to incorporate peasant agriculture into the colonial development agenda took precedence over idealistic concerns about preserving all forests of the colony.

Part of the administration's rationale for backing away from a thoroughgoing conservationism by the end of colonial rule was the conclusion that East African forestry provided no direct benefits to the metropole. One observer pointed out that Germany had no particular shortage of wood, and if outside sources were needed, it would be far more cost-effective to obtain timber from Russia or Austria-Hungary.[99] East African timber could not compete with already established trade networks that brought West African timber (mostly from non-German colonies) to Germany, and the need to pay transport fees through the Suez Canal increased costs to levels unacceptable to German timber consumers. Furthermore, German consumers were unfamiliar with how exotic East African woods could be used in industry. Scientific forestry had always linked conservation with profit, and the unclear economic benefits of East African forests for Germany led policy makers to water down colonial forest policy.

Despite the loss of a sense of urgency, policy makers did not completely abandon scientific forestry. While the 1909 forest ordinance allowed for forest use outside of declared reserves, it provided the rationale for a quick declaration of forest reserves where it was deemed valuable timbers should be conserved. Furthermore, the effort to find markets for East African timber continued, with South Africa as the most promising potential outlet.[100] Non-reserved Crown Land was also opened up for the industrial use of timber within German East Africa with modest oversight. In 1914 Governor Schnee requested that all district authorities identify valuable wood species that warranted government protection.[101] Emphasizing that there should be no obstacles to domestic use of forests outside of reserves, the government concentrated on exacting fees for the commercial exploitation of forests and required that young trees and a certain percentage of forest stands be protected from deforestation.

Forest policy evolved alongside the shifting colonial development agenda. Forest reserves represent that compromise. While enabling the administration to regulate forests on approximately 1 percent of the colony's land, forest policy recognized that the colonial administration was simply too weak to effectively police the forests of the entire colony according to a metropolitan vision. Whereas some quarter million foresters were employed in Germany at the turn of the century whose upkeep was "more than repaid by the sales of timber," in 1910 seventeen European foresters and eighty-five African forest wardens and police oversaw about 300,000 hectares of forest reserves, an extent that more than doubled by 1914.[102] Forest reserves at the same time served the development agenda by channeling villagers out of the forests into open areas where their labor power could be tapped for colonial pursuits, or where they could be pressured to grow cash crops. The use of forest reserves for development and social control in modern Tanzania is one of the ongoing legacies of German colonial rule.

Finally, the limited colonial vision of what constituted a forest was the result of an obsession with German-like landscapes. The roughly 250 forest reserves that had been demarcated by the end of German colonial rule privileged montane forests. The exception was the coastal mangroves, whose quick control by the state stemmed from an established fiscal value that predated German rule. However miombo woodlands, which constituted about half of the Tanzanian landscape and actually increased under German rule as a result of the consequences of population movements, colonial conquest, and the devastation of the pastoral economy, were not deemed worthy of the protections offered by scientific forestry. It was not until the aftermath of World War II that the British colonial government belatedly sought to correct this omission by increasing forest reserves from the 1 percent of the landscape demarcated under German rule to a target figure of 8 percent. Indeed, by the time Tanzania became independent in 1961, 14 percent of the landscape was classified as reserved forest, and the population removals inaugurated by German forest policy were about to reach a far greater magnitude.[103]

Notes

1. Hans G. Schabel, "Tanganyika Forestry under German Colonial Administration, 1891–1919," *Forest and Conservation History* 34 (July 1990): 130–141; Juhani

Koponen, *Development for Exploitation: German Colonial Policies in Mainland Tanganyika* (Hamburg: Lit Verlag, 1995), 529–536; T. Siebenlist, *Forstwirtschaft in Deutsch-Ostafrika* (Berlin: P. Parey, 1914), 7.
2. Siebenlist, *Forstwirtschaft*, 60.
3. German foresters publicized the figure of 26 percent, pointing out that German forest cover had increased in the nineteenth century owing to German forest management. M. Büsgen, "Forstwirtschaft in den Kolonien," *Verhandlungen des Deutschen Kolonialkongresses* (Berlin: Dietrich Reimer, 1910), 801–817, here 802; "The Forest Lands of Germany," *Indian Forester* 31, no. 6 (1905): 729–730.
4. Schabel, "Tanganyika Forestry," 138.
5. Gregory Barton, "Empire Forestry and the Origins of Environmentalism," *Journal of Historical Geography* 27, no. 4 (2001): 529–552, here 544.
6. William H. Rollins, "Imperial Shades of Green: Conservation and Environmental Chauvinism in the German Colonial Project," *German Studies Review* 22, no. 2 (1999): 187–213, here 187.
7. Roderick Neumann, *Imposing Wilderness: Struggles over Livelihood and Nature Preservation in Africa* (Berkeley: University of California Press, 1998), 19.
8. Neumann, *Imposing Wilderness*, 18.
9. Henry E. Lowood, "The Calculating Forester: Quantification, Cameral Science, and the Emergence of Scientific Forest Management in Germany," in *The Quantifying Spirit in the Eighteenth Century*, ed. Tore Frängsmyr, J. L. Heilbron, and Robin E. Rider (Berkeley: University of California Press, 1990), 315–342; Ravi Rajan, "Imperial Environmentalism or Environmental Imperialism? European Forestry, Colonial Foresters and the Agendas of Forest Management in British India 1800–1900," in *Nature and the Orient: The Environmental History of South and Southeast Asia*, ed. Richard H. Grove, Vinita Damodaran, and Satpal Sangwan (New Delhi: Oxford University Press, 1998), 324–371; James Scott, *Seeing Like a State: How Certain Schemes to Improve the Human Condition Have Failed* (New Haven: Yale University Press, 1998), 11–12. See also the paper by Michael Imort in this volume.
10. Peter Sahlins, *Forest Rites: The War of the Demoiselles in Nineteenth-Century France* (Cambridge, Mass.: Harvard University Press, 1994); Tom Scott and Bob Scribner, eds., *The German Peasants' War: A History in Documents* (Atlantic Highlands: Humanities Press International, 1991); Stephen P. Frank, *Crime, Cultural Conflict, and Justice in Rural Russia, 1856–1914* (Berkeley: University of California Press, 1999); Tamara Whited, *Forests and Peasant Politics in Modern France* (New Haven: Yale University Press, 2000).
11. Josef Mooser, "Property and Wood Theft: Agrarian Capitalism and Social Conflict in Rural Society, 1800–1850. A Westphalian Case Study," in *Peasants and Lords in Modern Germany*, ed. Robert G. Moeller (Boston: Allen and Unwin, 1986), 52–80; Reiner Prass, "Verbotenes Weiden und Holzdiebstahl: Ländliche Forstfrevel am südlichen Harzrand im späten 18. und frühen 19. Jahrhundert," *Archiv für Sozialgeschichte* 36 (1996): 51–68; Bernd-Stefan Grewe, "Shortage of Wood? Towards a New Approach in Forest History: The Palatinate in the Nineteenth Century," in *Forest History*, ed. M. Agnoletti and S. Anderson (Wallingford and New York: CABI, 2000), 143–159.
12. Germans served as inspectors-general of the Indian Forest Department from 1864 to 1917. Indra Munshi Saldanha, "Colonialism and Professionalism: A German Forester in India," *Environment and History* 2, no. 2 (1996): 195–219; Gregory Barton, "Keepers of the Jungle: Environmental Management in British India, 1855–1900," *Historian* 62, no. 3 (2000): 557–574, here 558; Nancy Lee Peluso, *Rich Forests, Poor People: Resource Control and Resistance in Java* (Berkeley:

University of California Press, 1994), 63, 65; Peter Boomgaard, "Forest Management and Exploitation in Colonial Java, 1677–1897," *Forest and Conservation History* 36 (January 1992): 4–14; Raymond L. Bryant, *The Political Ecology of Forestry in Burma 1824–1994* (Honolulu: University of Hawai'i Press, 1996).

13. Bryant, *Political Ecology*, 45–46; Jacques Pouchepadass, "British Attitudes Towards Shifting Cultivation in Colonial South India: A Case Study of South Canara District 1800–1920," in *Nature, Culture, Imperialism: Essays on the Environmental History of South Asia*, ed. David Arnold and Ramachandra Guha (New Delhi: Oxford University Press, 1996), 123–151; Madhav Gadgil and Ramachandra Guha, *This Fissured Land: An Ecological History of India* (Berkeley: University of California Press, 1993), 150–158; Ramachandra Guha, *The Unquiet Woods: Ecological Change and Peasant Resistance in the Himalaya* (Berkeley: University of California Press, 2000), 51.

14. Koponen, *Development for Exploitation*, 588–589; Thaddeus Sunseri, *Vilimani: Labor Migration and Rural Change in Early Colonial Tanzania* (Portsmouth, N.H.: Heinemann, 2001).

15. Bryant, *Political Ecology*, 79, 104; Nancy Lee Peluso, "A History of State Forestry Management in Java," in *Keepers of the Forest: Land Management Alternatives in Southeast Asia*, ed. Mark Poffenberger (West Hartford: Kumarian Press, 1990), 31; Paul Greenough, "Naturae Ferae: Wild Animals in South Asia and the Standard Environmental Narrative," in *Agrarian Studies: Synthetic Work at the Cutting Edge*, ed. James Scott and Nina Bhatt (New Haven and London: Yale University Press, 2001), 141–185; Anthony Reid, "Humans and Forests in Precolonial Southeast Asia," in Grove et al., *Nature and the Orient*, 106–126; Michael Mann, "Timber Trade on the Malabar Coast, c. 1780–1840," *Environment and History* 7, no. 4 (2001): 403–425.

16. Christopher Conte, "Searching for Common Ground: Reconstructing Landscape History in East Africa's Eastern Arc Mountains," in *Methods and Approaches in Forest History*, ed. M. Agnoletti and S. Anderson (Oxon and New York: CABI, 2000), 173–187.

17. This antithetical view of miombo was articulated with respect to Tabora district in Bundesarchiv-Berlin (hereafter BAB)/R1001/227, Charisius to Government, 16 January 1907, 4–6.

18. W. R. Rodgers, "The Miombo Woodlands," in *East African Ecosystems and Their Conservation*, ed. T. R. McClahan and T. P. Young (New York: Oxford University Press, 1996), 299–325.

19. Helge Kjekshus, *Ecology Control and Economic Development in East African History* (Berkeley: University of California Press, 1977).

20. "Die Urproduktion der Eingeborenen," *Berichte über Land- und Forstwirtschaft* (hereafter *BLF*) 1 (1903): 4.

21. Büsgen, "Forstwirtschaft," 804.

22. BAB/R1001/7680, Heyden to Caprivi, 11 April 1894, 27.

23. Eugen Krüger, "Die Wald- und Kulturverhältnisse in Deutsch-Ostafrika," *Deutsches Kolonial-Blatt* 5 (1894/95): 623–629. A more complete report is found in BAB/R1001/7680, Forstwesen in Deutsch-Ostafrika, 29–43.

24. BAB/R1001/7681, Götzen to Foreign Office, 8 March 1904, 111–112.

25. Krüger, "Wald- und Kulturverhältnisse," 626.

26. BAB/R1001/7681, Götzen to Foreign Office, 20 May 1904, 113–114.

27. A copy of the ordinance is found in BAB/R1001/7680, Forstwesen in Deutsch-Ostafrika, 3–5.

28. BAB/R1001/7680, Krüger report, 3 October 1894, 30b.

29. The Swahili version of the 1893 forest ordinance as applied to Rufiji district is

found in Tanzania National Archives (hereafter TNA) G8/19, untitled document 1581/94 (1894), 60–61.
30. The currency of German East Africa was the rupee, divided into sixty-four pesas. After 1904 pesas were substituted for a new currency of one hundred hellers to the rupee. One rupee equaled 1.33 marks.
31. BAB/R1001/7680, Liebert memorandum to district offices, 13 April 1898, 158.
32. L. A. Lewis and L. Berry, *African Environments and Resources* (Boston: Unwin Hyman, 1988), 136–138; R. de la B. Barker, "The Rufiji River," *Tanganyika Notes and Records* (1936), 10–16; Alexander Wood, Pamela Stedman-Edwards, and Johanna Mang, eds., *The Root Causes of Biodiversity Loss* (London: Earthscan, 2000), 309–336; Kjell Havnevik, *Tanzania: The Limits to Development from Above* (Uppsala: Nordiska Afrikainstituet, 1993).
33. TNA G8/588, Eberstein to Government, 25 May 1897; Prüssing, "Ueber das Rufiyi–Delta," *Mitteilungen aus den deutschen Schutzgebieten* (hereafter MaddS) 14 (1901): 106–113; Ziegenhorn, "Das Rufiyi-Delta," *MaddS* 9 (1896): 78–85.
34. Bruno Kurtze, *Die Deutsch-Ostafrikanische Gesellschaft* (Jena: G. Fischer, 1913), 106–107.
35. BAB/R1001/7680, Memorandum to District Offices and Sub-Stations, 5 December 1894, 52.
36. TNA G8/588, Eberstein to Governor, 6 May 1897.
37. BAB/R1001/7680, von der Decken to Foreign Office, 18 November 1898.
38. BAB/R1001/7723, Forest Administration in Rufiji, 22 April 1901, 167–170; Karl Grass, "Forststatistik für die Waldungen des Rufiyideltas," *BLF* 2 (1904–1906): 165–196.
39. BAB/R1001/7723, Stuttgart Chemical and Dye Works to Foreign Office, 28 December 1900, 87–90.
40. BAB/R1001/7722, German Consul, Johannesburg, to von Schuckmann, General Consul, Cape Town, 29 April 1898, 83–85.
41. TNA G8/529, Abgabe von Waldprodukten an andere Dienststellen, 1898–1915.
42. TNA G8/514, Grass report, 13 April 1900.
43. TNA G8/529, Grass to Government, 7 January 1903. In that year about six thousand rupees were paid as woodcutting and hauling wages out of a total cost of ten thousand rupees.
44. TNA G8/609, Foreign Office, Colonial Department to Imperial Government DOA, 31 May 1903.
45. TNA G8/609, Eckert report, 16 January 1904.
46. "Forstwirtschaft," *Jahresbericht über die Entwicklung der deutschen Schutzgebiete* (Berlin: Ernst Siegfried Mittler und Sohn, 1904), 30–31.
47. TNA G8/651, Land Commission, December 1904. By 1904 all mangroves along the Tanzanian coast were taken over by the state as forest reserves.
48. TNA G8/652, Kilwa Waldreservate "Naminangu," 1904–1914.
49. TNA G8/508, Denkschrift über die Forstwirtschaft in Deutsch-Ostafrika, 25 June 1905, 33–34.
50. Scott, *Seeing Like a State*, 13.
51. Rainer Tetzlaff, *Koloniale Entwicklung und Ausbeutung: Wirtschafts- und Sozialgeschichte Deutsch-Ostafrikas 1885–1914* (Berlin: Duncker und Humblot, 1970), 71–79.
52. Tetzlaff, *Koloniale Entwicklung*, 79.
53. TNA G8/508, Denkschrift über die Forstwirtschaft in Deutsch-Ostafrika, 25 June 1905, 33–34.
54. Tetzlaff, *Koloniale Entwicklung*, 71–73; Koponen, *Development for Exploitation*, 210–212.

55. TNA G8/653, Waldreservate Kilwa "Liwale," 1899–1909.
56. TNA G8/681, Waldreservate Rufiji "Tamburu," 1911–1912.
57. Martin Klamroth, "Beiträge zum Verständnis der religiösen Vorstellungen der Saramo im Bezirk Daressalam (Deutsch-Ostafrika)," *Zeitschrift für Kolonialsprachen* 1–3 (1910–1913): 37–70, 118–153, 189–223. See also T. O. Beidelman, *The Matrilineal Peoples of Eastern Tanzania* (London: International African Institute, 1967), 18; Marja-Liisa Swantz, *Ritual and Symbol in Transitional Zaramo Society with Special Reference to Women* (Lund: Gleerup, 1970).
58. Klamroth, "Beiträge," 49.
59. Klamroth, "Beiträge," 66–68.
60. Klamroth, "Beiträge," 192–193.
61. A.R.W. Crosse-Upcott, "Social Aspects of Ngindo Bee-keeping," *Journal of the Royal Anthropological Institute of Great Britain and Ireland* 86, no. 2 (1956): 81–108.
62. Crosse-Upcott, "Social Aspects," 87–88; Melissa Leach and Robin Mearns, *The Lie of the Land: Challenging the Received Wisdom in African Environmental Change and Policy* (Oxford: International African Institute, 1996).
63. Crosse-Upcott, "Social Aspects," 98.
64. The following discussion draws from Ziegenhorn "Das Rufiyi-Delta"; Havnevik, *Tanzania*, 146, 161–165, 169–170; Wood et al., *The Root Causes*, 309–336; R.E.S. Tanner, "Some Southern Province Trees with their African Names and Uses," *Tanganyika Notes and Records* 31 (1951): 61–70.
65. A.R.W. Crosse-Upcott, "Ngindo Famine Subsistence," *Tanganyika Notes and Records* 50 (1958): 1–20.
66. Lorne Larson, "A History of the Mbunga Confederacy ca. 1860–1907," *Tanzania Notes and Records*, nos. 81 and 82 (1977): 35–42.
67. "Bericht des Leutnants Z. S. Fromm über eine Rekognierungsfahrt nach dem Rufiji," *Deutsches Kolonialblatt* 4 (1893): 291–294.
68. Complaints about mustering workers for forest work were frequent. BAB/R1001/7723, Grass report, 22 April 1901, 167–170; BAB/R1001/7725, Haug report, 1908, 87–93.
69. TNA G8/888, Bewersdorf to Dar es Salaam District Office, 30 October 1906; TNA G8/589, Martin to Imperial Government, 20 February 1913.
70. I have advanced this argument in greater detail in "Reinterpreting a Colonial Rebellion: Forestry and Social Control in German East Africa, 1874–1915," *Environmental History* 8, no. 3 (July 2003): 430–451. On the Maji Maji rebellion generally see John Iliffe, *A Modern History of Tanganyika* (Cambridge: Cambridge University Press, 1979).
71. BAB/R1001/7776, Verordnung betreffend die Schonung des Wildstandes in Deutsch-Ostafrika, 17 January 1898, 56–57.
72. BAB/R1001/726, Westhaus testimony, 21 December 1905, 122b; Schultz report, 23 December 1905, 121.
73. TNA G8/677, Mpanga Forest Reserve, 26 August 1910.
74. TNA G8/632 Vikindu Forest Reserve, 28 January 1904; TNA G8/633 Massangania Forest Reserve, Land Protocol of 4 February 1904 at Mbindisi.
75. On the use of forests as havens during the war see Hans Paasche, *Im Morgenlicht: Kriegs-, Jagd-, und Reise-Erlebnisse in Ostafrika* (Berlin: C. A. Schwetschke und Sohn, 1907).
76. "Die Entwicklung Kilwas im Jahre 1908," *Deutsch-Ostafrikanische Rundschau*, 6 November 1909, 1.
77. BAB/R1001/7681, Götzen memorandum, 27 July 1904, 125–126.

78. BAB/R1001/7680, Liebert to Foreign Office, 19 April 1898, 153.
79. Verordnung, betreffend die Erhaltung von Privatwaldungen, 17 August 1908, Kaiserliches Gouvernement von Deutsch-Ostafrika, *Die Landes-Gesetzgebung des Deutsch-Ostafrikanischen Schutzgebiets* (Tanga and Dar es Salaam: Kaiserliches Gouvernement von Deutsch-Ostafrika, 1911), 587–588; Siebenlist, *Forstwirtschaft*, 54–56.
80. TNA G8/894, Otto to Colonial Office, 22 February 1908, 6.
81. TNA G8/894, Lambrecht to Imperial Government, 12 June 1908, 28–29; BAB/R1001/8190, Winterfeld to Otto, 18 April 1908.
82. BAB/R1001/8190, Rechenberg to Colonial Office, 1 November 1910; TNA G8/894, Otto to Colonial Office, 8 March 1908, 14.
83. TNA G8/894, Dernburg to Otto, 29 February 1908.
84. Waldschutz-Verordnung, 27 February 1909, *Landes-Gesetzgebung*, 588–591.
85. "Forstwesen," *Jahresbericht*, 53; Siebenlist, *Forstwirtschaft*, 7.
86. TNA G8/655, Waldreservat Tongamba, 7 October 1912.
87. TNA AB875, Tongamba Forest Reserve, Lancy to Chief Secretary, 20 May 1924.
88. TNA G8/651, Richter to Government, 6 January 1912; Kilwa District Office to Government, 11 December 1912.
89. TNA G8/589, Mohoro Forest Office to Government, 9 July 1914, 71–72.
90. TNA G58/4, Waldreservat Bunduki, October 1910.
91. TNA G11/1, Kilwa Strafbuch 1908–1909, 24, 102, 113, 194, 195, 232.
92. TNA G8/674, Waldreservat Kipo, 1908–1910.
93. "Sonderberichte der Forstverwaltung von Deutsch-Ostafrika für das Jahr 1909," *BLF* 3, no. 5 (1911): 294.
94. "Jahresberichte der Lokalforstbehörden," *Der Pflanzer* 8, suppl. 1 (1910–1911): 6.
95. TNA G8/589, Gouvernementsrat, Wald- und Jagdreservate und gesunde Eingeborenenpolitik, June 1913.
96. Tanner highlights the medicinal use of trees in, "Some Southern Province Trees," 61–70.
97. TNA G8/589, Utete District Officer to Government, 14 July 1913, 29–30.
98. TNA G8/505, Holtz to Imperial Government, 22 January 1909; TNA G58/98, Jahresbericht der Forstverwaltung Rufiyi 1910/11, 30 April 1911, 2–16.
99. BAB/R1001/668, Wirtschaftliche Erschließung der Waldungen der deutschen Schutzgebiete, May 1909, 174–175.
100. BAB/R1001/668, Renner to German General Consul, Cape Town, 15 August 1913, 224–225.
101. BAB/R1001/668, Runderlaß an alle Bezirksbehörden und Forstämter, 16 January 1914, 231–33.
102. Büsgen, "Forstwirtschaft," 811; "The Forest Lands of Germany," *Indian Forester* 31, no. 6 (1905): 729–730.
103. By 1961 reserved forests accounted for 14 percent of newly independent Tanzania, 89 percent of which were miombo woodlands. M. S. Parry, "Recent Progress in the Development of Miombo Woodland in Tanganyika," *East African Agricultural and Forestry Journal* 32 (January 1966): 307–315.

Part II
The Cultural Landscapes of Home

Chapter 5

Organic Machines
Cars, Drivers, and Nature from Imperial to Nazi Germany
RUDY KOSHAR

AUTO MANUFACTURERS PRODUCED one billion cars in the twentieth century. In view of the enormous impact this prodigious output had not just in the United States but throughout the globe, one would assume that scholarship would make the automobile a primary subject of investigation. But in fact, as Rudi Volti argued in 1996, "as a subject of scholarly inquiry the automobile remains vastly underexamined."[1] Even more surprising is the fact that within the extant literature on the automobile, research on the history of driving is limited.[2] An older automotive history generally favored accounts of inventors, manufacturers, designers, traffic planners, engineers, and workers rather than of users. Although the drivers' perspective was not entirely absent in classic studies such as James J. Flink's *America Adopts the Automobile*, the history of driving was generally taken as a given once everything else, from patterns of ownership to the formation of automobile clubs and the evolution of service facilities, was covered.[3] Since the 1980s, social historians in the United States have analyzed the experiences of women, farmers, and urbanites behind the wheel, but the cumulative amount of information conveyed by these studies is minor when compared to scholarship on the structures and organization of transportation systems, the development of the auto industry, patterns of state regulation, the history of car design, and other cognate topics.[4] Whereas more recent works by the American journalist Jane Holtz Kay and the German scholar Klaus Kuhm have specified the automobile's negative environmental and cultural effects, their ideological agendas lead them to overlook the nuanced history of human agents moving about and operating cars, while their focus on the present leads them to read contemporary crises into the past.⁵ Many billions of drivers all over the world have had a variety of experiences owning, operating, and identifying with their cars, yet their stories still constitute a tiny part of scholarly research on the automobile and an even more understudied topic in the social and cultural history of the twentieth century.

It is beyond the scope of this paper to speculate about the reasons for the relative paucity of scholarship on historical driving. But it is clear that inattention to drivers and driving practices results in a skewed historical narrative of the automobile. "A social history of the car that does not include the driver is a fetishized history," writes anthropologist Daniel Miller. "[Such a history] makes the critique of the car appear abstract and distant from the humanity in which it is involved.... The object and subject are set radically apart, and in much of the recent literature they appear mainly as antagonists where humanity is always the victim."⁶ Miller's critique should not be taken as a plea for a purely "agentic" view of the driver. After all, people driving cars find themselves in a wider field of power in which industry, the state, environmental groups, insurance agencies, and a host of other actors vie for influence. But even the most victimized drivers, whether they are Australian Aborigines⁷ or German Jews in Nazi Germany,⁸ have a degree of power, if only momentarily to escape insurmountable discrimination or oppression.

My general aim here is to sketch an argument about driving experiences and practices in Germany in the first third of the twentieth century, as part of a broader history of driving in Europe and the United States. Germany matters in the narrative of automotive history and motorized transport, of course, in part because German manufacturers and inventors such as Gottlieb Daimler and Carl Benz made the internal combustion engine commercially viable at the end of the nineteenth century, only to have French and American entrepreneurs take the necessary steps to create the modern automotive industry. After World War II, it was Germany that helped to shape the second major global surge in car production through the manufacture and marketing of the Volkswagen. Germany today remains one of the world's top three automobile producers, and its global reach was made painfully aware to the American auto industry in the creation of DaimlerChrysler, which was a culturally contentious takeover by the German car maker rather than a merger.⁹ If Germany's transformation into a leading motor vehicle producer was belated compared to the United States, France, or England, the automobile played an important cultural role in that society from the beginning of the automotive age.¹⁰ Whether as an object of desire or derision, the car put its stamp on Germany at a relatively early moment in time; German culture was motorized even if the society and economy were not. Germans imagined getting behind the wheel of a car well before many were able to do so.

My specific focus in the following pages is on how driving and representations of driving revealed attitudes about the interaction between human beings and natural environments. The initial idea was to use Leo Marx's evocative (and still much-used) metaphor of the "machine in the garden" as a point of entry.[11] But this concept proved to be inapplicable for two reasons. First, it derived both its metaphorical power and analytical purchase from an "externalist" view of the impact of technology on society. Marx analyzed how the literary imagination responded to the sudden intrusion of industrial development in American culture, allowing for little discussion of the historical experiences and social engagements of consumers and users. Second, the machine-in-the-garden metaphor was based on the idea that within American culture a persistent dichotomization of technology and nature operated through which the modern was always pastoralized, in "sentimental" or "complex" modes, as Marx used these terms.

Such a dichotomization was indeed present in German (and European) culture. Many German thinkers tried to overcome this dichotomization by conflating nature and technology in ways that were analogous to American responses, yet the conclusions they drew focused more often on industrial society's threat to the German Volk rather than on the clash between mechanical and natural environments per se.[12] Others, including those Jeffrey Herf described as "reactionary modernists," reconciled technology with Romantic ideas of nature and hierarchy, but they did so by rendering the modern in ahistorical terms, as a product of "fate." For many of them, the technological became "second nature": a product of social interaction and conflict became ontologized as cultural (and racial) essence.[13] No less important were those European thinkers and scientists who redefined nature for its machinic qualities, most notably in the notion of the "human motor" and the interdisciplinary discourse on "fatigue" as an impediment to labor power.[14] Recent research indicates there was tremendous variation in the response to industrial civilization in Germany, much more than scholars once maintained. Many Germans, from proponents of historic or nature preservation to advocates of technology, searched for "another modern" without either fully accepting liberal-capitalist notions of progress or resorting to reactionary modernism and fascism.[15] Building in part on this work, I argue that within German culture there was also the *potential* for acknowledging how the technical and the organic were intertwined in complex and historically specific ways without giving primacy to either pastoralism or technological determinism. The central question in the following pages is: how did this acknowledgment appear both as regards driving practices and ideas about the proper forms of driving in Germany?

Such questions remain the exception rather than the rule in the growing field of environmental history, in which the history of technology's impact as well as its social and cultural meanings have received systematic attention only sporadically. As Jeffrey Stine and Joel Tarr point out in a comprehensive review essay, there are "numerous superficial references to the intersections of technology and the environment" in scholarship, "but until recently this relationship has seldom constituted the principal focus of concern."[16] But Stine and Tarr also identify numerous important initiatives through which environmental history and the history of technology have converged with one another in creative ways. Moreover, recent scholarship has not entirely ignored the impact driving had on the environment. Studies of roads, landscapes, and cars in Germany, Austria, and Switzerland have accumulated over the past two decades.[17] Research has shown that German environmentalism held contradictory views on roads and cars. Deeply antimodern sentiments and rather traditional aestheticizing views of the landscape mixed with thoughtful legislative initiatives, impressive organizations, and practices based on the idea of adjudicating between development and preservation.[18] Scholarship on the German Autobahn is a well-developed subgenre here, as will be discussed. In American historiography, historians of the car's impact on the environment have also been active, though the amount of attention devoted to the experiences of drivers varies greatly.[19] A recent study by Paul Sutter demonstrates how anxiously important conservationists such as Aldo Leopold reacted to the motorized army of tourists that descended on American national parks in the first third of the twentieth century.[20]

Despite the richness of this scholarship, drivers most often remain hidden, or they appear as epiphenomena of large structures or big processes of planning, road building, theme park construction, suburbanization, transportation policy, automobile design, and (especially in the European context) politics and the state.[21] Individual agency and experience, and the rich variance of sensations and perceptions drivers had as they navigated through natural and urban spaces, get short shrift in such accounts, even when scholars acknowledge how the view through the windshield may have shaped prevalent definitions of nature in the first place. What is more, just as in Leo Marx's perspective, the relationship between the car and nature is explicitly or implicitly antagonistic, fraught with political and moral disablements, and deeply troublesome for an environmentally responsible worldview, however that may be defined.

The perceived antagonism between the automobile and nature is exemplified in J. R. McNeill's important study of twentieth-century environmental history in which the author discusses a "Motown cluster" of social, technical, and organizational innovations that created deleterious effects for the environment.[22] McNeill's roster of problems associated with the mass production of automobiles is by now well known. The roster now amounts to a kind of orthodoxy, not only among scholars but also in the wider culture of the United States, and it requires little elaboration here. What is significant, however, is that McNeill himself offers evidence that the environmental problems caused by the Motown cluster in most advanced industrial countries were temporary, and that things improved over the course of the twentieth century—though one often has to read the book against its major tenets to extract such information. Despite popular assumptions about the carnage caused by automobile travel on modern highways, statistical evidence indicated that car travel was safer than travel by horse,[23] and increasingly safer cars caused fatalities to decline dramatically when measured against total miles driven—even before vigorous state legislation required car makers to incorporate more safety features.[24] Much urban air in the United States, Europe, and Japan became dirtier but then cleaner throughout the course of the century thanks to state and industry initiatives.[25] McNeill discusses the environmental impact of the oil industry but he does not say much about critics over the course of the twentieth century who predicted a rapid depletion of fossil fuels even when known oil reserves were higher at the end of the twentieth century than at the beginning.[26] Being wrong so many times over such a long period should make any scholar wary of the claims advanced by those representatives of environmentalist history with strong political agendas—which is not to say McNeill should be numbered among these. In his discussion of air quality in Los Angeles, Athens, and other cities, McNeill writes, "Most citizens preferred driving cars and breathing smog to limits on driving and less smog." His intent was to underline the erroneousness of this choice, but one could just as easily marvel at the persistence of notions of individual rights, a heritage of modern liberal societies. Many environmental historians have neither trusted human agency nor defended individual liberty as a key priority, and their generally alarmist tone disallows more nuanced evidence demonstrating that, as Bjorn Lomborg writes, "Things are better—but not necessarily good."[27] It is unsurprising that environmental history treats driving as a disease, or ignores it altogether.

More than Marx's venerable metaphor or recent environmentalist ideology, Richard White's concept of the "organic machine" appeared to be a useful point of departure for addressing the car's itinerary through nature, though with the analytical signs modified to a substantial degree. White discussed the history of regulating and developing the Columbia River, where natural, political, economic, cultural, social, and technological forces interacted in often contradictory and paradoxical variations.[28] It is useful to think of drivers as being inserted in another kind of organic machine, a technological-commercial complex created by human labor, and encompassing cars, roads, auto dealerships, motels, repair shops, gas stations, suburbs, and strip malls—whole social infrastructures that engage, damage, or exploit nature, and that respond to nature's possibilities and limits in historically distinct ways. Whereas White's concept begins with a natural phenomenon and charts the processes whereby it became mechanized, my perspective begins with the man-made elements of an automotive culture and works toward natural environments. However, there is an additional variation on the concept that I would emphasize: within this machinic complex, "nature" refers not only to natural environments in the broadest sense but also to the organic nature of human operators and passengers themselves.[29] If cars and drivers are situated in a broader organic machine, then cars and the drivers interacting with one another also constitute organic machines in their own right whose meanings and itineraries reveal a deeply reciprocal, often conflict-laden dynamic of historical connectivities.[30]

Nature from behind the Wheel

In contrast to the United States between the world wars, when the automobile became fully enmeshed in everyday life, cars in Germany were still regarded mainly as vehicles for leisure driving. This is not to say that the utilitarian features of the car went unrecognized at this early moment in German automotive history. As they were in the United States and France, medical doctors who used their cars to make their rounds were among the earliest adopters of the new technology in Germany.[31] Even so, "gentlemen-drivers" (*Herrenfahrer*), often aristocrats or members of the wealthy upper middle classes (*Bürgertum*) who raced their cars as amateurs or tooled down public byways at high speeds, dominated the roads before World War I. But the roads were almost bare: there was only one motor vehicle per thousand population in Germany in 1914, a year in which the comparable number for the United States, England, and France was 17.8, 5.8, and 4.5 respectively. France was the European leader in car production, manufacturing more cars from 1902 to 1907 than all other European countries combined. It would retain this lead until 1924, when England, an avid importer of French autos, became the leading European car producer. But it was far behind the United States by 1913, when the American car industry accounted for an astounding 80 percent of global automotive production.[32]

Motorized transport proved its importance for the German military in World War I,[33] and the automobile was increasingly available for business and professional purposes among middle-class consumers by the 1930s.[34] The number of drivers' licenses issued increased by more than twelve-fold in the decade after 1920.[35] The transformation was even more telescoped than this suggests: in just four years, from 1925 to 1929, the share of traffic accounted for by horse-drawn vehicles fell from almost 55 percent of the total in 1925 to 26 percent, while the share for cars and motorcycles increased from nearly 35 to nearly 61 percent.[36] Even so, the density of automobiles varied considerably. Munich led the country in 1934 among German big cities with one motor vehicle for every twenty inhabitants. There was one vehicle for every thirty-one Berliners in the same year, but just one for every fifty inhabitants in Dortmund.[37] From a global perspective, Germany continued to lag behind, as there were 26.6 car owners per thousand inhabitants in 1939 compared to 236.9 in the United States and 57.3 in France.[38]

The syncopated spread of the car was unmistakable, then, but even when Adolf Hitler championed the "people's car," he had in mind not commuting or daily shopping trips but weekend outings in nature or motoring vacations for the working-class family.[39] A media project as much as it was an engineering feat, the Autobahn was designed to facilitate and enrich this entry into nature, as motorists were invited to cruise along sinewy, blue-gray bands of concrete that purportedly integrated technology and the environment in a truly *German* work of art.[40] In the United States, too, the automobile was an important tool in American tourism's engagement with the environment, but in Germany, where the rail system still accounted for the lion's share of commercial transport, the car's itinerary was shaped much more by leisure than by utility.[41]

What this situation meant for the evolution of driving practices may be briefly sketched in the following discussion of evidence culled from automotive writing in Imperial Germany (1871–1918), the Weimar Republic (1918–1933), and Nazi Germany (1933–1945). The first is an example of "road trip literature," a genre much less developed in Germany than in the United States but nonetheless still important for a history of European driving.[42]

The Impressionist writer Otto Julius Bierbaum was the first German-speaking author to write an account of a motoring trip in book form; all previous literature of this type had appeared in newspapers or car magazines.[43] Accompanied by his wife and a chauffeur, Bierbaum wrote of his "sensitive journey" (*empfindsame Reise*) through southern Germany, the Austro-Hungarian Empire, and Italy in 1902 in an Adler Phaêton. Bierbaum commented on forests, landscapes, architecture, hotels, and city life—the whole panorama of the organic machine that automobile use engaged in Europe. Significantly, he discussed the automobile not as a neutral instrument, but as a technical entity that both responded to and constituted human practice. This may be explained in part by the fact that the automobile was a much less reliable means of transportation in 1902 than in later years, and early automotive travel writers on both continents were compelled to comment on mechanical matters. Often, such writing pitted the car-driver against the elements of nature, as in popular accounts of the first transcontinental motor tours in the United States.[44] Moreover, having offered Bierbaum the opportunity to take the trip, the Adler firm was most interested in getting positive feedback on the car's performance. But this is not the whole story.

Bierbaum's account revealed important attitudes about relationships among the car, nature, and human users.[45] Bierbaum conducted his tour according to the motto "drive, don't race," thereby positioning himself against the early *Auto-Wildlinge*, or "automotive wild ones," who regarded the car as "an individualistic and ruthless speed machine." Referred to as "scorchers" in the United States and Britain, early hotrodders gained their inspiration not from the train but, as in the overwrought writings of the Italian futurist F. T. Marinetti, from the metaphors and figures of speech associated with early aeronautic culture.[46] These aristocratic sportsmen-drivers had already made their presence felt on European roads, causing many accidents and not a few deaths, and also arousing the ire of farmers and villagers, who hated the dust and noise of racing automobiles.[47] When Bierbaum reissued his auto-travelogue four years later, he repeated his advice in slightly modified form in the preface: "Learn to travel without racing." He argued that many motorists had taken his motto to heart, but that the "concept of sport" still dominated "automobilism."[48] Bierbaum's approach to motor touring required both slower (if still steady) speeds and more consistent acknowledgment of the people and places encountered by motorists. He had little tolerance for the "speed fanatics" who "have nothing to report about a trip in Italy other than the number of kilometers per hour they gobbled up."[49]

Harro Segeberg argues that Bierbaum's appeal for moderation was a "futile" gesture against an emergent auto industry that concentrated on racing and power as a way of making its products more appealing to the public. But he also notes that there were early motoring enthusiasts who advocated not speed per se but rather the individual's ability to regulate his tempo. This point is closer to the mark if one moves away from the literary evidence and focuses on social practice. For early motorists, the choice was never solely whether to speed or not to speed, except perhaps for the reckless few who embraced the "tumbling and incendiary violence" (Marinetti) of racing.

Rather, moderation and speed seemed to be two sides of the same coin. When the Austrian motorist Theodor von Liebieg took his summer tour of 1895, he not only drove his Benz "with the speed of the wind" but he also admired nature and the Reims cathedral, took lots of photographs, gave relatives and friends short rides, and drove in sync with bicyclists whom he met along the way.[50] His modus operandi was eclectic, as he mixed both the joy of speeding and the pleasure derived from sustained, moderate touring. Even so, Bierbaum's advocacy took on a more programmatic inflection as the driving culture evolved, at least judging from discussions in the car magazines. One writer in *Automobil-Welt* promoted "respectable driving" at slower speeds in 1910. In the same year, another early motorist, the head of a Magdeburg business firm, wrote in a letter to the NAG auto company published in the *Allgemeine Automobil-Zeitung* that on business trips he drove his NAG "Darling" at moderate speeds "so as not to be regarded as a wild driver." Three years later, Eduard Engler, writing in *Motor*, used Bierbaum's motto, and noted the general derision held by many drivers for speeders and irresponsible joy riders.[51]

Bierbaum insisted that the moderately driven car allowed one to perceive cultural and natural variations much more effectively than did the train, which Bierbaum criticized mightily throughout his account. "If one travels along in a touring car at a good pace, but nonetheless comfortably, as we did, then you notice that the great cultural steamroller [*Kulturwalze*] of the age had still not flattened out all differences," he wrote.[52] For the "sensitive" automobilist, modernity's alleged homogenizing effects were still far from complete. The diversity of landscapes and customs in German-speaking Central Europe were still much in evidence despite advances in transportation networks, rail lines, and urban amenities.

German thinkers such as the folklorist Wilhelm Heinrich Riehl and the geographer Friedrich Ratzel regarded the villages, farms, and small towns of rural society as part of nature, and therefore to emphasize the continuation of strong local variations, as Bierbaum did, was also to make an argument about the persistence of "authentic" natural environments.[53] In the strongest version of this viewpoint there was a tendency to see the city, the factory, and everything that went with them as a severe threat. In this view, modernity was capable of eviscerating age-old natural and cultural settings and replacing them with alienating, homogeneous, and inauthentic industrial-technological systems. But not all critics embraced this alarmist argument, and not all were uniformly opposed to technology, the city, and other elements of modernity. Some defenders of "homeland" culture, the devotees of the Heimat movement, argued that modern structures such as train stations, bridges, and motor roads could be inserted in natural environments with a minimum of damage, both physical and cultural. But generally, whether they advocated saving plants and animals or nurturing the aesthetic values of landscapes, the "homelanders" believed, much as their American counterparts did, that nature had a determinant influence on national or even racial character. In Germany, the Heimat movement had a strong regionalist and anti-centralist bias, shaped in no small part by Riehl's writings. But even when they were defending a local or regional setting, they regarded their work in a broader cultural-national context. To protect nature against the cruder defilements of industrial-technical advance was to somehow preserve the authentic character of the regionally nested and internally diverse Volk.[54] To adapt to modernity was to undertake a kind of preemptive strike, whether as a full mobilization or a more dispersed action, against the more threatening and homogenizing tendencies of industrialization and urbanization.

In assessing the car as a positive cultural value, Bierbaum set himself up against the stronger version of the aforementioned Heimat argument, and he offered a significant qualitative twist to the weaker one. For Bierbaum, recognition of authenticity was facilitated not hindered by a technological mechanism, the car, whose attributes gave it a purchase on nature that the train could never offer. The automobile was not only to be tolerated as long as it did not make a ruckus in nature; rather it was to be embraced as a path to and through the natural world. In addition, by linking auto-driving with heightened recognition of local cultural difference, Bierbaum addressed the homelanders' most basic value as he insisted that the result of interaction between nature and the machine was both pleasurable and productive of cultural improvement. Perhaps because his argument engaged contemporary debates at a number of angles, Bierbaum felt he had to reiterate his point in *Mit der Kraft*, in which he wrote: "No, the world is not yet so miserably alike everywhere as it is at train stations and their environs."[55]

Awareness of the motorist's environs put the emphasis not on the destination alone, but rather on the pleasurable experience of passage itself and on the autotelic motion of car travel. What was it exactly that made automobile travel so appealing? Aside from the ability to view passing landscapes and peoples, Bierbaum wrote of the "exhilaration of movement" (*Bewegungsrausch*) for its own sake, nowhere more delightfully felt than when motoring along a country lane or a small village street.[56] "Passage," Eric Leed reminds us, "may dissolve inherited containments and be experienced as a cure of the diseases and attributions acquired in place."[57] The therapeutic tradition of car travel was already established by the time Bierbaum wrote, but it should be emphasized because it reflects Bierbaum's idea that the car and nature (or more specifically, human nature) existed in a complimentary rather than contentious or exploitative relationship. Whereas some saw such movement as just another symptom of the "age of nervousness," others were willing to emphasize its soothing or ameliorative effects, sometimes with respect to the bicycle, but increasingly in connection with automotive travel.[58] Moving along in the car had the added effect of allowing one to hear the sound and to feel the vibration of the engine or, as Carl Benz once stated evocatively, "to feel the marvelous language that the gear wheels talk when they mesh with one another."[59] This was hardly only the rapturous car-talk of the father of the modern automobile, for one could hear such language in more ordinary accounts, as in the 1913 piece on a Harz motor tour in which the author wrote, "The roads are good, the weather beautiful, and soon the powerful machine sings the song we have come to love."[60]

One can find numerous traces of this perspective in ensuing decades throughout Western car cultures as well. We find it in England between the wars, where driving is celebrated as an antidote to depression or a spur to alertness and creativity.[61] It persists in the United States in many different settings, not least in the literature on long-distance touring from roughly 1915 to the late 1920s, when the reliability of cars and the slow improvement of roads had taken some of the earlier uncertainty out of driving but the bodily pleasure afforded by motoring was still experienced as something novel and exhilarating.[62] In the 1950s we find it in George R. Stewart's account of his travels on U.S. 40, a highway that once stretched from Atlantic City to San Francisco. "There are, too, those vague sensations that we call kinesthetic," wrote Stewart, a professor of English. "All the rivers of fresh air coursing over the face. The pressure backward with acceleration . . . and the continual joggling from the springs, doubtless good for the digestion and the nerves and the general well-being, reminiscent perhaps even of the joggling of the child within the womb."[63] Of course, the tradition of pinpointing the pleasure of movement, the sound and feel of the engine, indeed, the entire spectrum of sensations association with motoring "passage," gains expression innumerable times in hotrodding cultures in a variety of social milieux. This tradition fits firmly within the discourse of the "human motor," which Anson Rabinbach so effectively analyzes as the constitutive metaphor in modernity's invention of "labor power."[64] But in this case the goal was not ever greater productivity, at least not directly, but rather the pleasure of experiencing how the machine, the human driver, and nature existed in harmony, replicating each other's rhythms, movements, and patterns. It was consistent with the logic of the organic machine whereby, as White writes, "the emphasis was on mimicking and not conquering nature."[65]

Bierbaum's championing of the car over the train was analogous to the walker's claim that travel by foot was preferable to mechanized transport. Walking as an end in itself had centuries-old roots in Europe, but at the time Bierbaum wrote, many Germans advocated walking as an antidote to the alleged alienating effects of train travel or mass tourism.[66] From the perspective of much travel literature of the time, the good bourgeois "pedestrian," his slow but knowledgeable gaze fixed on natural beauty or cultural monuments, achieved a level of cultural understanding that was closed to denizens of the weekend bus tours or to the working-class travelers in third- and fourth-class train cars.[67] Against the sightseeing of Cook's Tours and the Stangen travel agency's group excursions, the bourgeois peripatetic, often shaped by the advice of the Baedeker handbook, appeared to be the superior alternative.[68] Measured walking of this kind had its quasi-populist elements in turn-of-the-century Germany as well, as youth groups advocated wandering throughout the countryside, away from machines and the bustle of the modern city. When such advocacy combined with cultural criticism of modern civilization, as happened in the small but influential *Wandervogel* movement, it often included a strong racist element, as Jews became identified with capitalism and the commercialization of leisure time, including leisure travel. For participants in such practices and discourses, peripatetic offered a more organic relationship to nature and culture, just as for Bierbaum automobile travel engaged the individual in the external world in a manner that train travel made impossible. Above all, walking and slow automotive promenading through nature shared the potential to link the individual and the community in mutually reinforcing ways rather than falling prey to either the constricting collectivism of the train or the unbridled individualism of reckless automotive racing. The speeding train or racing automobile presented comparable threats to nature's ability to mediate identities in this manner.

Driving through the scenic Vienna Woods on the way from the Austrian capital to Munich in April 1902, Bierbaum wrote "We took our time and often stopped to gaze about leisurely."[69] An extended stay in Vienna seemed to put the car in rhythm with the tempo of the culture, as "the Adler assumed its most beautiful four-stroke time, and it appeared as if the motor was happy to be able to work again."[70] Anthropomorphic themes reappeared in variations throughout the account.[71] Reunited with the car after touring Venice, Bierbaum and his wife greeted the Adler "like an old friend." The chauffeur had "visited" the car several times during their Venice stay, equipping it with new tires and inspecting it for mechanical problems. "This personal relationship to the object is no doubt pleasant and not at all a burden," wrote Bierbaum. "It is even a kind of love, and it is satisfying to nurture it." The author went on to say that women in particular could have great empathy for mechanical artifacts "because they understand better than men to treat even objects with sensitivity."[72]

The full implications of this statement must be left for another time, but it is important to note that the argument of women's natural inclination to take a nurturing stance toward the car provides rich material for investigating the "sex of things" as relates to automotive history.[73] It also suggests that the tried and true gender divide associated with the automobile—whereby the man took care of the car's mechanical needs while the woman concerned herself with "superficial" items such as the color of the body and upholstery—was open to a potential for contradiction. After all, if Bierbaum posited "woman" as a more sensitive auto-user, then why could this argument not extend to the care and maintenance of the engine, brakes, and other mechanical components? In this case, the relation between (human) nature and the car issued into broader questions about gender stereotypes and their effects in the social practices of automobility.

In Bierbaum's account, the car not only gave the motorist the opportunity "to converse pleasantly"[74] with natural scenery, but also created an interactive and even loving relationship between the organic and the inorganic, between the human user and the machine. This was a far cry from the message of the *Zivilisationskritiker*, those who railed against the devastating effects of industrialization on German culture and nature. But it was also inassimilable to the discourse of someone like Walter Rathenau, the philosophically minded head of German General Electric whose musings on technology and the German "soul" were enormously revealing for German understandings of modernity on the eve of World War I.[75] Rathenau embraced technology, but he did so anxiously, seeking ways to assert the power of German culture and bourgeois ethics over modern industrial society. Bierbaum's road trip points to a more reciprocating and relaxed attitude toward the machine that was nonetheless not without a critical purchase on the negative effects of modernization as evidenced in train travel or the devil-may-care auto-sportsmen. Above all, Bierbaum loved cars, stating that those who criticized the automobile were incapable of "fantasy," and insisting that to drive a car was potentially to "de-philistinize" humankind.[76]

Making Peace with the Machine

The writings of the journalist, novelist, and travel writer Heinrich Hauser provide examples of a related perspective from the period between the wars. Eclecticism was the watchword of Hauser's oeuvre, which included a novel,

Brackwasser, that won the Gerhard Hauptmann Prize in 1929; a history of the Opel firm; a travelogue and film recording of a voyage he made from Hamburg to Talcahuano, a Chilean port city; an account of his experience as a farmer in Missouri during World War II; a controversial commentary on the American Occupation of Germany; and a stint as the editor of the illustrated *Stern* magazine.[77]

Hauser also wrote about driving cars. In 1928 Hauser penned *Friede mit Maschinen*, or *Peace with Machines*, a primer on industrial culture in Germany.[78] Hauser wrote to overcome what he saw as the educated middle classes' resistance to technology, a quality he attributed in part to the persistence of their allegiances to classical learning and their romanticizing visions of the natural world. Teaching his readers about the manufacture and operation of automobiles was an important part of the effort to get Germans to live productively with modern technology. Paradoxically, this project was made even more difficult by the evolution of the machines themselves. Unlike the type of automobile Bierbaum used, the cars of the 1920s functioned so well—Hauser used the English term "foolproof" to describe them—that they seemed to alienate the driver from their mechanical nature.[79] Such a critique was not new, of course; one can find it already before World War I in both Europe and America, often in conjunction with the argument that automobiles had become "feminized" through the use of closed bodies (which became particularly widespread in the 1920s), automatic ignitions, and even body styles that substituted faddishness for functional integrity.[80] These transformations had made "motoring . . . curiously more amenable to woman than to man," in the words of one British observer.[81] Hauser did not make this connection, but the association circulated widely in Germany and elsewhere. Nor did Hauser go so far as to refer to the contemporary driver as an "automaton," as one prewar observer did, although again the critique was implicit.[82]

Hauser's work was influenced in part by the developing tradition of "reactionary modernism," which, in the prolific writings of Ernst Jünger, combined romanticism and aggressive, technological visions derived from the "front experience" in World War I. Many thinkers railed against technology's dehumanizing effects in Germany after the Great War, but Jünger, far from being a technological determinist, argued that technology should become the object of a "magical realism" in which rationality and mystery were wedded in new constellations appropriate to the time.[83] Fascinated with machines, Jünger maintained that human potential lies in the freedom to mobilize one's energies within the technologized *Gestalt*, or form, of the historical epoch.[84] Against the widespread charge that Jünger was a nihilist who longed for a kind of technological apocalypse, Thomas Nevin argues persuasively that he was optimistic, in a darkly austere way, about humankind's ability to bear the burdens of a machinic future.[85]

What is more, Jünger foresaw the eventual "fusion of difference [*Verschmelzung des Unterschiedes*] between the organic and mechanical world."[86] This implied that ideas about harmonization of nature and technology, based as they still were on dichotomous thinking, were ultimately irrelevant, since the world was tending toward a complete evisceration of the distinctions on which such notions rested. As for the practice of driving cars, Jünger's conclusions suggested not the human being working in syncopation with car and nature, not even the human motor, but rather the idea that the machine clothed the body as a kind of exoskeleton. To the degree that this organic machine envisioned a cyborg-like entity, it revealed the startlingly futuristic component of Jünger's thought.[87] But it also suggested a less synergetic vision than that deployed in the idea of the organic machine; for Jünger, the fundamental issue was not the reciprocal imitation of the organic and the inorganic, but the explosion of human qualities and their reallocation *inside* a machinic form. As for automobile design, we see cognate ideas in exotic concept vehicles such as General Motor's Hy-wire.[88] This creation has garnered attention from environmentalists because of its hydrogen fuel-cell technology, but it also features a "skateboard" concept allowing drivers to snap numerous bodies onto its chassis, almost as if they were wearing a different skin to adapt to the environment.

Hauser's small book also drew more broadly from *Neue Sachlichkeit*, or New Objectivity, whose advocates sought a "matter-of-fact" (*sachlich*) and basically positive relationship to the commercial and technical artifacts of a society basking in the glow of misleading prosperity before the Great Depression. The automobile fit this project well because, as Helmut Lethen writes, "the central topos of new objectivity literature is 'traffic.'" "Traffic," or *Verkehr*, could mean circulation, exchange, movement, or modern urban transportation including trains, streetcars, and automobiles. Urban traffic enabled "cool conduct" based on a new appreciation of reflex actions, as a driver responds to traffic signals and the actions of other drivers. Overcoming "resistance" through habitual behaviors such as driving created a context for freedom and for concealing the inner tensions that were part and parcel of the modernist sensibility. Responding to traffic matter-of-factly, coolly, also foregrounded the "superiority of the supra-individual system that directs the behavioral forms."[89] Even so, Hauser was trying to depict the "humane" in the automobile.[90] But in so doing he aimed for more than serving industry's needs by calming the nerves of bourgeois consumers, as Lethen argued.[91] Instead, he was making a point about how organic and machinic elements interacted (or should interact) in German culture. He thereby carried forward Bierbaum's project, though under different historical conditions.

Hauser often anthropomorphized the automobile. The car he would use for his first driving lesson was introduced to the reader almost as if it was a prized racehorse, with its "vibrating motor," glistening paint, and high radiator. One recalls the scenes from Walther Ruttmann's 1927 film *Berlin: Symphonie einer Großstadt* in which shining automobiles are wheeled out of garages like noble animals. Here Hauser drew on the tradition of the gentlemen-drivers, whose key leitmotifs were derived not only from prewar aeronautics but also from a much older tradition, that of equestrian culture.[92] Just as the medieval knight identified closely with his loyal horse, so the modern driver treated his automobile as a "trusted steed," as the Austrian driver Theodor von Liebieg had referred to his Benz in 1895.[93] The driving experience was rendered in naturalistic metaphors, as Hauser described the entry into a "canalized stream" once his instructor drove him into the flow of traffic. Learning to drive was a lesson in the history of reciprocity between humankind and machines. Hauser observed that the blue leather seats of the instructor's car were worn like a much-used club chair, that the metallic luster of the emergency brake lever and gearshift were clouded by the sweat of countless human hands, and that the clutch, accelerator, and brake pedals were "deeply worn away down to the brass."[94]

In this account, the car functioned as a kind of mobile memory site of the everyday in which countless, anonymous actions by drivers and passengers left their traces. This perspective will be unsurprising to the numerous car enthusiasts who have treated automobiles as foci of nostalgia, childhood dreams, and romantic evocations of first love.[95] But it is intriguing that it has rarely been considered in academic scholarship, even despite the fact that a whole generation of writing on European and American memory cultures has come to an end, and a new phase of work on memory has begun. Only as an exception has the history of memory taken the car as a serious object of analysis.[96]

Earlier in the book, Hauser compared pre–World War I automobiles and 1920s models, arguing that the former "looked more human."[97] He participated in a broader critique here, evident in both Britain and Germany. The idea was that cars had become more uniform and less imitative of the human beings that operated them, which meant that the cars had become less distinctive and unique or, as in the case of discussion that went on around the 1928 Berlin auto show, less reflective of national differences.[98] Design features aside, even the newer vehicles took on organic qualities, though only with human use, according to Hauser. "Every one of these autos has very distinct and personal characteristics, a thoroughly individual face [*Gesicht*]," he wrote of late model used cars, "a car receives its appearance only through use, just as a human also gains an appearance through life."[99] From a narrower perspective, one can see how such an observation anticipated the Nazi concept of a "comradeship with the machine."[100] In this view, the machine world was available for manipulation by the collectivity, whose ultimate goal was the aggressive expansion of national power.[101] National solidarity entailed the complete mobilization of machinic resources, which, in the form of cars, motorcycles, planes, and other technological wonders, achieved the status of völkisch comrades, or perhaps cyborg adjuncts of the racial community. Operating in such a context, the machines not only were products of racial "know-how"; they also took on the character of the Volk that used them.

But a transnational view suggests another perspective. Hauser attributed his view not to National Socialism but to American influences: Buster Keaton films gave cars individual personality traits, whether funny or threatening, and in big cities in the United States, police were trained to recognize minute differences between autos to help them hunt down car thieves. Paradoxically, the only country in which the mass production of automobiles had become a reality also gave the car an individuated personality through human usage, according to the German journalist. There was no shortage of commentators in the United States who would have agreed with Hauser, of course. America's embrace of the car was based on ideas of individuation and consumer preference. To ask "how do buyers feel about today's cars," as one American car magazine did, was to evoke an entire culture of individualized interaction with machines.[102] The British automotive journalist Owen John also took Hauser's position on the effects of car use, but he stated that whereas American car culture had become more homogenized, cars and drivers in England still evidenced individual traits and idiosyncracies.[103] Neoconservatives such as Jünger or Oswald Spengler criticized processes of individuation as a typical excrescence of (usually) American consumer society. In contrast, Hauser embraced the development, and he saw driving automobiles as an important social practice in which individuation was objectified in the material environment. We become better individuals (and better citizens?) by driving cars, Hauser claimed, and the cars themselves bear the traces of this process of self-improvement through usage. What is more, in contrast to the National Socialist aestheticization of a "modern product culture," including cars and other technical artifacts, Hauser consistently stressed the way in which quotidian practice and human agency situated automobiles in the broader culture.[104]

If this deeply reciprocal relationship occupied much of Hauser's discourse on humanity and technology, then so too did the Autobahn in the 1930s.

Envisioned and partially planned before Hitler's rise to power, the Autobahn project under the Nazis resulted in the building of some 3,800 (out of a projected 7,000) kilometers of high-speed, limited-access highway. In the Autobahn we have an example of an organic machine in which technology and nature appear to have been fused through National Socialist politics. Celebrating the new superhighways, hundreds of writers, artists, and filmmakers echoed regime propagandists' (and not a few environmentalists') claims that the Nazis had finally overcome the deep chasm in German history between the machinic and the natural. Scenic bridges (some nine thousand of them!), undulating strips of gray concrete, quiet rest areas, roadside plantings, the absence of advertising, hills and forests—all suggested the triumph of this nature-friendly vision. But in fact the inorganic had subdued nature, and Nazi politics demanded that motorists recognize their "Aryan" heritage by passively consuming the lovely vistas through the windshields of their cars.[105] Moreover, environmentalist perspectives had a purchase on Autobahn planning only when they corroborated functional and financial objectives.[106] This is not to say that functionalist planning's victories were inevitable: Thomas Zeller demonstrates that conflicts ensued between engineers who favored a driver-oriented vision through which Autobahn vistas were appropriated as a commodity, represented by Fritz Todt, and those who had a more conservative understanding of the landscape rooted in Heimat tradition, for whom Alwin Seifert stood as a key spokesman.[107]

Though full of praise for Hitler's "genius," Hauser's commentary on the Autobahn again suggests the potential for an alternative, or perhaps for an interstitial reworking of Autobahn propaganda. Writing in 1936, Hauser proposed that the Autobahn necessitated a new driving practice, "automotive wandering," which in its intent and focus was not unlike Bierbaum's road trip project from more than three decades before.[108] Automotive wandering required slower speeds and a "more thoughtful driving experience." It necessitated smaller, more agile cars that would allow drivers to pull off the roads, to pass more quickly, and to be more in control of their machines. No less significantly, it highlighted the need for more sensitive interactions with the environment. A Sunday drive in Germany reflected the abysmal state of motoring in that country, according to Hauser. People still parked on curves, threw trash on the roadside, and undertook their motoring tours with too much equipment—everything from tents and canoes to gramophones. They should be learning from the Indian or the soldier, who knew how to make the most out of just a little equipment. Motorists were also taking too much food with them. To pack food for a longer journey was economical, observed Hauser, but this practice also robbed one of experiencing local cuisine and customs. "Eat and drink everywhere what the farmer eats and drinks," he counseled.

Writing about Autobahn design, William Rollins asks the question "whose landscape?" to pinpoint important differences between the goals of the Nazi party, technologists, and environmentalists.[109] Hauser might have insisted on adding drivers to the list because he outlined a mode of travel in which the Autobahn landscape opened up a potentially more sensitive relationship between humans and the environment than that envisioned by the political patrons or engineers and builders of "Hitler's roads." The moderation of speed was an important issue in configuring a driver's agenda, to be sure, as it had been for Bierbaum. But the discourse on speed had become riskier and more fraught with political controversy at the time Hauser wrote. Not only had advocates of a radicalized technologism made their activist claims. Jünger wrote of machines and technology that "there is no way out, no sideways or backwards; what is much more important is to increase the force and the speed of the processes in which we are involved."[110]

But the building of the Autobahn had forced the issue in more immediate terms. Auto magazines were full of articles stumping for cars and engines that not only went faster but that could sustain high speeds for longer periods of time. Faster cars not only more quickly got the occupants to their destination, they were also more efficient in terms of fuel costs and driver's effort, or so it was claimed.[111] Most German cars were either too ponderous for such purposes, or too underpowered. Autobahn driving caused premature wear and tear and mechanical breakdowns in such cars, prompting drivers to find ways to make their cars more "Autobahn ready."[112] German as well as foreign automotive magazines contained complaints about Germans driving too slowly on the Autobahns.[113] Among the numerous "Autobahn sins" reported by a *Motor-Kritik* correspondent was the slow driver who misunderstood the purpose of the fast lane.[114] One British traveler noted in 1936 there had been lots of trouble on Autobahns caused by people not pulling in fast enough after passing, thus "baulking faster cars behind."[115] The Aubobahn necessitated precisely the kind of acceleration the radical technologists called for, but many drivers of the 1930s appear not to have learned this lesson, or to have resisted it.

Ironically, it would be Hitler, Germany's most famous non-driving car enthusiast, who would put the brakes on speed discourse. Hitler had done everything he could to promote motorization, not only championing the Volkswagen and the roadbuilding scheme, but also doing away with the license fee for new cars and motorbikes, and even abolishing mandatory driving school attendance.[116] His regime had promoted speed, efficiency, and production—in rhetoric if not always in fact. In automotive culture, the state pumped resources into racing, and the Mercedes and Auto Union teams achieved enormous success in international events in the 1930s.[117] Even so, months before the outbreak of World War II, things had changed. Hitler complained to an audience at the Berlin Automobile and Motorcycle Show on 17 February 1939 that since the start of the Nazi regime, German road fatalities nearly equaled the number of deaths resulting from the Franco-Prussian War of 1870–1871, though he failed to say why this was a relevant comparison. In fact, more than 240,000 accidents were recorded in Germany in 1938, a year in which the index of accidents caused by speeding had risen to 106.3 (in 1936 it had been 100). Germans mistakenly assumed that average Autobahn speeds should exceed 120 kilometers per hour, whereas the Führer argued that 80 kilometers per hour was much safer and more reasonable. Hitler branded speeders as enemies of the Volk, and he promised the state's swift retribution against those who had not eliminated themselves already by careless driving on the Autobahn.[118]

Historians of modern Germany would be quick to contextualize Hitler's concerns about Autobahn driving as a product of his preoccupation with military preparedness, the careful overseeing of raw materials, and propaganda needs. They would be correct to do so, but only partially. The German leader's anxious diatribe against accidents and speeding also reflected and furthered a growing conversation in German society about the car accident as a cultural form of modern automotive environments. German car commentators had discussed automotive accidents for some time in relation to insurance, traffic regulation, police practice, drivers' education, speed limits, and many other topics.[119] In the United States and France, as recent scholarship has demonstrated, a flurry of writing, of a literary, forensic, and policy-oriented nature, grew up around the automotive accident as the twentieth century wore on.[120] Hauser's anti-speeding critique, like Bierbaum's, was also an effect of this wider concern. For Hauser, to discover more productive relationships between humankind and machine was also to slow down and avoid the increasingly alarming drama of car accidents that was being played out on the new superhighways. Hitler added ammunition to this argument, but for reasons that were secondary to Hauser's and other's concerns. It was not the first time that everyday needs became conflated with the Nazi regime's military and genocidal goals; or that "normalcy" was mobilized to meet terror's agenda; or for that matter, that Hitler used words with which the populace could identify even though he meant something very different.

By getting Germans to drive more sensitively and economically, Hauser hoped to strengthen the national community. In Nazi ideology, this concept was based on biologistic categories. National community was not based on an act of political assertion but rather on racial characteristics; Germans were not members of the nation for what they did but rather for what they were. Nature and culture were suffused with the traces of German racial being, just as historically evolved life-worlds constituting the Volk were honed by constant interactions with natural environments. An admirer of Hitler, Hauser nonetheless seemed to defy this understanding. He married a Jewish woman and wrote several critical articles in the late 1930s. When he fled Nazi Germany to live temporarily in America, he only grated more on the regime. But not for this reason alone was his writing significant.

Hauser drew on a tradition in which the automobile was situated as a point of interaction among nature, human users, and technology. Along with Bierbaum and numerous devotees of German car cultures, he appears to me to be among the most articulate representatives of a position that was finally inassimilable to all the discourses German cultural history scholarship has done so much to identify, whether antimodern critique of civilization, reactionary modernism, New Objectivity, environmentalism, or Nazism. The authors discussed here neither raged against the machine nor embraced technology to prepare for bloody wars to come. They neither fetishized the car as a "pure" technical artifact nor rendered it in unmediated organic forms. They neither reduced technology to "second nature," nor treated nature as an adjunct to man's technical progress. Instead, they tried, if only imperfectly, to see cars and drivers together as organic-machinic complexes embedded in larger sets of historical (and therefore changeable) relationships.

The foregoing suggests that the notion of the organic machine, in the double sense I have used the term, may have been an important determinant of the automotive culture, at least in the first half of the twentieth century. Above all, the proponents of this view argued that driving was a process of community-building and nation-building. Germans became who they were not due to their racial characteristics, but rather in part by how they drove. In a society increasingly shaped by the exigencies of mass mobility and communications, the car, a vehicle of pleasure and an insertion point of human action in both culture and nature, was a platform of societal belonging. Not passive consumerism or unthinking spectatorship, but rather active, self-conscious driving should shape Germans' use of the car. Driving therefore became an important constituent of civil society, which in Germany had been wracked by war, eviscerated by political conflict, and mobilized for what became a genocidal conflagration. Yet a way of driving had been established that Nazism was unable fully to mobilize or capture, a way of driving in which the relationship between the car and nature was not one-sided or exploitative but mutually reinforcing and complementary. If a truly mass culture of automobility evolved in the two Germanys only in the 1960s and 1970s, its cultural precedents—and in particular its varied ways of conceptualizing relationships between cars, drivers, society, and nature—were evident before World War I.

This leads me to conclude with a number of questions: Did popular discourse do a better job of envisioning the automobile's deeply malleable relationships to both the natural world and the human actor than either most German thinkers or the state did? If popular discourse, or the substantial part of it constituted by car-talk, "got it" in Germany or Europe, then can we make the same assertion about the United States, where the triumph of the automobile was earlier and more complete? If we offer an affirmative response to these questions, then what are the implications for present-day public debate, in which a deeply ahistorical anti-car rhetoric carries the day? If in modern industrial and civil societies we are what we drive, what happens when shortsighted policies become so preoccupied with the environmental costs of driving that they lose sight of the other civic contributions of the car?

Notes

1. Rudi Volti, "A Century of Automobility," *Technology and Culture* 37, no. 4 (October 1996): 663–685, here 663.
2. An important exception is Virginia Scharff, *Taking the Wheel: Women and the Coming of the Motor Age* (Albuquerque: University of New Mexico Press, 1991); see also H. F. Moorhouse, *Driving Ambitions: A Social Analysis of the American Hot Rod Enthusiasm* (Manchester: Manchester University Press, 1991). For a recent study from European historiography that devotes significant attention to the history of driving: Sean O'Connell, *The Car in British Society: Class, Gender, and Motoring, 1896–1939* (Manchester: Manchester University Press, 1998). For a fuller statement of this point with reference to European social history, see my "On the History of the Automobile in Everyday Life," *Contemporary European History* 10, no. 1 (2001): 143–154.
3. James J. Flink, *America Adopts the Automobile, 1895–1910* (Cambridge, Mass.: MIT Press, 1970).
4. Michael L. Berger, *The Devil Wagon in God's Country: The Automobile and Social Change in Rural America, 1893–1929* (Hamden, Conn.: Archon Books, 1979); David A. Kirsch, *The Electric Vehicle and the Burden of History* (New Brunswick, N.J.: Rutgers University Press, 2000); Ronald Kline, *Consumers in the Country: Technology and Social Change in Rural America* (Baltimore: The Johns Hopkins University Press, 2000); Clay McShane, *Down the Asphalt Path: The Automobile and the American City* (New York: Columbia University Press, 1994); Scharff, *Taking the Wheel*.
5. For a recent example, Jane Holtz Kay, *Asphalt Nation: How the Automobile Took Over America and How We Can Take it Back* (Berkeley: University of California

Press, 1997); a recent German study by Klaus Kuhm, *Das eilige Jahrhundert: Einblicke in die automobile Gesellschaft* (Hamburg: Junius, 1995), proposes to analyze driving as a "sickness," but actually contains precious little discussion of the *experience* of operating an automobile.
6. Daniel Miller, ed., *Car Cultures* (Oxford: Berg, 2001), 9.
7. Diana Young, "The Life and Death of Cars: Private Vehicles on the Pitjanjatjara Lands, South Australia," and Gertrude Stotz, "The Colonizing Vehicle," both in *Car Cultures*, ed. Daniel Miller, 35–57, 223–244.
8. It is extraordinary to find how much time Victor Klemperer devotes to describing his experiences owning and operating a used Opel in his celebrated *I Will Bear Witness: A Diary of the Nazi Years, 1933–1941* (New York: Random House, 1998).
9. David Waller, *Wheels on Fire: The Amazing Inside Story of the DaimlerChrysler Merger* (London: Hodder and Stoughton, 2001).
10. Christoph Maria Merki, *Der holprige Siegeszug des Automobils, 1895–1930. Zur Motorisierung des Straßenverkehrs in Frankreich, Deutschland und der Schweiz* (Vienna: Böhlau, 2002); Barbara Haubner, *Nervenkitzel und Freizeitvergnügen: Automobilismus in Deutschland 1886–1914* (Göttingen: Vandenhoeck und Ruprecht, 1998); Reiner Flik, *Von Ford Lernen? Automobilbau und Motorisierung in Deutschland bis 1933* (Cologne: Böhlau, 2001).
11. Leo Marx, *The Machine in the Garden: Technology and the Pastoral Ideal in America* (Oxford: Oxford University Press, 1964).
12. George L. Mosse, *The Crisis of German Ideology: Intellectual Origins of the Third Reich* (New York: Grosset and Dunlap, 1964). See also Joachim Wolschke-Bulmahn, "The Nationalization of Nature and the Naturalization of the German Nation: 'Teutonic' Trends in Early Twentieth-Century Landscape Design," in Joachim Wolschke-Bulmahn, ed., *Nature and Ideology: Natural Garden Design in the Twentieth Century* (Washington, D.C.: Dumbarton Oaks Research Library, 1997); Joachim Wolschke-Bulmahn and Gert Gröning, "Nationalistic Trends in Garden Design in Germany during the Early Twentieth Century," *Journal of Garden History* 12 (1992): 73–80.
13. Jeffrey Herf, *Reactionary Modernism: Technology, Culture, and Politics in Weimar and the Third Reich* (Cambridge: Cambridge University Press, 1984).
14. Anson Rabinbach, *The Human Motor: Energy, Fatigue, and the Origins of Modernity* (Berkeley: University of California Press, 1990).
15. Thomas Rohkrämer, *Eine andere Moderne? Zivilisationskritik, Natur und Technik in Deutschland, 1890–1933* (Paderborn: Schöningh, 1999); "Antimodernism, Reactionary Modernism and National Socialism. Technocratic Tendencies in Germany, 1890–1945," *Contemporary European History*, 8, no. 1 (March 1999): 29–50. See also Rudy Koshar, *Germany's Transient Pasts: Preservation and National Memory in the Twentieth Century* (Chapel Hill: University of North Carolina Press, 1998); Thomas Lekan, *Imagining the Nation in Nature: Landscape Preservation and German Identity, 1885–1945* (Cambridge, Mass.: Harvard University Press, 2004); Kevin Repp, *Reformers, Critics, and the Paths of German Modernity: Anti-Politics and the Search for Alternatives, 1890–1914* (Cambridge, Mass.: Harvard University Press, 2000); William Rollins, *A Greener Vision of Home: Cultural Politics and Environmental Reform in the German Heimatschutz Movement, 1904–1918* (Ann Arbor: University of Michigan Press, 1997).
16. Jeffrey K. Stine and Joel Tarr, "At the Intersection of Histories: Technology and the Environment, Updated with a Postscript by the Authors," *Technology and Culture* 39, no. 4 (1998): 601–640; updated at http://www2.h-net.msu.edu/~environ/historiography/ustechnology.htm (accessed 17 January 2005).

17. Georg Rigele, *Die Großglockner Hochalpenstraße: Zur Geschichte eines österreichischen Monuments* (Vienna: WUV Universitätsverlag, 1998); *Die Wiener Höhenstraße: Autos, Landschaft, Politik in den dreißiger Jahren* (Vienna: Turia und Kant, 1993); Walter Zschokke, *Die Straßen der vergessenen Landschaft: der Sustenpaß* (Zurich: gta Verlag, 1997).
18. Karl Ditt, "Nature Conservation in England and Germany 1900–1970: Forerunner of Environmental Protection?" *Contemporary European History* 5, no. 1 (1996): 1–28; Michael Wettengel, "Staat und Naturschutz 1906–1945: Zur Geschichte der Staatlichen Stelle für Naturdenkmalpflege in Preußen und der Reichsstelle für Naturschutz," *Historische Zeitschrift* 257, no. 2 (1993): 355–399.
19. See for example, David M. Wrobel and Patrick T. Long, eds., *Seeing and Being Seen: Tourism in the American West* (Lawrence: University Press of Kansas, 2001). This collection includes Marguerite S. Shaffer, "Seeing America First: The Search for Identity in the Tourist Landscape," which contains substantive discussion of auto-tourists' experiences, as well as David Louter, "Glaciers and Gasoline: The Making of a Windshield Wilderness, 1900–1915," in which travelers' sense of their surroundings receives relatively little sustained analysis.
20. Paul Sutter, *Driven Wild: How the Fight Against Automobiles Launched the Modern Wilderness Movement* (Seattle: University of Washington Press, 2002).
21. See for example, Kenneth T. Jackson, *Crabgrass Frontier: The Suburbanization of the United States* (New York: Oxford University Press, 1985), which offers much analysis of the car and the suburb, but only incidental information about how people drove cars, what their experiences were, and what they said about driving. More recently on the suburbs, see Adam Ward Rome, *Bulldozer in the Countryside: Suburban Sprawl and the Rise of American Environmentalism* (Cambridge: Cambridge University Press, 2001).
22. J. R. McNeill, *Something New Under the Sun: An Environmental History of the Twentieth-Century World* (New York: W. W. Norton, 2000), 296–324.
23. McNeill, *Something New*, 311, n. 23, buries the point in a footnote as he writes "cars, incidentally, were much safer . . . than horses by 1925," but how incidental is this information?
24. For example, the fatality rate per 100 million vehicle miles traveled in the United States declined from 40.59 in 1908, the highest total recorded, to 10.83 in 1939. See U.S. Census Bureau, "Mini-Historical Statistics," *Statistical Abstract of the United States* (Washington, D.C., 2003), 77.
25. McNeill, *Something New*, 50–83.
26. Bjorn Lomborg, *The Skeptical Environmentalist: Measuring the Real State of the World* (Cambridge: Cambridge University Press, 2001), 121–126.
27. Ibid., 4.
28. Richard White, *The Organic Machine* (New York: Hill and Wang, 1995). It is impossible here to explore the connections between White's concept and broader Euro-American traditions of thinking through relations between the organic and the inorganic, as discussed most notably in Rabinbach's *Human Motor*.
29. The term is used by John Urry, *Sociology beyond Societies: Mobilities for the Twenty-first Century* (London: Routledge, 2000).
30. It may be noted that John Brinckerhoff Jackson's discussions of "humanized" and "synthetic" landscapes, which include cars, trucks, and roads as important constituents, are relevant to the conceptual background of this piece; see his *A Sense of Place, A Sense of Time* (New Haven: Yale University Press, 1994).
31. Peter Voswinckel, *Arzt und Auto: Das Auto und seine Welt im Spiegel des Deutschen Ärzteblattes von 1907 bis 1975* (Münster: Murken-Altrogge, 1981); Angela Zatsch,

Staatsmacht und Motorisierung am Morgen des Automobilzeitalters (Konstanz: Hartung-Gorre, 1993), 400–404.
32. Flik, *Von Ford Lernen?*, 288, table 1.8; James J. Flink, *The Automobile Age* (Cambridge, Mass.: MIT Press, 1988), 19, 25.
33. M. Schwarte, ed., *Die Technik im Weltkriege* (Berlin: E. S. Mittler, 1920).
34. Heidrun Edelmann, *Vom Luxusgut zum Gebrauchsgegenstand: Die Geschichte der Verbreitung von Personenkraftwagen in Deutschland* (Frankfurt am Main: Verband der Automobilindustrie, 1989); Haubner, *Nervenkitzel und Freizeitvergnügen*; Kurt Möser, "World War I and the Creation of Desire for Automobiles in Germany," in *Getting and Spending: European and American Consumer Societies in the Twentieth Century*, ed. Susan Strasser, Charles McGovern, and Matthias Judt (Cambridge: Cambridge University Press, 1998), 195–222; Wolfgang Sachs, *For Love of the Automobile: Looking Back into the History of our Desires* (Berkeley: University of California Press, 1992).
35. Reichsverband der deutschen Automobilindustrie, ed., *Tatsachen und Zahlen aus der Kraftverkehrswirtschaft 1938* (Berlin: Union Deutsche Verlagsgesellschaft, 1939), 116.
36. Paul Wohl and A. Albitreccia, *Road and Rail in Forty Countries. A Report Prepared for the International Chamber of Commerce* (London: Oxford University Press, 1935), 125.
37. Statistisches Reichsamt, *Statistisches Jahrbuch für das Deutsche Reich* 53 (Berlin: Verlag der Reimar Hobbing, 1934), 171, table VI. B. 3.
38. Flik, *Von Ford lernen?*, 288.
39. Hans Mommsen (with Manfred Grieger), *Das Volkswagenwerk und seine Arbeiter im Dritten Reich* (Düsseldorf: ECON Verlag, 1996), 179–202.
40. Edward Dimendberg, "The Will to Motorization: Cinema, Highways, and Modernity," *October* 73 (Summer 1995): 91–137, esp. 94–116; William Rollins, "Whose Landscape? Technology, Fascism, and Environmentalism on the National Socialist Autobahn," *Annals of the Association of American Geographers* 85, no. 3 (1995): 494–520; Erhard Schütz and Eckhard Gruber, *Mythos Reichsautobahn: Bau und Inszenierung der "Straßen des Führers" 1933–1941* (Berlin: Christian Links, 1996); Rainer Stommer, ed., *Reichsautobahn: Pyramiden des Dritten Reiches* (Marburg: Jonas Verlag, 1982); Thomas Zeller, "'The Landscape's Crown': Landscape, Perceptions, and Modernizing Effects of the German Autobahn System, 1934 to 1941," in *Technologies of Landscape: From Reaping to Recycling*, ed. David E. Nye (Amherst: University of Massachusetts Press, 1999), 218–238. For one of many contemporary accounts: Kurt Schuder, *Granit und Herz: Die Straßen Adolf Hitlers—ein Dombau unserer Zeit* (Braunschweig: G. Westermann, 1940).
41. Hal K. Rothman, *Devil's Bargain: Tourism in the Twentieth-Century American West* (Lawrence: University Press of Kansas, 1998); Warren James Belasco, *Americans on the Road: From Autocamp to Motel 1910–1945* (Baltimore: The Johns Hopkins University Press, 1979); Wrobel and Long, *Seeing and Being Seen*; Sutter, *Driven Wild*.
42. Roger N. Casey, *Textual Vehicles: The Automobile in American Literature* (New York: Garland Publishers, 1997); Kris Lackey, *RoadFrames: The American Highway Narrative* (Lincoln: University of Nebraska Press, 1997); Ronald Primeau, *Romance of the Road: The Literature of the American Highway* (Bowling Green, Ohio: Bowling Green State University Popular Press, 1996).
43. Otto Julius Bierbaum, *Reisegeschichten: Yankeedoodle-Fahrt. Eine empfindsame Reise im Automobil*, rev. ed. (Munich: G. Müller, 1906); for Bierbaum's life and oeuvre, William H. Wilkening, *Otto Julius Bierbaum: The Tragedy of a Poet. A Biography*

(Stuttgart: Akademischer Verlag H. D. Heinz, 1997). On German car magazines and other topics related to early car culture in Germany, see Barbara Haubner, *Nervenkitzel und Freizeitvergnügen. Automobilismus in Deutschland, 1886–1914* (Göttingen: Vandenhoeck und Ruprecht, 1998); on car magazines, see Ulrich Kubisch, *Das Automobil als Lesestoff: Zur Geschichte der deutschen Motorpresse 1898–1998* (Berlin: Staatsbibliothek zu Berlin—Preußischer Kulturbesitz, 1998).

44. See Curt McConnell, *Coast to Coast by Automobile: The Pioneering Trips, 1899–1908* (Stanford: Stanford University Press, 2000); for an early account of long-distance touring in Austria, Germany, and France that has much commentary on the mechanical ins and outs of the car, see Theodor von Liebieg, *Benz-Reise*, unpublished manuscript, 1895, in Daimler-Benz Archiv Stuttgart.
45. For Kuhm, *Das eilige Jahrhundert*, 28, Bierbaum's account regards the automobile as a means of "control over nature" (*Naturbeherrschung*); I find this analysis to be exaggerated.
46. Harro Segeberg, *Literatur im technischen Zeitalter: Von der Frühzeit der deutschen Aufklärung bis zum Beginn des Ersten Weltkriegs* (Darmstadt: Wissenschaftliche Buchgesellschaft, 1997), 248, 250; Eugen Weber, ed., "The Foundations of Futurism ["Manifesto of Futurism," 1909]," in *Paths to the Present* (New York: Dodd, Mead, 1960).
47. Christoph Maria Merki, "Die 'Auto-Wildlinge' und das Recht: Verkehrs(un)sicherheit in der Frühzeit des Automobilismus," in *Geschichte der Straßenverkehrssicherheit im Wechselspiel zwischen Fahrzeug-Fahrbahn und Mensch*, ed. Harry Niemann and Armin Hermann (Bielefeld: Delius Klasing, 1999), 51–73; Harry Niemann and Armin Hermann, "Den Fortschritt bremsen? Der Widerstand gegen die Motorisierung des Straßenverkehrs in der Schweiz," *Technikgeschichte* 65, no. 3 (1998): 233–253, and *Der holprige Siegeszug des Automobils 1895–1930*, 257–264. The strength of popular opinion in prewar Germany against *Autoraserei* (careless, fast driving on public roads) is mentioned in Joachim Radkau, *Technik in Deutschland: Vom 18. Jahrhundert bis zur Gegenwart* (Frankfurt am Main: Suhrkamp, 1989), 300, and now more systematically explored—though perhaps also exaggerated—in Uwe Fraunholz, *Motorphobia: Antiautomobiler Protest in Kaiserreich und Weimarer Republik* (Göttingen: Vandenhoeck und Ruprecht, 2002).
48. Otto Julius Bierbaum, *Mit der Kraft: Automobilia* (Berlin: Bard, Marquardt, 1906), 2. This edition included not only a lightly revised version of the 1902 travelogue, but also several addenda.
49. Bierbaum, *Mit der Kraft*, 328.
50. Von Liebieg, *Benz-Reise*.
51. Felix Schlagintweit, "Eine Frühjahrsreise im Automobil," *Automobil-Welt* 3, no. 35 (2 September 1910): 1543–1546; Firma S. Pr. & Cie, Magedeburg, "Darling," *Allgemeine Automobil-Zeitung* 11, no. 40 (7 October 1910): 55; Eduard Engler, "Auto-Reisen—einst und jetzt," *Motor* 13, no. 1 (January 1913): 32–39.
52. Bierbaum, *Reisegeschichten*, 261.
53. On Riehl and Ratzel, see Mosse, *The Crisis of German Ideology*, 16–23.
54. Celia Applegate, *A Nation of Provincials: The German Idea of Heimat* (Berkeley: University of California Press, 1990); Alon Confino, *The Nation as a Local Metaphor: Württemberg, Imperial Germany, and National Memory* (Chapel Hill: University of North Carolina Press, 1997); Ditt, "Nature Conservation in England and Germany"; Koshar, *Germany's Transient Pasts*, 21–22, 26–29.
55. Bierbaum, *Mit der Kraft*, 334.
56. Bierbaum, *Reisegeschichten*, 250.

57. Eric J. Leed, *The Mind of the Traveler: From Gilgamesh to Global Tourism* (New York: Basic Books, 1991), 80.
58. Joachim Radkau, *Das Zeitalter der Nervosität: Deutschland zwischen Bismarck und Hitler* (Munich: Carl Hanser, 1998), 206–207.
59. Cited in Raymond Flower and Michael Wynn Jones, *100 Years on the Road: A Social History of the Car* (New York: McGraw-Hill, 1981), 12.
60. M. I. Oppenheimer, "Eine Harzreise im Automobil," *Motor* 1, no. 7 (July 1913): 47.
61. H. S. Linfield, "Winter Motoring: Not Only Worth While in Itself, but it Tends to Brighten What is Otherwise Apt to be a Dull Period," *Autocar* 62, no. 1734 (25 January 1929): 153–155; J.M., "The Reason for Motoring: Benefits to Body and Brain Accruing to the Fortunate Owner of a Car," *Autocar* 62, no. 1754 (14 June 1929): 1174.
62. Effie Price Gladding, *Across the Continent by the Lincoln Highway* (New York: Brentano, 1915); Dallas Lore Sharp, *The Better Country* (Boston: Houghton Mifflin, 1928); Jan Gordon and Cora J. Gordon, *On Wandering Wheels: Through Roadside Camps from Maine to Georgia in an Old Sedan Car* (New York: Dodd, Mead, 1928).
63. George R. Stewart, *U.S. 40: Cross-Section of the United States of America* (Boston: Houghton Mifflin, 1953), 29; see also Thomas R. Vale and Geraldine R. Vale, *U.S. 40 Today: Thirty Years of Landscape Change in America* (Madison: University of Wisconsin Press, 1983).
64. Rabinbach, *The Human Motor*.
65. White, *The Organic Machine*, 62.
66. This was by no means a uniquely German phenomenon. For examples from English culture, see Anne D. Wallace, *Walking, Literature, and English Culture: The Origins and Uses of Peripatetic in the Nineteenth Century* (Oxford: Clarendon, 1993); Frank Trentmann, "Civilization and its Discontents: English Neo-Romanticism and the Transformation of Anti-Modernism in Twentieth Century Western Culture," *Journal of Contemporary History* 29 (1994): 583–625.
67. Motor bus lines began operation in Germany in 1903; ten years later there were 367 lines operating over more than 6,800 kilometers. See Paul Wohl and A. Albitreccia, *Road and Rail in Forty Countries* (London: Oxford University Press, 1935), 129.
68. Rudy Koshar, *German Travel Cultures* (Oxford: Berg, 2000), 34–35.
69. Bierbaum, *Reisegeschichten*, 269.
70. Ibid.
71. On anthropomorphic themes in the history of car-talk, see Ingrid Piller, *American Automobile Names* (Essen: Blaue Eule, 1996); "Extended Metaphor in Automobile Fan Discourse," *Poetics Today* 20, no. 3 (1999): 493–498.
72. Bierbaum, *Reisegeschichten*, 299.
73. See Victoria de Grazia (with Ellen Furlough), eds., *The Sex of Things: Gender and Consumption in Historical Perspective* (Berkeley: University of California Press, 1996), which, it must be noted, ignores the relationship between gender and the automobile.
74. Bierbaum, *Reisegeschichten*, 299.
75. Rohrkrämer, *Eine andere Moderne?*, chap. 5.
76. Bierbaum, *Mit der Kraft*, 327, 329.
77. On Hauser's post-1945 career, see Michael Ermarth, "The German Talks Back: Heinrich Hauser and German Attitudes toward Americanization after World War II," in *America and the Shaping of German Society, 1945–1955*, ed. Michael Ermarth (Providence, R.I.: Berg, 1993).

78. Heinrich Hauser, *Friede mit Maschinen* (Leipzig: Reclam, 1928).
79. Hauser, *Friede mit Maschinen*, 3.
80. See Engler, "Auto-Reisen."
81. Owen John, *The Autocar-Biography* (London: Iliffe and Sons, Ltd., 1927), ix–x; see also, "On the Road: Does Motoring Have a Tendency to Destroy Individuality?" *Autocar* 62, no. 1735 (1 February 1929): 208–209.
82. Karl D. Helfenberg, "Aus den Kinderjahren des Automobilismus," *Motor* 1, no. 7 (July 1913): 57–61, here 58.
83. This is Herf's argument in *Reactionary Modernism*, 105; for the view endorsed by this author, see Rohkrämer, *Eine andere Moderne?*, 301–338.
84. Marcus Paul Bullock, *The Violent Eye: Ernst Jünger's Visions and Revisions on the European Right* (Detroit: Wayne State University Press, 1992), 140–141.
85. Thomas Nevin, *Ernst Jünger and Germany: Into the Abyss, 1914–1945* (Durham, N.C.: Duke University Press, 1996), esp. 138–140.
86. Ernst Jünger, "Der Arbeiter," in *Ernst Jünger: Sämtliche Werke*, vol. 8 (Stuttgart: Klett-Cotta, 1981), 181.
87. See for example Chris Hables Gray, *Cyborg Citizen: Politics in the Posthuman Age* (New York: Routledge, 2001).
88. Frank Markus, "GM Hy-wire," *Car and Driver* 48, no. 9 (March 2003): 77–79.
89. Helmut Lethen, *Cool Conduct: The Culture of Distance in Weimar Germany* (Berkeley: University of California Press, 2002), 26, 29.
90. Hauser, *Friede mit Maschinen*, 3.
91. Helmut Lethen, *Neue Sachlichkeit 1924–1932: Studien zur Literatur des "Weissen Sozialismus"* (Stuttgart: Metzler, 1975), 68–71, which fails to mention Hauser's discussion of driving.
92. Merki, *Der holprige Siegeszug des Automobils*, 257–264.
93. Von Liebieg, *Benz-Reise*.
94. Hauser, *Friede mit Maschinen*, 40.
95. One of the best U.S. examples in this genre is the magazine *Special Interest Autos*, published by *Hemmings Motor News*, but see also *Collectible Automobile* and the more "scholarly" *Automobile Quarterly*.
96. For a recent, and much-welcomed, example from German historiography: Erhard Schütz, "Der Volkswagen," in *Deutsche Erinnerungsorte*, vol. 1, ed. Etienne François and Hagen Schulze (Munich: C. H. Beck, 2001).
97. Hauser, *Friede mit Maschinen*, 23.
98. E.J.A., "Berlin!," *Autocar* (16 November 1928): 1144–1149. In this account, only the Mercedes appeared to retain styling characteristics that reflected its origins in a distinctly national car culture.
99. Ibid., 23–24.
100. Schuder, *Granit und Herz*, 11.
101. Wolfgang B. von Lengercke, *Kraftfahrzeug und Staat: Ein Versuch* (Heidelberg: K. Vowinckel, 1941).
102. See Thomas G. MacGowan, "How do Buyers Feel About Today's Cars . . . Today?," *Automotive Industries* (16 July 1938): 86–90, 94, 140–144.
103. Owen John, "On the Road: Does Motoring have a Tendency to Destroy Individuality?," *Autocar* (1 February 1929): 208–209.
104. The term is used by Peter Reichel, *Der schöne Schein des Dritten Reiches: Faszination und Gewalt des Faschismus* (Munich: Carl Hanser, 1991).
105. Thomas Zeller, "Landschaften des Verkehrs. Autobahnen im Nationalsozialismus und Hochgeschwindigkeitsstrecken für die Bahn in der Bundesrepublik," *Technikgeschichte* 64, no. 4 (1997): 323–340; see also Zeller, " 'The Landscape's Crown.' "

106. Wettengel, "Staat und Naturschutz," 392–393.
107. Thomas Zeller, *Straße, Bahn, Panorama: Verkehrswege und Landschaftsveränderung in Deutschland von 1930 bis 1990* (Frankfurt am Main: Campus, 2002), 158–164, 187–194.
108. Heinrich Hauser, "Autowandern, eine wachsende Bewegung," *Die Straße* 3, no. 14 (July 1936): 455–457; Rudy Koshar, "Germans at the Wheel: Cars and Leisure Travel in Interwar Germany," in *Histories of Leisure*, ed. Rudy Koshar (Oxford: Berg, 2002), 215–230.
109. Rollins, "Whose Landscape?"
110. Ernst Jünger, *Der Arbeiter. Herrschaft und Gestalt* (Hamburg: Hanseatische Verlagsanstalt, 1932), 194, as cited in Rohkrämer, *Eine andere Moderne?*, 324.
111. Albert Liese, Walter Ostwald, Alfred Rothweiler, "Vergleichsfahrten auf Reichsautobahn und Reichsstrasse," *Die Straße*, Sonderdruck, 10 (1937); Kurt C. Volkhart, "Leichtbau und Autobahn: Erfahrungen eines Praktikers," *Automobiltechnische Zeitschrift* 40, no. 2 (25 January 1937): 27–28.
112. Hans Bahr, "Mein Kraftwagen auf der Reichsautobahn," *Die Straße* 3, no. 13 (July 1936): 415–416; letters to the editor nos. 8109 and 8251, *Allgemeine Automobil-Zeitung* 38, no. 1 (2 January 1937): 20, 21.
113. "Gedanken auf Reichsautobahn," *Allgemeine Automobil-Zeitung* 37, no. 21 (22 May 1937): 652; "Erfahrungen auf den Reichs-Autobahnen," *Der Motorist: Presse- und Nachrichtendienst* 12, no. 389 (28 July 1937).
114. C. Volkhardt, "Autobahnsünder," *Motor-Kritik* 18, no. 17 (September 1938): 623–627.
115. Harold Nockolds, "Cross-Channel Holiday," *Autocar* (UK) 77, no. 2130 (28 August 1936): 392.
116. Dietmar Fack, "Das deutsche Kraftfahrschulwesen und die technisch-rechtliche Konstitution der Fahrausbildung 1899–1943," *Technikgeschichte* 67, no. 2 (2000): 111–138.
117. Alan Henry, *Mercedes in Motorsport* (Sparkford: Haynes, 2001), 32–68; von Eberan, "Der Auto-Union-Rennwagen im europäischen Rennsport," *Automobiltechnische Zeitschrift* 40, no. 13 (10 July 1937): 329–333.
118. Max Domarus, ed., *Hitler: Speeches and Proclamations 1932–1945*, vol. 3 (Wauconda, Ill.: Bolchazy-Carducci, 1997), 1476–1479; on accidents: *Tatsachen und Zahlen 1938*, 113–114.
119. Rolf Schäfer, "Die tödlichen Verkehrsunfälle in den Jahren 1929–1934" (Ph.D. diss., University of Heidelberg, 1935); Helmut Lossagk, *Sinnestäuschung und Verkehrsunfall* (Berlin: Franck Verlag, 1937); Wa. Ostwald, "Die neue Zeit," *Automobiltechnische Zeitschrift* 42, no. 4 (25 February 1939): 85. For recent scholarship on accidents in Germany, see Merki, *Der holprige Ziegeszug des Automobils*, 346–371; see also Fraunholz, *Motorphobia*.
120. Mikita Brottman, ed., *Car Crash Culture* (New York: Palgrave, 2002); Kristin Ross, *Fast Cars, Clean Bodies: Decolonization and the Reordering of French Culture* (Cambridge, Mass.: MIT Press, 1998).

Chapter 6

Biology—Heimat—Family

Nature and Gender in German Natural History Museums around 1900

SUSANNE KÖSTERING

REFLECTING UPON NATURE means grappling with culturally constructed notions of nature. This process, in turn, involves analyzing, overturning, or upholding historically constituted constructs in order to reorganize them in new or different ways over and over again. Numerous social institutions have taken on this task, one of the most enduring being the museum. It links objects, spaces, and people and, in applying categories from the present, uses a historical or natural past congealed into collections as a marker of the future. The museum is thus engaged in an ongoing process of interpretation: collecting, ordering, filing, researching, and presenting are all unending contributions to the constructs of "nature" and "culture." Simultaneously, the museum serves as an index of cultural change and a major point of departure for researching such change. Fluid portrayals of nature in museums are one way of gauging how modern societies have conceived of "nature" and thought about the fate of both human communities and the nonhuman environment.

This ongoing process of constructing and interpreting nature underwent a paradigm shift in German natural history museums around 1900. During the era of the Second German Reich from 1871 to 1918, natural history collections and research centers had evolved into public education agencies that transmitted new ideas about the natural world, developing concepts that continue to shape our perceptions of nature today. The massive rationalizing, disciplining, and efficiency-enhancing interventions in nature that accompanied Germany's rapid industrialization in these years engendered novel biological and ecological representations that broke away from traditional natural history collections based on taxonomic classifications of nature. This search for innovative images of nature fit for the present and the future corresponded with the perception that industrialization had decisively changed the natural world. Just as fin-de-siècle natural scientists, writers, artists, and philosophers sought new syntheses between nature and culture, so did museum curators search for modern ways to represent nature in their public displays.

By 1900, there were about 150 natural history museums in the German Reich. Some fifty of these—primarily major municipal and state agencies for the most part—shaped the way museum curators assembled natural history collections for years to come. A dozen of them developed innovative presentation concepts during the imperial era and accordingly played a leading part in the overall German museum reform movement.[1] Prominent among these were the Zoological Museum at the University of Kiel, the Natural Science Museum at the University of Berlin, the Altona Museum, the Hessian State Museum in Darmstadt, and the Local Natural History Museum in Leipzig, all of which relocated between 1881 and 1912 and thereafter ran innovative exhibitions on their new premises. During this period of change, museum curators reconceived collections that had hitherto been arranged according to taxonomic categories along biological/ecological lines. This reorganization involved displaying animals in a lifelike manner within imitation natural surroundings. Instead of merely imparting scientific findings, museum directors strove to present images of nature that were "close to reality," even as ever larger parcels of the natural landscape and the traditional cultural landscape were disappearing.

These changes were a significant turning point in the development of natural history museums. In the late eighteenth century, these museums had originated from the collections of dukes and bourgeois societies. With the creation of a German national state in 1871, natural history museums became an essential component of bourgeois cultural patronage and identity. Starting in 1870, the collections' profile changed due to the influx of natural objects from German colonies. However, representing the diversity of the world's fauna, flora, and the lithosphere remained the dominant mission, as it had been before. The order of natural objects on the basis of Linneaus's systematic biology—the taxonomic categories of kingdom, phylum, class, order, family, genus, and species—remained the leading principle for the representation of nature. Even Darwin's evolutionary theory did not change this order of representation, at least not at first. However, the differentiation of biology into different subfields as a result of the spread of evolutionary theory brought about a paradigm shift in natural history displays. This paradigm shift favored animal geographic and ecological approaches and turned animals' ways of living into the central focus of research. Ecological thinking penetrated the descriptive natural sciences in the second half of the nineteenth century, focusing scientists' attention on the environmental conditions that shaped different species' subsistence and reproduction patterns as well as the relationships between species in a given spatial context.

The German scientist Ernst Haeckel popularized such ecological thinking in Germany and coined the term "ecology" as early as 1860, without, however, pursuing this approach further.[2] Instead, Karl Möbius, the director of the natural history museum in Hamburg, made the term usable for the sciences and established it in research and for museums; Haeckel's role was thus that of a mentor for generations of museum leaders. Before the turning point around 1850, representations of nature in natural history museums all over the world had solely relied upon the taxonomic categories of natural history. After this paradigm change, different approaches of natural-historical representation coexisted, including ones based on taxonomic, biogeographical, and ecological schema, which could not easily be reconciled. In the United States, Sweden, and Germany, group representations of animals appeared in nineteenth-century natural history museums that showed the diversity of species within their natural topographical and floristic habitats. Other countries soon followed this trend, which satisfied the desire to make natural historical knowledge accessible to broad sectors of the populace, since visitors considered ecological approaches more vivid and memorable than glass cabinets filled with systematic arrays of animals.

Since it is impossible to deal with the entirety of museal representations of nature in Imperial Germany in this paper, I will focus my attention on this shift from taxonomic to biogeographical or ecological display across Germany. Many specific museums deserve further scholarly attention; one could mention here the decidedly animal-geographic exhibitions in Münster and Darmstadt, which displayed the contrast between colonial and domestic fauna, or museums such as the one in Frankfurt am Main, which devoted themselves explicitly to popularizing evolutionary theory.

The biological/ecological turn in natural history collections around 1900 reconfigured not only the underlying scientific principles of these exhibits. Indeed, while most scholars of Germany's natural history museums have analyzed this shift in how museums represent nature solely in terms of the history of science, this chapter has a different goal: to demonstrate that the paradigm shift in natural museum display around 1900 reflected new social and cultural, rather than scientific, priorities. This process involved modification of exhibition practices, the architectural and spatial arrangements in museums, and the composition of museum patronage. The production of display collections was thus an open-ended process of representation that constantly engendered newness.[3] Seen in this light, the natural history museum of 1900 was a laboratory for developing and testing new cultural constructs of "nature." Just like a laboratory, however, the museum was no self-enclosed cosmos but was shaped instead by its historical and social context.

Two themes dominated the representation of natural history museums at the time: *Heimat* (homeland) and family. The regional homeland collections were transformed into spaces for the portrayal of homeland, a move that employed the museum to naturalize politically contingent territorial borders. In a similar vein, biological images that fixed gender roles within the family came to dominate natural history museum displays, where the homeland symbolized the nation's peaceful home and animals prepared in a lifelike fashion were grouped into family units. Natural history museums in other countries displayed similar nationalist and social concerns. Germany was by no means unique in the socially constructed quality of its natural history museums; the intense focus on spatial perception and gender roles was one of degree rather than kind. In the following analysis, I examine how Heimat ideals and family models dovetailed with the portrayal of nature in museums and how, in the process, museum collections were modified, realigned with one another, and gave rise to newness.

From "Fatherland" to "Homeland" in Natural History Museums

After the founding of the German Reich in 1871, many natural history museums in Germany augmented their general collections, which were at that time still arranged in accordance with the Linnaean taxonomic system, with a separate display of fauna from their own regions.[4] The administrative subdivision of the Reich into states and provinces served as the basis for collection areas. Thus, for instance, in Stuttgart the animal "fatherland" was Württemberg, in Dresden it was Saxony, in Karlsruhe it was Baden, and in the capital Berlin it was Germany. The collection of Hessian animals at the Grand-Ducal State Museum in Darmstadt contained exclusively animals that had been found or killed on the territory of the Grand Duchy of Hessen and not, by contrast, from the neighboring Prussian province of Hessen. Provincial museums within Prussia, such as in Münster, and some university collections, such as the Zoological Museum at the University of Breslau, presented provincial faunae—in these cases the Westphalian and Silesian animal kingdoms, respectively.

Classification of the animal world, therefore, was based not only on natural science categories but also on politically designated borders. One example of such political reorganization of the natural world was the natural history museum in Berlin, which assembled a fatherland collection comprising "German" animals despite the fact that it was not a national Reich museum or even a major regional museum, but simply one of several Prussian university museums. Many collections in municipal or association-owned museums were likewise founded on the symbolic appropriation of political spheres of influence; urban environs became informal reference areas as well. The main criterion for the demarcation of collection areas in such cases was accessibility, rather than the actual geographical extent of animal populations.[5] The zoological "fatherland" collections remained committed to the goal of documenting their area's fauna as exhaustively as possible for a longer period than did the general collections. A large and diverse array of wildlife in the collection symbolized the power of the state, province, or municipality. The fatherland collections were also unique in their early adoption of scenic dioramas that depicted the lives and pursuits of domestic animals. The fatherland departments thereby became the catalysts for the biological turn in museum representation, for during the course of the late nineteenth and early twentieth century these sections evolved into Heimat departments that literally brought regional identity to life in scenic re-enactments of local animals hunting for food, looking for mates, and raising their young.

The evolution into Heimat collections resulted from the adaptation of biological principles to museum display. In the period around 1900, the discipline of biology was concerned with the evolution of life and focused on how animal lifestyles corresponded with the possibilities and limitations imposed by their habitat in their "struggle for survival." Biology gave special consideration to the physical living conditions of animals, the changes that their bodies underwent as a result of seasonal variations and age, and the behavioral traits that they displayed in different situations. Because biological concerns could be applied to daily life, the science was increasingly popular in amateur circles and at schools, where teachers played a seminal role as pioneers of biological research and in the popularization of its findings.

This new biological approach debuted at the Natural Science Museum in Berlin in May 1889 in the "Birds and Mammals of the Fatherland" collection. This collection showed birds in flight, in the process of capturing their prey, and with their offspring in nests, rather than arranged neatly behind glass cabinets according to phylum, order, or family. Other museums quickly followed Berlin's lead in switching to biological displays: in 1896, the Hamburg Natural History Museum presented animals from the Lower Elbe region, while in that same year the Museum for Natural Science, Ethnology, and Commerce at Bremen ran a show entitled "Character Images from the Local Animal

World" and the Museum of the Senckenberg Nature Research Society in Frankfurt presented biological portrayals of animals from the Taunus region.[6] The Altona Museum, which opened in 1901, based its entire public zoological collection on the portrayal of biological phenomena.[7]

The museum dedicated the largest exhibition space to dioramas with meter-high groups of animals. Visitors encountered three elks being attacked by wolves, boars preparing for a tussle, and foxes, badgers, polecats, does, and bucks searching for food. Birds and insects were also presented in large-scale, three-dimensional compositions that included the birds with their nests and eggs and the insects in various stages of metamorphosis, the plants in which they lived and those that they ate, and the predators that stalked them. With this novel approach, the Altona Museum's natural history section became the prototype for "biological display" in the Reich.[8] The Altona Museum also pioneered the use of the term *Heimat* to refer to its biological displays, a designation that gave the museum's political setting—the once-contested Prussian province of Schleswig-Holstein—a sense of rootedness in the natural landscape. The portrayal of extinct animals, moreover, was intended to revive a mystical link between the region's people and their natural surroundings. The museum thus naturalized the contours of Heimat and offered a vision of political permanence amid rapid social and economic change.

The organizers of the first German Museum Conference, held at Mannheim in 1903, celebrated biological collections as symbols of the reformatory power of natural science museums and, by extension, of Wilhelmine society's capacity for renewal. The conference heralded the breakthrough of the biological turn; in the years thereafter, almost all natural history museums turned their fatherland collections into sections devoted to the homeland by setting up biological installations. Though most of the museums, including Altona, continued to order their biological groups according to taxonomic criteria, animals were no longer arranged in stiff poses devoid of movement or pegged in rows inside glass cases.[9] Categories such as phylum, order, and family helped to organize the displays, but vivid biological portrayals of animal interactions commanded visitors' attention. In this sense, natural history museums' scientific mission of documenting the fauna of the fatherland moved backstage while "life," packaged according to the social and political categories of the Heimat, entered the limelight.

Heimat as Environment: Ecological Concepts in Museums

The metamorphosis of taxonomic fatherland collections into biological portrayals of Heimat was fully complete with the incorporation of the discipline of ecology that occurred in the early twentieth century. In 1908, the Berlin zoologist Friedrich Dahl, building on the concept of "biocoenosis" (life community) devised by Karl Möbius in 1877, devised the ecological model of the "biotope," which formed the basis for new types of natural history museum displays.[10] The Leipzig Local Natural History Museum, which opened in 1912, was the first natural science museum in Germany to replace a Linnaean-based approach with an ecologically grounded portrayal of urban environs as a complex system of ecologically defined natural areas. In these ecological displays, biological animal groups were grouped and portrayed according to shared conditions of life, rather than according to taxonomic criteria; indeed, the staff transferred entire habitats from the surrounding region to the museum to ensure the quality of verisimilitude. The Leipzig Natural History Museum was also the first in Germany to identify and define an urban catchment area as a natural environment, adopting natural science criteria to understand the ecological character of a metropolitan area.

Behind the public collection in Leipzig was an innovative system of research into the ecology of the homeland that preceded any similar academic exploration of this topic. In this case, the demand for realistic portrayal shaped and was in turn shaped by the research process. Under the stewardship of Richard Buch, museum supporters, teachers, and amateur researchers working on a volunteer basis processed the findings of such museum-driven research into an integrated depiction of Leipzig's urban basin and recreation areas as a complex system of interlocking biotopes. The museum depicted Leipzig as the center of a circle with a twenty-five-kilometer radius known as the "Leipzig Lowland Bight," a geographical fiction that supposedly formed an extension of the North German Lowland Bight. The Heimat movement, which had reached a peak in its popularity around 1900, shared in the process of research and interpretation for the new museum. The movement's involvement reflected the recreational value of Leipzig's surrounding woodlands, meadows, rivers, and lakes, which Heimat clubs viewed as the foundation of homeland feeling, and which were easily accessible to city dwellers via public transportation. Leipzig was thus part of a larger eco-Heimat, a shared place of social interaction that transcended class distinctions and provided a font of local and regional identity.[11] The museum's ecological depiction of the homeland thus collapsed distinctions between the social and natural spheres, as museum designers used scientific categories to define *Heimat* as the web of ecological relationships that surround and permeate an urban entity. Though the phrase "environment" had yet to be used at that time, *Heimat* had broadened its meaning to include the environment: the human home and its natural surroundings.

Animal Preparation and the Construction of Nature and Gender in Museum Displays

Just as regional Heimat collections attempted to naturalize territorial arrangements that had emerged in the newly created German nation, so did animal displays try to naturalize bourgeois gender and family arrangements that had developed during the industrial revolution. Immediately after the German Reich was established in 1871, the concept of gender in the natural history museum was radically revised. This change had its origins in the work of animal preparators, rather than in scholarly studies. Preparators' works were grounded in scientific and aesthetic norms, and they gauged the success of their work by its scientific accuracy as well as its realism. Socially constructed views of nature and gender nonetheless played a crucial role in how preparators arranged animal figures for display and presented their work to the public.

The most celebrated and influential nineteenth-century animal preparator in Germany was Philipp Leopold Martin (1815–1885), who played a significant role in redefining the nature-culture relationship in animal displays. Most importantly, Martin was responsible for transforming taxidermy into dermoplasty in the preparation of animal figures. The standard method of preparing animals since the eighteenth century had been taxidermy, a process that involved preserving the skin of the dead animal in one piece as far as possible and stuffing it with vegetable material. Dermoplasty, by contrast, sacrificed the external integrity of the skin in favor of building up the animal's body from the inside. Martin opted for the radical step of sectioning the animal's skin once he had measured the deceased creature and recreating it as a life-size model using wood and straw or clay and gypsum. He then covered the artificial body with the various bits of skin and sewed them back together. This switch from taxidermy to dermoplasty was more than just a technical refinement in animal preparation; it was a wholesale paradigm shift that paralleled the shift from descriptive taxonomy to holistic biology/ecology in fatherland collections. Whereas taxidermists traditionally saw themselves as having the task, like natural scientists, of documenting animal species, dermoplasticians were active fashioners, and hence more akin to sculptors. Etymologically speaking, after all, they were "shapers of skin."[12]

In the preface to his *Natural History in Practice* (*Praxis der Naturgeschichte*, especially part 2, *Dermoplastik und Museologie*, published in Weimar in 1870), Martin described his innovation as being born of the changing relationship between nature and culture in his time. "Wherever advancing culture supplants or actually eliminates primeval nature, all thinking people are seized by the earnest desire to learn as much as they can about the essence of the natural entity in its original composition. Our ever-changing culture has preserved some of this [primeval nature] though subjecting it to different conditions, has already supplanted much of it and has, indeed, destroyed a very great deal."[13] In the second edition of the book (1880), he intensified his criticism of modern industrialism, which, he argued, is not based on "life" but has "erected its hollow structure on dead husks" and called for a reorientation of society toward the elemental force of nature. Martin was no nostalgic dreamer, however; rather, he sided with "progress," which in his view could be attained in a fresh synthesis of nature and culture: "Nature steels our strength, but our artificial life enfeebles it."[14] The element of nature in modern culture needed strengthening. He believed that the goal of natural history museums was to compensate for the loss of primeval nature, but criticized them as being ill-equipped to accomplish this task. "Our natural history collections, which ought to be archives of the entire subject, offer little comfort in this regard due to their one-sided, solely systematic presentation. It almost seems as if the specimens in our collections merely serve to corroborate the prevailing systematics, when in fact they should lead the way in this broad field [of taxonomy]."[15] Martin thus criticized the natural history museum's dependence on the increasingly narrow approach of taxonomy, and, by extension, the accelerated professionalization and specialization of scientific research. He believed that museums had a duty to popularize scientific knowledge. He wrote, "As with the natural sciences as a whole, natural history needs to be treated as a popular science, which serves all people, since every human being makes up a part of this greater natural whole."[16]

Martin himself was self-taught, which was not unusual in his profession at the time. He acquired his skill during research trips, by studying collections of specimens, and by consulting sets of instructions from the field experience of practitioners. Beginning in 1851, he worked at the Zoological Museum at Berlin University, holding one of the first museum posts in Germany. He experienced the consequences of narrow professionalization of scientific collections for animal preparation firsthand. Whereas Hinrich Lichtenstein, the museum's first curator, gave Martin wide latitude for the portrayal of animals, his successor Wilhelm Peters, who entered office in 1857, instructed Martin to prepare all animals in a uniform, head-on position. Peter's wish reflected the systematician's wish to be able to arrange the individual exemplars of an order or family randomly and in this way keep the collection taxonomically state-of-the-art. But Peters's direction conflicted utterly with Martin's vision of displaying animals in natural, realistic poses. He resigned from the Berlin museum in 1859 and went to Stuttgart, where he worked as a liaison between the State Natural History Collection, the city's zoological gardens, and a museum that he set up himself, albeit unsuccessfully. His résumé thus echoed the tensions between professionalization and popularization that characterized nature studies as a whole in the mid-nineteenth century.[17]

Martin did not cite the midcentury change in scientific perspective from systematics to evolutionary biology as the reason for innovating animal portrayal, despite acknowledging it and using it to support his case: "The objects that strict science presents us with in its long galleries and cabinets are dead—and thus boring and fatiguing, whereas biological display has an impact on the observer that is as exciting as it is educational. Among rigorous scientists, the former style of presentation has long been recognized as a deplorable one-sidedness and, at least partially, some efforts have been made to move closer to a biological view."[18] Martin explicitly drew upon colleagues, including Hermann Ploucquet, while the only formal scientist whom he accepted was the zoogeographer and Darwin popularizer Gustav Jaeger.[19] One can assume that Martin read the most influential nineteenth-century nature writers in formulating his ideas on dermoplasty and biological animal display, including Charles Darwin's theory of evolution, Wilhelm Heinrich Riehl's *Natural History of the German People*, and Alexander von Humboldt's biogeographical work, from which Martin may have learned about biological holism. But Martin did not explicitly acknowledge either zoology or biology as intellectual influences, but instead referred to "life" itself, a loose construct that could refer both to the "natural" reality of animals and the social reality of human beings—or both at once. This indicates that changes in animal preparation did not result primarily from scientific debates, but instead reflected evolving social ideas about the relationship between nature and culture in modern times and the role of the museum in representing this changing relationship.

A clear indication of the influence of social practices in shaping museum representation was Martin's reliance on the family unit as the key to dermoplasty. "If we are agreed that, in a real and not a systematics-related sense, the family has to be the center-point of a species in the collection as it is in real life, then we suddenly find ourselves at the threshold of a major new field, at the beginning of evolutionary history," he wrote in his comprehensive work *Natural History in Practice*. "Once we have the family as the epitome of a given natural whole before us, its dwelling place belongs in the picture as well, and all of this must, of course, yield an image of family life. Faithful images from the family life of animals, therefore, need to be and to become the aim of our future collections of stuffed animals."[20] Martin was not referring to the taxonomic category of "family," which had become the lowest unit portrayed in many natural history collections in the second half of the nineteenth century.

What he meant instead was the family as a social category: father, mother, children. This "family," in his eyes, only became fully constituted by being placed in the "home," hence Martin felt it to be crucial that "its dwelling place" be shown. "Burrows, nests, and dens are finite things quite inseparable from the life-history" of the animal, he stated, and had to be included in the portrayal.[21]

In stark contrast to these peaceful family images, Martin described a second category of animal groups: animal fights. "Family groups, which will invariably form the main thrust of a zoological collection, are all about peaceful togetherness in a restricted space, whereas these [fighting] groups often demand a greater extension and richer features, since hunting and desire in the wild unfold not only in woodlands, fields, steppe, desert, and water but also in the air. To be able effectively to portray an antelope being chased by a lion or leopard, therefore, we must add a painted desert as a backdrop."[22] Using the Humboldtian division of the Earth into zoogeographical regions as the basis for his ideas, Martin recommended faunal scenes from the polar seas, the Alps, the jungle, the savannah, the prairie or steppe, and the coast.[23]

The juxtaposition of animal families and animal fights, which were spatially configured in the home and the outside world, mirrored the "separate spheres" that characterized emerging gender relations in the nineteenth century.[24] Martin's images from the family lives of animals aped the division of roles in the bourgeois nuclear family, which was considered the benchmark of a "normal family" with "normal" behaviors adapted to social requirements. He assigned the male and female animals differing roles and positions in the displays to portray the idealized division of labor and cooperation thought to dominate both the human and natural worlds.[25] This polarization of male and female animal roles naturalized socially constructed human gender ideals and suggested a harmonious complementarity between male and female characteristics. The focus lay less on the contrast between male and female animal specimens than on the bourgeois family system founded upon these opposites.

Martin placed the authentic portrayal of animals within their burrows, dens, and nests at the center of natural historical display. Social portrayals of this type were not new to natural history collections, as they had been used since the eighteenth century to add color to aristocratic and bourgeois collections. Martin, however, made the family scene the "center-point of a species, in the collection as in life" and hence the point of departure for all representations of nature in museums. The portrayal of the "family life of animals," organized according to masculine and feminine gender roles, became not *one* but "*the* task for our future collections of stuffed animals" (emphasis added). In the future, wrote Martin in 1870, "the family group will, therefore, form the main task of our portrayal, because through it we can cause a whole chain of natural correlations to unfold before the eyes of the beholder ... Future generations will thank us for having put an end to the jaded, anachronistic systematics approach by means of a portrayal that will never age."[26] Martin argued that the dermoplastic family groups also offered museum directors a more objective, lifelike portrayal of animals, a factor that enhanced the popularity of his techniques in late nineteenth-century museums. Like the switch from taxonomy to biology/ecology, therefore, the replacement of taxonomic categories with animal families and zoogeographical regions represented a paradigm shift in natural history display that paralleled broader social changes in the German family structure.

Martin's search for naturalistic portrayal of animal specimens had always been the goal of animal preparation, yet the actual guidelines for achieving such realism were amazingly fluid in the nineteenth century, a reflection of the varying constructions of gender that prevailed at different points in time. Preparation instructions from the 1830s, for example, emphasized the use of animals to visualize emotional states vicariously. Animal preparators accepted that animals had "passions like people, fewer in number but of a far more tempestuous nature."[27] The central categories for achieving accurate portrayal were "positions" and "poses." Position meant the animal's normal stance, such as sitting, lying, or standing, while pose was a momentary, psychologically determined movement that articulated fear, anger, or love.[28] In such displays, female animals represented motherly love, males the longing to hunt, a reflection of early nineteenth-century social views on masculinity and femininity.

The portrayal of animal emotions declined in significance toward the middle of the nineteenth century, when preparators began to focus instead on positions and poses as manifestations of moral virtue. A key exponent of this approach to animal preparation was the famous bird collector Christian Ludwig Brehm. In his widely read preparation instructions, he adjudged birds that hold themselves erect, maintain a smooth, orderly plumage (like a uniform), and survey the world with a piercing gaze to be noble, while he held a dim view of those that bent their necks and tended their feathers in a "slovenly" way.[29] In his eyes, the noblest birds were the condors: "They stand quite erect, *like a man* [emphasis added], their wings close to their bodies, a splendid crest on top, all their feathers tightly pressed in, their necks slightly withdrawn, their heads resting majestically on their necks, and with protruding, marvelously gleaming eyes."[30] Other bird species paled in comparison; as Brehm's verdict on the vulture stated: "They do stand vaguely erect, but theirs is not a noble posture. Their wings extend too far above where their breast begins, enveloping them almost like a cloak ... and are borne in such a slovenly way that the second layer of flight feathers often overlies the first and does not create a smooth surface."[31] Brehm's judgments on animal nobility also reflected mid-nineteenth-century notions of gender, for while male bird specimens could achieve perfection in form, he often implied that stooping habits and "ungainly" forms were characteristic of the female sex.[32]

In 1845, a book on preparation appeared that jettisoned the focus on emotions or moral virtues in favor of scientific categories. In an "Appendix on the Customs of Mammals," for example, the Munich curator Alexander Held classified the various genera of animals according to seventeen posture variants, which he declared to be fixed "characters" of a given genus.[33] Martin adopted Held's concept of characteristic, rehabilitating the vultures Brehm had found so reprehensible by arguing: "The bird's slack habit and the curving of its neck are characteristic features of this taxon [bearded vulture], a correct portrayal which is generally very much lacking in collections."[34] According to Martin, positions and poses were references to life, to a being's inner nature, rather than to temporary feelings or anthropomorphic categories of virtue. The portrayal of "really visible forms," he wrote, would strive to portray the "visible, true-to-nature musculature" and evoked an "indicative expression of life," an approach Martin viewed as an expression of a new, objective depiction of animals.[35]

This new "objectivity" was an important argument in favor of the dermoplastic portrayal of animal groups. Dermoplasticians claimed that their animal displays were rooted in natural facts, rather than the application of human values to the animal kingdom. Their new method of working outward from the inside articulated the desire to represent animals in an era of accelerated transformation of the natural world. In their eyes it was not enough to describe and to create a static image of animals; they believed that dynamic motion offered deeper insights into the reality of animals in the wild. They argued that animals ought to be shown in motion, though this did not mean random movements, but rather the characteristic movements that typified their particular genus. Martin believed, however, that these characteristic movements would be articulated best in family scenes, without realizing that he had effectively substituted yet another sociocultural category—the bourgeois family—in his quest for a wholly objective portrayal of the natural world as a whole.

In Martin's animal family scenes, the notion of gender complementarity predominated, a reflection of the late nineteenth-century belief in the "natural" division of labor between the sexes. At the 1903 Museums Conference in Mannheim, for example, the bulk of the animal displays showed animal families, predominantly mothers with young, inside or within close vicinity of their dens, burrows, or nests (e.g., "Vixen suckling her young in the foxhole"), often teaching the immature animals how to catch prey. Where animal couples were shown, it was in the context of stalking prey, looking for a nesting place, or attending to their young. The supposed natural division of labor by gender appeared repeatedly in such scenes. "While the female [coot] remains on the nest," noted one display, "her already fledgling, delightfully marked young go off into the water with their father."[36] The male "heads of family" were frequently portrayed procuring or defending their home. "A group portraying our deer in summer-time" a scene produced at the Altona museum in 1901, contained the following elements: "The powerful buck protects, the young buck knocks its antler against black dog-wood switches . . . and boisterously paws the ground with its hooves, causing the characteristic patch marks to appear. The doe in the foreground is turning round towards the 'whimpering' young buck while the other animals calmly continue grazing . . . a lovely scene."[37] The biological groups set up at Hamburg's Natural History Museum provide a similar insight into the naturalization of animal gender roles, where gender-neutral titles often masked the reproduction of gender stereotypes that occurred in such scenes. "Deer in early winter" depicted a doe with two fawns in the undergrowth, while the buck kept a lookout, "Wild boar in mid-summer" showed a sow with her young and, in the background, a powerful male, "Noble deer in autumn" portrayed a female with fawn following a buck approaching the wallow, while the young pricket clambered up an incline.[38] Opinions seemed to differ as to the relative intelligence of the two sexes: a mountain cock group at the Hanover School Museum portrayed a mountain cock with its plumage fanned out as a sign of warning to two hens that a fox was creeping up, while the Provincial Museum in Hanover reversed the roles, showing the hens giving the warning.[39] Despite such minor differences in the characterization of gender roles, the family remained the focal point of animal display throughout the imperial era, a testament to the power of human social constructions to shape and be shaped by shifting conceptions of the natural world.[40]

The portrayal of family life in animal displays served the didactic mission of natural history museums in the Wilhelmine era's rapidly industrializing and urbanizing society. Dermoplastic animal families served as normative, accessible models of bourgeois family life; they helped to represent, define, and negotiate gender relations and the allocation of gender roles in Germany's society. Museum curators balanced scenes of male conflict with portrayals of family harmony, mirroring and, in turn, reifying the separation of the competitive work life from the domestic sphere in Victorian society as a whole. Biological displays addressed people without any scientific background from all social strata, including the growing number of working-class and petty bourgeois families that frequented museums around 1900. Patrons could easily make sense of such images because they drew on familiar concepts and offered means of identification. Museum officials hoped that scenes of natural harmony and family tranquility would provide useful social models for this increasingly broad audience.

Such true-to-life animal groups also helped to compensate for the vanishing natural world by giving museumgoers the feeling that they were surrounded by a landscape still intact and alive. Biological scenes drew visitors' attention to animals species threatened by industrialization and urbanization, such as the lynx, the wild cat, and the beaver, by making them the heroes of the displays. Such scenes enabled audiences to reestablish proximity to such animals and temporarily forget their alienation from nature. Yet at the same time dermoplastic animal scenes, along with the increasing popularity of regional biological/ecological displays helped to domesticate nature, transforming it into an amenity for city dwellers. After 1900, the "biology—Heimat—family" theme found its way into natural history museums across Germany, the benchmark for the portrayal of nature writ large.

In the decades before World War I, therefore, natural history museums drew upon nonscientific sources—particularly the social constructs Heimat and family—to represent nature to urban audiences. Museums increasingly sought a flexible balance between disseminating scientific knowledge and developing realistic, accessible displays. Without sacrificing their scientific credentials, they shifted away from their traditional focus on taxonomy and taxidermy in response to accelerated changes in both the natural environment and modern social life that accompanied industrialization. Dermoplasticians such as Martin, for example, argued that museums of the industrial age should no longer be mere repositories of taxonomic knowledge, but compensate for the disappearance of nature with displays that represented the vitality and dynamism of the natural world. Museums represented nature as a source of knowledge, an aesthetic pleasure, and an intact, organic entity—capable of existing without humanity yet affording it protective shelter and offering examples of domestic harmony. In relying on the idea of the homeland (or Heimat) and the nuclear family to organize zoogeographical and animal displays, moreover, the museum helped to reify existing territorial arrangements and gender roles as natural and timeless. Using biological categories to describe regions and urban environs as biotopes or to envision the family as elemental core of animal life gave these concepts a scientific underpinning that they previously lacked. In addition, a gendered vision of nationhood emerged, with a female Heimat serving as the emotional counterpart to the male fatherland.

Notes

1. This essay contains some findings from my dissertation entitled "Natur zum Anschauen. Wissenspräsentation im Naturkundemuseum des deutschen Kaiserreichs, 1871–1914," which was published as *Natur zum Anschauen. Das Naturkundemuseum des deutschen Kaiserreichs, 1871–1914* (Cologne: Böhlau, 2003).
2. On the development of Haeckel's ecological approach to biology, see Robert C. Stauffer, "Haeckel, Darwin, and Ecology," *Quarterly Review of Biology* 32 (1957): 138–144. On Haeckel's influence in the development of plant autoecology, the study of individual plant-environmental interactions, see Eugene Cittadino, *Nature as the Laboratory: Darwinian Plant Ecology in the German Empire, 1880–1900* (Cambridge: Cambridge University Press, 1990).
3. Hans-Jörg Rheinberger et al., eds., *Räume des Wissens. Repräsentation, Codierung, Spur* (Berlin: Akademie Verlag, 1997).
4. Kurt Lampert, "Die naturhistorischen Museen," in *Die Museen als Volksbildungsstätten: Ergebnisse der 12. Konferenz der Zentralstelle für Arbeiter-Wohlfahrtseinrichtungen*, ed. Zentralstelle für Arbeiterwohlfahrtseinrichtungen (Berlin: Heymann, 1904), 20–27.
5. See, for example, *Jahrbuch des Mannheimer Vereins für Naturkunde* (1853): 7; *Guide to the Museum of the Senckenberg Natural Research Society at Frankfurt am Main* (Frankfurt am Main: Knauer, 1896).
6. *Zur Geschichte des Städtischen Museums für Natur-, Völker- und Handelskunde zu Bremen* (Bremen: Heilig und Bartels, 1928), 28; *Guide to the Museum of the Senckenberg Natural Research Society at Frankfurt am Main* (Frankfurt am Main: Knauer, 1896), 104–106.
7. Otto Lehmann, *Festschrift zur Eröffnung des Altonaer Museums. Zugleich ein Führer durch die Sammlungen* (Altona, 1901); Otto Lehmann, *Guide through the Altona Museum*, 5 pts. (Altona [1915?]) and "Der Erweiterungsbau des Altonaer Museums," *Museumskunde* 13/14, nos. 3/4 (1917/1918): 93–134. See also Uwe Claassen, "Denkmäler des Volkstums: Zu einem biologistischen Objektverständnis in einer kulturgeschichtlichen Museumsabteilung," in *Symbole: Zur Bedeutung der Zeichen in der Kultur*, Thirtieth German Ethnology Congress in Karlsruhe 25–29 September 1995, ed. Rolf Wilhelm Brednich and Heinz Schmitt (Münster: Waxmann, 1997), 399–407.
8. In 1912/1913, the Altona museum's public collection was redesigned in the course of extension work. A comprehensive geological section was added together with a section entitled "Function and Form in Nature." Altona Museum Archive File No. 21.7.16.3: Natural Science, Fauna and Flora.
9. Ferdinand Pax, *Führer durch die Schausammlungen des Zoologischen Museums in Breslau*, 4th ed. (Breslau: Korn, 1925).
10. Friedrich Dahl, "Grundsätze und Grundbegriffe der biocönotischen Forschung," *Zoologischer Anzeiger* 33, no. 11 (1908): 349–353. See also Lynn Nyhart, "Civic and Economic Zoology in Nineteenth-Century Germany: The 'Living Communities' of Karl Möbius," *Isis* 89, no. 4 (1998): 605–630.
11. Celia Applegate, *A Nation of Provincials: The German Idea of Heimat* (Berkeley: University of California Press, 1990).
12. Philipp Leopold Martin, *Die Praxis der Naturgeschichte*, vol. 2, *Dermoplastik und Museologie oder das Modelliren der Thiere und das Aufstellen und Erhalten von Naturaliensammlungen* (Weimar: Voigt, 1870), 74.
13. Martin, *Die Praxis der Naturgeschichte*, vol. 2 (1870), vii–viii.
14. Philipp Leopold Martin, *Die Praxis der Naturgeschichte*, vol. 2, *Dermoplastik und Museologie*, 2nd ed. (Weimar: Voigt, 1880), vii.

15. Martin, *Die Praxis der Naturgeschichte*, vol. 2 (1880), vii–viii.
16. Ibid.
17. Andreas Daum, *Wissenschaftspopularisierung im 19. Jahrhundert: Bürgerliche Kultur, Naturwissenschaftliche Bildung und die deutsche Öffentlichkeit, 1848–1914* (Munich: R. Oldenbourg, 1998).
18. Philipp Leopold Martin, *Die Praxis der Naturgeschichte*, vol. 1, *Taxidermie oder die Lehre vom Präparieren, Konservieren*, 2nd ed. (Weimar: Voigt, 1886), 137.
19. Martin, *Die Praxis der Naturgeschichte*, vol. 2 (1870), 5, 11.
20. Ibid., 6.
21. Ibid., 69.
22. Ibid.
23. Ibid., 71.
24. Karin Hausen, "Die Polarisierung der 'Geschlechtscharaktere'—Eine Spiegelung der Dissoziation von Erwerbs- und Familienleben," in *Sozialgeschichte der Familie in der Neuzeit Europas*, ed. Werner Conze (Stuttgart: Klett, 1976), 363–393.
25. When Martin portrayed animals hunting, he accorded males a higher position than females, often in extravagant poses. In the 1860s, he produced a hunting lion couple for the Detmold Museum that was described as follows in the 1920s: "The male is poised menacingly on a rock, whilst somewhat lower down we see the lioness leaping after an antelope." Quoted in Otto Werth, *Die zoologische Sammlung des Lippischen Landesmuseums. Ein Führer* (Detmold: Meyersche Hofbuchhandlung, 1922), 12. See also Rainer Springhorn, "150 Jahre Lippisches Landesmuseum Detmold," *Heimatland Lippe* 78, nos. 8/9 (1985): 287–314; Springhorn, "Die frühe Phase der naturhistorischen Sammlung des Lippischen Landesmuseums. Stiftungen aus der Zeit von 1835–1891," *Lippische Mitteilungen aus Geschichte und Landeskunde* 54 (1985): 85–107.
26. Martin, *Die Praxis der Naturgeschichte*, vol. 2 (1870), 66.
27. W. Boitard, *Die Kunst, Tiere auszustopfen und Pflanzen und Mineralien aufzubewahren* (Quedlinburg: Gottfried Basse, 1835), 157–158.
28. Georg Pistorius, *Anleitung zum Ausstopfen und Aufbewahren der Vögel und Säugetiere. Aus eigenen Erfahrungen und sachkundigen Männern geschöpft* (Darmstadt, 1799), 126.
29. Christian Ludwig Brehm, *Die Kunst Vögel als Bälge zu bereiten, auszustopfen, aufzustellen und aufzubewahren. Nebst einer kurzen Anleitung, Schmetterlinge und Käfer zu fangen, zu präparieren, aufzustellen und aufzubewahren* (Weimar: Voigt, 1842), 83–84.
30. Ibid.
31. Ibid.
32. Brehm disparaged earth-bound birds, about which he said there was "something ungainly and common" due to their "slack" plumage posture. Brehm, *Die Kunst*, 86.
33. Quoted in Walter Jülicher, "Alexander Held. Ein Vater der Präparatoren und seine Naturgeschichte," *Der Präparator* 34, no. 4 (1988): 337–342, here 340.
34. Martin, *Die Praxis der Naturgeschichte*, vol. 2 (1880), 4; Martin, *Atlas zur Praxis der Naturgeschichte* (Weimar: Voigt, 1886), chart 7, fig. 4.
35. Martin, *Atlas zur Praxis der Naturgeschichte* (Weimar: Voigt, 1886), 21; Martin, *Die Praxis der Naturgeschichte*, vol. 1, 3rd ed. (Weimar: Voigt, 1886), 131, 136.
36. "Zooplastische Tiergruppen für volkstümliche Museen, entworfen und ausgeführt von Heinrich Sander, Köln a. Rh.," in Exhibition Catalog from the Twelfth Conference of the Centralstelle für Arbeiter—Wohlfahrtseinrichtungen, 21–22 September 1903 in Mannheim (Berlin: Heymann, 1903), 5–12.
37. Otto Lehmann, *Festschrift zur Eröffnung des Altonaer Museums, zugleich ein Führer durch die Sammlungen* (Altona, 1901), 19.

38. Similar family images were also set up in the Hanover School Museum, among them a "deer group"—protective buck, recumbent doe, two punctated kids; a fox group—father fox has caught a hare and is eagerly "awaited by his family outside their den." W. Wehrhahn, "Das Städtische Schulmuseum in Hannover, Part 3: Die Zoologische Sammlung," *Hannoversche Schulzeitung* 35, no. 26 (1899): 203–204.
39. *Illustrierte Rundschau* 8 (1912): 125–127.
40. Preparation instructions retained the division of dermoplastic groups into family groups, fighting groups, and zoogeographical groups introduced by Philip Leopold Martin and incorporated the range of examples proposed by him almost unchanged. M. Selmons, *Handbuch für Naturaliensammler*, vol. 1, *Praktische Anleitung zum Fangen, Züchten, Konservieren und Präparieren von Naturkörpern, sowie zur Einrichtung von Sammlungen* (Berlin: E. A. Böttcher, 1907), 60.

Part III
The Politics of Conservation

Chapter 7

Indication and Identification

On the History of Bird Protection in Germany, 1800–1918

FRIEDEMANN SCHMOLL

A<small>ROUND</small> 1900 a programmatic debate over the general principles of nature conservation reached its climax within German bird protection. Conservationists asked: Why should birds be protected? Why do animals living freely need the preserving care of man? What are the reasons why nature should be protected by culture from culture? One pole of this debate was based on a utilitarian point of view. Hans von Berlepsch (1857–1933), a leader of the bird protection movement, justified his motive for protecting birds with an argument grounded in the "Practical Enlightenment" tradition. He considered nature to be a pseudomechanical and, therefore, adjustable system in which birds were beneficial supporters for man's material interests:

> Bird protection is not only leisure pursuit, a passion arising out of ethical and aesthetic motives—it does not stem, therefore, from admiration for the birds' song or from the endeavor to beautify and enliven nature—but in the first place, bird protection is exclusively an economical question of most eminent significance. Bird protection is a measure for which we humans have material and great pecuniary use. Bird protection wants to protect and breed the birds useful to man on account of their immediate necessity.[1]

In Berlepsch's anthropocentric view, birds deserved protection first and foremost because they served the needs of human beings, rather than having intrinsic value of their own.

The counterargument in this debate about the aims of bird protection rejected an instrumental valuation of nature. Important protagonists of bird protection such as Ernst Hartert (1859–1933) looked at birds as creatures worthy of protection for aesthetic, emotional, or ethical reasons. Hartert wrote: "After all, the question why we actually want to provide birds with protection is not as easily settled as it is usually done. . . . We indeed have the courage to say that we want the birds themselves, that we want to protect them for their own sake, that we don't want nature to become one-sided because of our pouches and our 'development.'"[2] In Hartert's view, bird protection was an ethical affair; even birds traditionally deemed "harmful" to crops and songbirds were vital to the overall balance of nature and therefore deserved protection. From this ethical perspective, birds had an intrinsic value beyond human needs.

Although Berlepsch's utilitarian perspective dominated discussions of bird protection during the nineteenth century, Hartert's ethical view was gaining ground in the years around 1900. This ethical standpoint emphasized that not only "beneficial" birds were to enjoy human care, but also those formerly classified as "harmful." The institutional formation of the bird protection movement integrated these and other reasons for the conservation of nature: utilitarian ones based on theories derived from agriculture and forestry, aesthetic and emotional ones based on a passion for birds, ethical ones adopted from animal protection, and ecological ones borrowed from scientific ornithology. Bird protectionism's ability to embrace these different perspectives enabled the movement to become a popular force for social integration.

In this chapter I want to reconstruct the different motivations for bird protection and show how they reflected the different cultural meanings attached to nature around 1900. During the institutionalization of conservation movements in the final third of the nineteenth century, bird protection was among the earliest, best organized, and most popular environmental movements. In the German Empire, a society very concerned about difference and distinction, bird protection brought people together in societies and institutions who otherwise would have had no contact with each other. Nature conservation allowed new social networks to form: animal-loving aristocrats mixed with the sensitive bourgeoisie, the lower middle class, including farmers moved by utilitarian thinking, joined proletarian members of the Friends of Nature (Naturfreunde), anarcho-syndicalist drop-outs of the political Left found common ground with racist spokesmen of the political Right.

The reason for the popularity of bird protection and its status as a vanguard environmental movement stemmed from the manifold ways in which birds could be appreciated. Birds fulfilled needs and longings for an (emotional and aesthetic) identification with the natural world, on the one hand, and for (ecological) indication of human impacts on that world, on the other. Birds, in comparison with other animals or plants, enjoyed a special cultural status in the imagination of modern man. As the ornithologist and nature writer Curt Floericke (1869–1934) noted: "No other class of animals has won man's pleasure and affection to such a high degree as has the class of birds . . . It is the bird's beauty, its grace, its innocence, its admirable flight and, above all, its magnificent song, which captivates us so much. Under these circumstances it was only natural for man to give his support early to the weal and woe of his feathered friends, for him to be struck by the idea of bird protection."[3] Due to their abilities and characteristics, birds were better suited than other animals and plants to serve as cultural symbols and sources of emotional identification. These "feathered friends" were a suitable medium for projecting the values and aspirations of the liberal, Protestant bourgeoisie of Central and Northern Europe.[4] Through the binoculars of the ornithologist they embodied the very ideal of the bourgeois society: they are clean and spotless, diligent and orderly. Their sexual life is brief and discreet apart from the emotional production of courtship display and song. It is nonetheless exceedingly effective as far as the success of the efforts of reproduction is concerned. The birds' social existence is filled with a harmonious sense of family and self-sacrificing love for children. They are freedom loving yet socially oriented. They are open-minded and have proven themselves to be cosmopolitans when migrating to the south for the winter. But their love for their native land appears to be even greater since they regularly return to their traditional breeding grounds come springtime. Birds build their nests even more skillfully than architects build their homes, and they sing more beautifully than opera tenors—and they even don't have to practice. Thus, on the one hand birds are wild and remain alien and strange, but at the same time they are exceedingly suitable for anthropomorphism and for the construction of emotional relations.

Birds are, however, also bio-indicators through which complex ecological changes may be observed and documented. Anyone with an interest in birds, whether layman or scientist, learns much about ecological interconnectedness. Through birds, questions about nature—its endangerment and possibilities for its preservation—became an issue for not only elite discussion, but also popular mobilization. Discoveries about birds provided evidence about the state of the natural environment and therefore material for comments about the development of civilization. Since knowledge about birds was knowledge about the state of nature in general, it became easier to criticize social processes and the pace of historical change associated with industrialization.

The rapid growth of book titles handling themes of bird protection from the 1860s onward signaled that this issue was no longer reserved for specialists and scientists but was a concern for everyone.[5] The question concerning the welfare of the bird kingdom became a question of public welfare and interest. The moving forces in the public discussion on nature were civic organizations, which tried to establish nature protection as the objective of social action against competing—mostly economic—interests. As institutions of social self-organization they articulated their claim to participate in the shaping of social processes. These civic organizations and societies, which mainly drew from the educated middle class, however, had another objective: the self-education of their members, who came into contact with like-minded people through the cultivation of their interests and passions in nature study within a widespread national and international network. Therefore, occupation with birds was a topic of individual experience as well as of social communication.

After local societies for bird protection and for the friends of birds had been founded on a broad basis, the first institution that operated in the whole German Empire, the German Organization for the Protection of the Bird Kingdom (Deutscher Verein zum Schutze der Vogelwelt) was established in 1875. It was an organization, though, whose membership quickly stagnated at around 1100. In 1899, on the initiative of entrepreneur's wife Lisa Hähnle (1851–1941), the Society for Bird Protection (Bund für Vogelschutz) was founded. This organization brought together individuals with different backgrounds and philosophies about bird protection, including scientific ornithologists, scientific amateurs, animal ethicists, and those who embraced the aesthetic-emotional traditions of the connoisseurs of birds. Due to the low fees a rapid growth in the number of members occurred, which reached its peak at 41,323 in 1914. The society developed into a modern lobby organization concerned with the interests of nature. It coordinated public relations work with mass media, including modern forms such as postcards, photos, and films, and committed itself to politics by pursuing legislative initiatives. On a local level members pursued not only education in natural history, but also practical bird protection by creating copses, building nesting boxes, and providing winter feed. Society members safeguarded ecologically valuable areas as nature reserves through purchasing or leasing land. Additionally, the society vigorously fought against the national and international trapping of birds for the feather industry and food purposes. The bird protection movement thus pursued two interrelated tasks. On the one hand, the movement of bird protection articulated a claim to political participation. On the other hand, as a value-oriented reform movement, it broached questions of individual lifestyle and tried to establish a moral attitude toward nature.

Natural Balance and Harmonic Order:
Birds as Useful Helpers in the Household of Nature

The first groups of people to think systematically about bird protection were natural historians and forestry theorists during the Age of Enlightenment. The background for this interest in bird protection was the rationalization of forestry in Central Europe in the late eighteenth century. The supposed shortage of wood, national indebtedness, and the regional onset of the industrial revolution incited reforms in forestry in the eighteenth century aimed at an economically more effective use of the material resource of wood. One consequence of the conversion of extensively utilized mixed woodlands into intensively utilized monocultures was their ecological instability: at the end of the eighteenth century, complaints mounted about the massive damage to the forest caused by insects in certain regions.[6]

To counteract such damage, forestry scientists such as Johann Matthäus Bechstein (1757–1822) called for the protection of birds as a means of biological pest control. "The conservation of insectivorous birds is a major contraceptive," wrote Bechstein. "What an immense number of harmful insects dwelling in forest and wood is not being exterminated in the span of one year by the chaffinches, the titmouse, the mistletoe—and song thrushes, the blackbirds, the robins, the redstarts, and the warblers! . . . Therefore, the most rigorous bans against the catching of birds . . . are to be issued."[7] Bechstein's analysis meant that living birds were of greater material use than dead birds in a pot killed for food purposes. By using the bluetit as an example, a bird that was mainly hunted in Thuringia, Johann Friedrich Naumann (1780–1857) claimed that the material exploitation of birds was a morally reprehensible action: "Their meat is a pleasant fare; they will be, however, so extraordinarily beneficial to us by demolishing a tremendous number of harmful insects that it is a sinful thing to kill such beneficial birdlife for the sake of such a small tasty bite."[8] Thus, some local decrees and regional laws in the German small states aimed at the protection of "beneficial" species of birds as early as the end of the eighteenth century.

Models of nature as a system of interlocking parts that had to be kept in balance served as the basis for such instrumentalist thought. Authors within the physico-theological and Enlightenment traditions regarded nature as a harmonic order in which humans had to keep the balance between civilization and nature as paternalist housekeepers. According to Bechstein, the advancing development of human culture with its intrusions into nature tended to disturb this harmonic balance: "Thus, the sophisticated man has unforeseeably extended his rule over the Earth with his advances in culture and he dares to disturb that balance with regard to his own interest."[9] If man regarded himself nonetheless as a "sensible ruler and well-appointed housekeeper of the visible nature," then he would have to intervene in its household and regulate it, asking the question: "To what extent do I have to turn and set that clock in order to intervene as little as possible and without wantonness on the course of nature in favor of my own interest?"[10] In Bechstein's view, then, human beings were stewards of the great systems of nature.

The useful function of birds facilitated their cultural re-valuation and changed the relations between man and bird. As ordinary people began to see birds as helpers, they could no longer imagine killing their friends in order to have them for dinner or serve as a trophy afterward. Since friendship is about a pure, personal, and emotional relationship on a voluntary basis, at least as far as the underlying idea is concerned, beneficial birds deserved not only considerate treatment and care for their "work" but also personal acknowledgment. According to the Polish ornithologist Casimir Graf von Wodziki (1816–1889), birds were "feathered friends": "It is consequently the responsibility of the forester and the agriculturalist to become acquainted with these friendly allies nature supplies them with and to treat them as good friends, to raise them, to spare them, to care for them, and to nurse them: this should happen instead of the widespread habit of looking upon them as unwelcome guests, and therefore even hunting and killing them."[11] From this perspective, the potential for developing an emotional partnership beyond the material relation between man and bird emerged. Instrumentalization and sentimentalization do not stand in contradiction to each other, but instead form two facets of one and the same possible relation between man and bird.

The argument about birds' "usefulness" provided, for a long time, plausible legitimacy for their protection. After Bechstein there followed numerous natural historians or theorists of forestry, such as Emil Adolf Roßmäßler (1806–1867), Harald Othmar Lenz (1798–1870), Julius Theodor Ratzeburg (1801–1871), or Constantin Wilhelm Lambert Gloger (1803–1863), who made observations about the protection of birds on a systematic basis. Even after 1870, the "utility principle" proved to be a plausible argument for establishing the movement during its first phase of popularization and institutionalization. The instrumental function of birds in nature's household was the center of commitment, though less in the national context than within the framework of international attempts at cooperation. This position was very important, for example, at the International Conventions of Farmers and Foresters in Vienna in 1873, at the Second Budapest International Convention of Ornithologists in 1891, or for the Paris International Agreement on the Protection of Birds Useful to Agriculture (Übereinkunft zum Schutze der für die Landwirtschaft nützlichen Vögel) in 1902.[12]

The anthropocentric idea of a "relation of accomplices" between man and bird certainly also implied the status of "harmfulness" for other species of birds and raised the question how they were to be treated. The criteria of "usefulness" and "harmfulness" divided the world of birds up into good and evil; they parted into species either worth preserving or exterminating. As the love for "beneficial" birds grew, by this same logic, so did the hate for "harmful" birds expand. Thus, at the end of the nineteenth century, there were still numerous efforts to exterminate "harmful" animals. The first local societies for bird protection, for example, were still carrying on the persecution of "harmful" species before 1900, offering paid rewards for the killing of raptors, sparrows, or shrikes. As late as 1899, Hans von Berlepsch dedicated a whole chapter of his standard work *The Complete Bird Protection* (*Der gesamte Vogelschutz*) to the "Extermination of the different enemies of the birds worthy of protection," wherein he offered concrete instructions on how to hunt down these enemies. It was necessary, according to Berlepsch, to "declare war without mercy" on cats, martens, ravens, starlings, and shrikes.[13] He considered the sparrow to be the main enemy, for which he worked out a full-blown program of extermination. Berlepsch based this program on the premise that the sparrow was endowed with "a particularly powerful sexual drive," that the males could be easily distinguished from the females, and that males outnumbered the females. Berlepsch reasoned that the production of a libidinal surplus of that drive would necessarily result in the self-destruction of the populations of sparrows by the means of mass rapes: "If, now, we intensify this unhealthy ratio of the sexes even more by exterminating solely the female sparrows, I am convinced that the time will come when the few surviving females will be annihilated by the males altogether. Of course, that would be tantamount to the doom of the whole species."[14] Berlepsch was not an outsider; his book ranked as one of the bestsellers of contemporary literature on bird protection.

Diagnoses of Decline: On the Perception of Nature in Field Ornithology

Another group of people that analyzed the decline in the bird kingdom and promoted their protection was the fowlers. Due to their experience in tracking down birds, those who practiced fowling became some of the most well-known ornithologists of the nineteenth century. As one historian of bird sciences noted, "The ornithological research has mainly been fed by two sources here at its beginning: on the one hand, by the joy of the hunt and of bird-catching, and by the passion for birds on the other."[15] Outdoor observation, systematized in fowling and field research, and usually pursued by autodidacts, was the basis of ornithology in the nineteenth century. Johann Andreas (1744–1826) and his son Johann Friedrich Naumann, for example, were farmers; Christian Ludwig Brehm (1787–1864) was a pastor.

The fowlers were the first to recognize that the structural change in agriculture resulted in an impoverishment in the diversity of species and a drop in bird populations. In 1849 Johann Friedrich Naumann retrospectively reported about the fowlers living in his hometown Köthen in Anhalt around 1800: "Already fifty years ago the complaint about the decrease in bird population was commonplace among these people, and although they deemed their snaring measures improved, they spotted and caught by far not as many birds as their ancestors had actually spotted and caught. Hence, one fowling floor after the other died with its owners . . . Soon fowling won't be worth its endeavors"[16] For Naumann, the "heightened culture," by which he meant the intensification of agriculture and forestry, transformed the rich textures of the traditional countryside into a modern agrarian wasteland, devastating hedges, marshes, moors, and meadows. "All too certain is it [the decline] a result of the increase in human population and human needs, of the rise in industry and of a profitable employment of soil."[17] As evidence for the aesthetic and ecological impoverishment of cultural landscapes, Naumann mentioned the conversion of varied meadows into "monotonous arable lands," the "now almost fashionable clearance of wild copses, hedges, and isolated woods in order to reclaim land for agriculture," as well as the draining of marshes and lakes.[18] Thus, the experience and systematic observation by the first field ornithologists provided information on ecological changes in the environment. Their diagnosis for the decrease in bird population was another force that brought the issue to the forefront around 1900.

Aesthetics and Morality: The Popularization of Bird Protection around 1900

At the beginning of the 1870s the ornithologist Ferdinand von Droste-Hülshoff (1841–1874) noted that the protectors of the bird kingdom were about to gain cultural and social dominance over the users: "Now it has become different. In town and country one has to search out the ancestral fowlers with a Diogenes' lantern. . . . In their place all sorts of educated gentlemen and jovial regulars gather monthly around a long table and, having a pint of beer, deliberate with appropriate solemnity whether a house should be built for this or that family of sparrows. These are the gentlemen of the municipal society for bird protection, the everlastingly vigorous foster fathers of all birds." Droste-

Hülshoff also described the sensitive members of societies for animal protection and against cruelty to animals, "who would rather hang a street urchin than let a sparrow go hungry." Noting that state governments were increasingly issuing decrees banning the killing and catching of birds as well as the removal of eggs from the nests of a large number of birds, he wrote, "Times have changed, considerably changed."[19]

Droste-Hülshoff thus indicates that aesthetic and ethical arguments were gaining ground in the late nineteenth century, as the union of scientific ornithology, animal protection, and the passion for birds in popular societies brought about a massive critique of utilitarian reasoning for bird protection. Karl Theodor Liebe (1828–1894), for example, criticized economic reasons for protecting birds at the Second Budapest International Ornithologists' Convention in 1891, stating: "For the most part the utility principle prevails. In our opinion, however, the discussion about questions of bird protection is extraordinarily aggravated on account of it." In Liebe's view, the motives for bird protection also needed to encompass a "compassion for birds," since "the appearance of the bird living free rightly appealed to the sense of beauty inherent in every man and that the motive for the protection of birds is derived from that."[20] Those who protected birds also seemed to be better persons; whoever assumes responsibility for nature had a better moral position, because this signaled a high cultural stage of development in line with the cultural evolutionary thinking of that time.

While they differed in their grounds and justifications for their bird protection activity, the competing wings of the movement for bird protection were nonetheless united by their common images of nature and their common patterns of interpretation. Ethically and aesthetically motivated protectors of birds like Liebe also embraced models of nature as a harmonious system of order. Liebe argued that nature should be preserved in its integrity in spite of humanity's "continual struggle for our existence and for our culture." Condemning the willful destruction of nature, Liebe asserted that we have the "duty to preserve nature in its integrity . . . Additionally we must not forget that we don't stand merely opposite nature with our culture, but that we move inside of the same and are a part of the same. Therefore, the animals and the plants are our fellow beings and we must respect them as such." Liebe thus argued that we have a moral obligation toward animals and plants. "No animal is originally either genuinely beneficial or harmful," he noted; "each one has been shown to have a place in nature's big household, where it is pleased by its own presence and contributes its own share to the preservation of the big, beautiful whole. Animals only become beneficial and harmful when they come into pleasant or unpleasant contact with man and his culture."[21] While sharing the utilitarians' vision of nature as a self-regulating system, Liebe used this conception to formulate an ethical justification for conservation that insisted on the intrinsic value and integrity of nature.

Both wings of the bird protection movement were influenced by romantic myths of origin and purity. Their ideal was a harmonic order based on a belief in design in nature, an order they believed was being destroyed by humans. In opposition to the nineteenth century's largely progressive view of history, bird protectionists described historical change as a process of decline, degeneration, and loss. In their glorification of pristine nature and diagnoses of decline, bird protectionists shared many ideological affinities with the nature conservation (Naturschutz) and Heimat protection (Heimatschutz) movements that were also emerging around 1900. Like these movements, bird protection transformed pristine nature to a sacred icon,[22] and touted nature's moralizing effects on the "wonderful organism of a whole popular personality."[23] Bird protection also shared the interests of nature conservationists and Heimat protectionists in the work of Wilhelm Heinrich Riehl (1823–1897), a conservative cultural commentator who is commonly seen as the founding father of German folklore studies, and who argued that nature and national character were intertwined. Only by retaining the bond between a people and its national origins, he asserted, could the nation remain healthy and vital in the long term. "There is something terribly frightening for every natural person's fantasy," he wrote, "in the thought of seeing every little spot burrowed over by the hand of man; it is especially loathsome, however, to the German mind."[24] Among the different peoples of Europe, he believed, Germans maintained a special tie to nature. In his eyes, only original, pristine, and pure nature enjoyed the status of real nature and thus deserved preservation. Like Riehl, many bird protectionists bemoaned the progress of "civilization" as a tale of disaster and decline, a digression from the original state of "naturalness." As Liebe noted, "Where cultural man sets his foot, his step destroys nature's naturalness with its sublime wildness and authoritative harmony; and what freshly blossoms out of the ruins is not the old nature anymore, it's just countryside dismembered by geometric lines, by margins, by ditches, and by paths [creating] a boring caricature of that original magnificent creation of nature."[25]

Though Germany's nature conservation and Heimat protection movements also viewed Riehl as their intellectual forefather, the debates among bird protectors were in general far less culturally pessimistic than these other movements. Though bird protectionists bemoaned the aesthetic disfigurement of the countryside, particularly the transformation of the richly diverse preindustrial landscape into dull monocultural tracts, their analysis was for the most part free of conservative ideology. In addition, bird protectionists were far more willing than adherents of nature conservation or of Heimat protection to concede that humans could have a beneficial impact on the natural environment. Their observations of the bird kingdom revealed that humans' cultural shaping of preexistent nature had in many cases created new ecological niches for a great variety of species of animals and plants. In 1921, for example, Otto Schnurre (1894–1979) noted: "In the last decades, complaints of varied kind from the side of conservation and bird protection have come to the fore about the desolation of local nature by human culture. No one is going to deny the legitimacy of these complaints. The melioration of the bogs, the regulation of the rivers, the monotonization of the forests, and suchlike have removed the ground from a very characteristic flora and fauna."[26] Schnurre, though, simultaneously disabused his readers of the myth of the great variety of unspoiled nature. "It is precisely the culture so reviled by the nature conservationists," Schnurre noted provocatively, "to which we owe the abundance of birds in our gardens and parks. The relative lack of birds in the original woods stands out sharply against it . . . Only where man intrudes, sets up settlements, creates a varied landscape does an ornis with a large number of species present itself."[27] In Schnurre's analysis, then, Germany's great abundance of bird species was a result of human intervention; birds and humans had coevolved to form the cultural landscape on which both were dependent.

Because bird protectionists accepted the possibility of positive human impact on bird populations, they were much more willing than conservationists to create the conditions necessary for bird propagation, rather than simply lament the loss of pristine nature. Their practical work aimed above all at compensating for human intrusion upon nature and the landscape with measures of balance. In answer to the biological and aesthetical impoverishment of natural environments—the disappearance of hedges and marshlands, the intensification of agriculture use by re-parceling land, or the felling of ancient trees—bird protectionists planted copses for bird protection or hung nesting boxes. Their credo was that it was acceptable to replace the small-scale destruction of habitats and nesting places with artificial alternatives in order to bolster bird populations in city and countryside alike.

Though it is not possible to speak of a uniform understanding of bird protection within the movement, but rather competing basic trends, this diversity of aims enabled bird protectionists to gain a large number of members and some legislative success. After attempts to standardize local and regional protective decrees had repeatedly failed since the early 1870s, bird protectionists worked successfully to ensure passage of the Imperial Bird Protection Law after heated discussions in 1888. Bird protectors hoped that the law could at least compensate for the loss of habitat caused by modern agriculture and forestry. The law stated that "the changed operation of agriculture, by the removal of breeding grounds (such as hedges and trees) carried out on a large scale" had decimated populations of birds, which is why "all the greater care has to be taken to diminish as much as possible the intrusions of killing and catching in order to at least preserve the current stock."[28]

The Imperial law was innovative for its time in going beyond utilitarian reasons for protecting birds. "Apart from the reasons of usefulness which can be considered first," the law stated, "the aesthetic and moral considerations upon which the endeavors established in the popular consciousness are based certainly deserve attention."[29] The framers of the law admitted that there was no basis that was either objective or anchored in the regularities of nature for the separation into "beneficial" and "harmful" species. These categories were neither measurable nor fixed by general criteria; they were socially constructed, defined by the interests of various interest groups such as farmers, fishermen, or foresters. Nonetheless, the legislation was still based largely on the material criterion of "utility" and denied "harmful" species such as raptors, stranglers, or crows protection.[30] From that point of time onward, this classificatory separation was continually met with criticism. When the bird protection law was amended in 1908, additional species of birds were given protected status and certain methods of hunting were restricted for ethical reasons, even though the utilitarian classification into "beneficial" and "harmful" species still remained. Still, the Imperial Bird Protection Law, which attempted to define new legal relations between humans and groups of animals, reflected the ever-increasing social status of bird protection in the Second Empire. The law was one of Germany's earliest nature protection statutes and enshrined bird protection as the avant-garde of the country's budding nature conservation movement. The state of the natural environment—its biological state as well as its aesthetic appearance—had become a parameter of public welfare.

Armed with state support and growing social esteem, the bird protection movement soon moved beyond its critique of mechanized agriculture and forestry and began a systematic campaign against the human persecution of birds. The movement's leaders realized that land use issues were only partly suitable for creating emotional bonds between humans and the endangered bird kingdom. Through public campaigns against the hunting of birds for food or the use of birds in the fashion industry, bird protectionists promoted a new, uncompromising moral stance toward the natural world.

Idealism versus Materialism: Birds' Corpses on Women's Heads

One issue that demanded an ethically unambiguous relationship toward the bird kingdom, and therefore toward nature as a whole, was the conflict about the use of bird feathers in the fashion industry.[31] Around 1900, numerous organizations mobilized against the killing and use of bird feathers for hat decorations or other accessories and labeled these practices, following the lead of popular writer Carl Georg Schillings (1866–1921), as "mass murder," a "tragedy of civilization," and "concomitants of barbaric life" in the modern age.[32] In his essay "Humans and Earth" (*Mensch und Erde*), Ludwig Klages (1872–1956), a critic of progress and an apocalyptic philosopher, reported that about 300 million birds each year were being cruelly sacrificed for the sake of human vanity. "An unparalleled orgy of devastation has taken its hold of humanity," he lamented. "'Civilization' bears the feature of an unleashed desire to kill, and the wealth of the Earth is withering in its poisonous stench. That's how the fruits of 'progress' look!"[33] Carl R. Hennicke (1865–1941) detailed the carnage in the 1898 *Book of Bird Protection* (*Vogelschutzbuch*). In that year alone, at least 1,538,738 herons had been killed for fashion purposes in Venezuela. In 1902, the importation of feathers into Paris had amounted to 703,300 kilograms, rising to 1,039,300 kilograms in 1911. Of the 1911 total, 245,400 kilograms had come from the feathers of "noble" birds: herons, ostriches, and birds of paradise.[34]

Not only the number of birds destroyed but also the cruelty of the killing methods roused the ire of bird advocates. Hennicke's article reported that the plumage of hummingbirds was skinned alive in order to preserve the colors of the feathers for a longer time. While the feather industry disputed the truth of this claim, it did not comment upon reports that herons were being shot right in the middle of their brooding act. Societies such the International Women's Association for the Protection of Birds (Internationaler Frauenbund für Vogelschutz) or the Society for Bird Protection (Bund für Vogelschutz) spoke out against this debasement of nature, while the popular natural history journal *Kosmos*, the cultural reform organization Dürerbund, and the German Merchants Association (Deutscher Käuferbund) designated the use of birds for fashion purposes as an unequivocally unethical treatment of nature.

Bird protectionists' critique of the fashion industry's use of bird feathers inevitably made female consumption a central object of debate; the result was a gendered understanding of personal ethics. German commentators contrasted the fashion practices of "heartless and thoughtless demimondaines in Paris" with the "decent" German woman, who was assumed to value simplicity over extravagance and "naturalness" over urban sophistication.[35] For Carl Georg Schillings, the slaughter of birds for such needless vanity was plain proof that the cultural evolution of everyday life had not kept up with the technical development of civilization. "A lady, a maiden of fine, noble inclination," he argued, "cannot possibly show herself with a dead, dismembered bird attached to her head, with a bird's corpse, once she has been informed about the state of affairs. A large amount of today's feather hats have a painful effect, in the true sense of the word, on a nature-loving, noble, and decent feeling person. The fact that ladies wear hats with bird's corpses or wings . . . can be interpreted as a striking evidence that our times, which are so highly developed scientifically . . . [and] technically, cannot claim that praise for itself in the cultural respect."[36] When it came to reforming everyday life, in other words, Schillings, like many of his contemporaries, assumed that women were the primary consumers of superfluous goods. By turning away from such goods, however, they would show that the German cultural nation was idealistic and close to nature, in contrast to the materialism and superficiality of Romance civilization.

Against such useless exploitation of nature, bird and nature conservation societies roused the public. They agitated in the press and in lectures, targeting people's environmentally destructive lifestyle and personal conscience. Public figures such as the dowager queen Alexandra of Great Britain, who banned the use of feather decoration in her court, were held up as examples to be emulated by ordinary people. Members of the societies for bird protection were asked to convince people in their immediate social network—relatives, friends, church members, and other clubs—to stop using bird feathers. The Society for Bird Protection distributed "lists of the renunciation of feathers," in which women were able to publish their declaration not to use feather decoration. The debate on feather usage for the fashion industry thus proved appropriate for demanding moral responsibility from humans toward nature.

Culinary Morality: On the Creation of a Modern Food Taboo

The use of birds for food was another consumption issue that bird protectionists used to separate the friends and foes of nature, the idealists from the materialists.[37] Songbirds had been an integral part of noble, bourgeois, and country cooking for centuries in Central Europe. Until well into the nineteenth century, the hunting and consumption of larks, thrushes, titmouse, and other species of songbirds was perfectly natural in Central Europe, albeit shaped by regionally available species and tastes. "Leipziger Lerchen" (larks of Leipzig) were still considered a delicacy, while "Thüringer Meisensuppe" (Thuringian titmouse soup) was a local specialty at this point in time. The hunting of "Krammetsvögel" (thrushes) was customary in the whole of Germany. According to hunting statistics, about 1.2 million specimens had been caught in Prussia in 1898 alone, yet this large number was probably only a fraction of the birds actually shot.[38] Recipes for preparing larks or thrushes would have been found in any cookbook around 1900.

As a rule, hunting with nets, snares, lime switches, or on the fowling floor was practiced seasonally and as a second occupation; noble bird hunting enjoyed only small prestige in comparison to these other hunting practices. The mass catching of birds by farmers had a large economic significance, especially in regions of economic shortage. "Many a poor day laborer tries to earn his bread with it, and many a rich man obtains the most beautiful delight of mind with it," Johann Andreas Naumann noted in 1789.[39] Such economic advantages were still evident in the mid-nineteenth century; as Christian Ludwig Brehm noted in 1855: "The catching of birds is very advantageous; it supplies us with a very healthy and tasty meal, which is even more highly esteemed in the country, the less one has the opportunity in many places to obtain fresh meat without high costs. For that reason as well fowling is important for the farmer, and it is also a not insignificant line of business in many areas."[40]

The bird protection movement, however, estimated the aesthetic and emotional value of living birds higher than the material and nutritional value of dead ones and pursued the systematic extinction of hunting and eating practices that had been considered "natural" for centuries. The campaign had its origins in late eighteenth-century campaigns against cruel hunting practices, gathered momentum as various observers pointed out the usefulness of live birds for agriculture and forestry, and ended with the "extinction" of songbirds in German kitchens around 1900. The critique of hunting birds for food had important class dimensions, for members of the "sensitive" and the self-conscious educated bourgeoisie spoke up most vehemently to end such practices against the "decadent" aristocracy and the "primitive" and "cruel" peasantry.

German bird protectionists also portrayed the issue of bird consumption as a measure of a people's degree of cultural evolution and a characteristic of a country's national-cultural identity. The Germans as a "cultural people who have become sensitive" would respect the life of birds out of a special love for nature, claimed Ludwig Reinhardt in 1912. "Different are the hard-hearted Latin peoples," he wrote, "who not only have not taken offense to bloodshed and cruelty to animals since the time of the Roman Empire, but who rather enjoy these practices. They put those small plucked... corpses on the market.... Actually it is a shame that such a delicacy is still tolerated in a usually so advanced cultural nation."[41] Defenders of birds disparaged Italian fowling practices and anthropomorphized birds' suffering by referring to their capture and killing as "disgusting mass murders" or a "war of destruction" while describing their tormentors as "bloodthirsty, uncultured Italians."[42] In Germany, the bird protection movement fought against bird catching with everything they had. The 1888 Imperial Bird Protection Law established closed seasons for hunting, banned the hunting of certain species, and forbade the use of particularly cruel capture techniques. The amended 1908 bird protection law expanded the list of protected species and outlawed additional hunting techniques, including the infamous "line of springs" (*Dohnenstieg*).

By 1900, therefore, bird protectionists had succeeded in making songbirds an impossible meal; eating songbirds became a moral scandal, a social taboo. Taboos are guides to social interaction and regulate human behavior. They mark cultural and moral norms—in this case, norms of dealing with nature. Within the space of a few decades, the taste for fried larks, titmouse, or finches vanished; songbirds were no longer identified as appropriate sources of food. Except for a brief period during the starvation years of World War I, the consumption of songbirds from that point on aroused disgust and loathing, the impossibility of using them for food anchored deeply in the psychic structure of modern Central Europeans. The moral gain of renouncing birds obviously had become higher than the material return from their consumption. The establishment of this food taboo is a clear indication of the cultural value birds had achieved by the turn of the century. The ethos of protection was internalized as a culinary morality that shaped Germans' very appetite.

Preservation and Extermination: On the Ambivalence of Natural Relations in the Modern Age

The state of the natural environment had indisputably become a major issue of public welfare in Wilhelmine society by 1900. The second industrial revolution had spurred the formation of numerous societies and institutions dedicated to the protection and the preservation of nature and landscape. With its diffuse combination of utilitarian, sociopolitical, ecological, ethical, and aesthetic motives, bird protection proved to be avant-garde among the movements of nature conservation. Bird protectionists distinguished themselves from their counterparts in the nature conservation and Heimat protection movements in important ways. The nature conservation movement sought to defend a range of individual relics of nature such as "strange" trees, unique rock formations, and, later, more complex habitats or aesthetically valued landscapes against intrusions of civilization. Heimat protection organizations tried hard to preserve a number of scenic or historically significant landscapes against the leveling forces of the modern age. Bird protection, by contrast, committed itself to an exclusive group of animals: the birds.

Despite this narrow focus, bird protection was better able to see the connection between the protection of the avian world and nature as a whole, with birds serving as indicators of broader ecological relationships. The other streams of conservation developed in the context of historicism and sought to set aside witnesses to Germany's historical-cultural and natural past. Such "museumification" was in no way concerned with the resolution of modern environmental problems, but functioned instead to secure a historically certain origin and national identity in an age of rapid change. The movement of bird protection, on the other hand, placed far greater emphasis on the need to shape the landscape consciously, with far less attention to bemoaning the loss of supposedly primordial nature or traditional cultural landscapes. Bird protectionists sought ways to compensate for ecological changes that were negative for bird populations in what they accepted to be a culturally shaped environment. This environment was not perceived to be an entity independent of man but as the result of human history.

Bird protectionists' concentration on aesthetic arguments cannot be interpreted as escapism, a flight into an apolitical inwardness, or antimodernism. By criticizing social change on aesthetic grounds, they raised thoroughly rational issues about modern societies' self-image: How should the environment in which people live be? Bird protectionists proposed that the beauty of nature and the sensuous perception of ecological and aesthetic variety enriched human life. Their observations of the decrease and even extinction of particular species provided insights about the ecological integrity of nature and raised pertinent questions about more complex ecological problems that fueled speculation about the material and biological limitations of progress. If particular species should be preserved, then, following the insight into the relations between organisms and their wider environment, more complex ecological systems and habitats also had to be preserved. These ecological insights gave a scientific basis for feelings of human guilt in environmental destruction. As one nature conservationist noted, "Man does not change nature's equilibrium without punishment, as it has developed in that battle of a thousand years fought between the races, the genera, the species, and the individuals among flora and fauna. If in recent times pests have appeared here and there in such a quantity that it amounts to catastrophes . . . then man should not only examine the obvious causes for why the pests got out of control, but also deliberate whether that catastrophe is not, in fact, an inevitable consequence of our whole cultural activities."[43] With respect to ecological models of balance, the historical process of modernization could thus be interpreted as an offense against a nature independent of humans and, consequently, as an unmistakable sign of the decline of culture and the triumph of a civilization based purely on materialistic values.

These questions about human culpability for nature's destruction forced bird protectionists to develop a new morality toward nature, one that accepted nature's intrinsic value, rather than solely its usefulness for human beings. Nature conservationists in general were developing new discursive strategies to vilify unbridled environmental destruction; nature conservation and Heimat protection organizations, for example, adapted the sexually and religiously charged vocabulary of nature's virginity to describe the effects of modern civilization on the natural environment. In this rhetoric, nature was construed as a formerly innocent and sacred subject full of integrity, while modern ways of handling and using nature, which satisfied only material interests, were elevated to the status of moral offenses. Conservationists portrayed industrial abuse, the opening up of landscapes for tourism, the rationalization of ecosystems for agriculture and forestry, and other forms of exploitation as "mutilation," the "rape of the landscape," the "defiling of the landscape," or the "maltreatment of nature" to underscore this anthropomorphic and gendered formulation of nature's vulnerability.[44] Bird protection was not immune to such emotionally charged language, but in general restricted itself to a comparably sober and rational description of biological and aesthetic change. There was a widespread agreement in bird protection circles that industrial progress, which turned traditional cultural landscapes into biologically and aesthetically monotonous environments, could not be stopped. The question of how human beings could shape this progress to retain nature's vitality was paramount in their organizational strategies.

Bird protectionists also distinguished themselves from traditional "natural monument protection" (*Naturdenkmalpflege*) and Heimat protection in their international orientation and institutionalization. Although most bird protectionists were concerned primarily with saving and preserving nature within Germany, the movement did not limit its activities to national borders, and concern about the entire world of birds led to extensive correspondence and cooperation between German bird protection organizations and their counterparts throughout Europe and the United States. The development of bird protection as a reflection of a diffuse mixture of utilitarian, ethical, and aesthetic rationales for nature protection, on the one hand, and its amateur-scientific orientation, on the other, was in no way limited to Germany. Indeed, one could find such contradictory bourgeois cultural attitudes toward nature and bird protection to varying degrees throughout modern, Western, Protestant-influenced lands of the late nineteenth and early twentieth centuries.[45] The unique ability of bird protection to synthesize scientific, utilitarian, aesthetic, ecological, and ethical interests thus elevated the movement to an avant-garde role in the development of comprehensive nature and environmental protection in Europe and North America.[46]

Bird protectionists' struggle to save Germany's bird populations from the ravages of industrial modernity also exemplifies the paradoxes that characterize modern societies' relations with nature in general. The same culture that civilized and subjugated nature discovered a different and neglected aspect of it, which it classified as worth protecting and preserving. On the one hand, modernity demanded a seemingly boundless exploitation and instrumental subjugation of nature; on the other hand, modern "cults of nature" professed boundless admiration for the bird kingdom and other parts of the environment deemed scientifically unique or aesthetically pleasing. Why did the catching and killing of birds hold such an enormous potential for emotional agitation around 1900, while at the same time other animals, such as pigs and cattle could be treated like mechanical devices or industrial goods? Germany, like other modern societies around 1900, organized its relations with nature at the extremes of the poles of instrumental use, on the one hand, and non-material appreciation, on the other: in other words, between the poles of extermination and preservation. Whereas the relations at the instrumental pole led to the objectification, disenchantment, and mechanization of nature, those at the opposite side created re-enchantment, aestheticization, and quasi-sacral idealization of nature. The desire for re-enchantment grew in tandem with the rationalization, juxtaposing the dominant utilitarian means-end relationship with the sacred "other" of nature: its beauty, its magic, its secrets.

Both developments were thus inextricably intertwined. Technical and economic modernization was only possible when the memory of lost relations with nature was simultaneously present, be it an isle of memory in the shape of a natural reserve or a rare species of birds saved from extinction. Birds underwent a symbolic transformation within the framework of this process: they turned from objects of material use into subjects with intrinsic value, "feathered friends" who enjoyed human protection. This new appreciation only increased as knowledge that numerous species of birds were becoming extinct became widespread. Nature in the modern age thus divided into two separate and exclusive spheres that precluded the possibility for real reconciliation. Preservation and extinction emerged as elements of one and the same process within the unfolding logic of modern civilization.

The bird protection movement, with its ability to integrate scientific, aesthetic, and ethical interests, functioned as an agency of cultural modernization. Birds seemed predestined to spark humans' desire to protect nature because they, like no other animal group, were suitable not only as (ecological) objects of indication, but also as emotional-aesthetic objects of identification. Though utilitarian, scientific, and aesthetic rationale for bird protection helped to mobilize the movement during the course of the nineteenth century, around 1900 a new understanding of birds' intrinsic value emerged that called for an integrated, ethical understanding of the avian kingdom, indeed of nature in general. As one conservationist wrote: "Let us not always ask pettily and greedily: what is the use of this bird or that bird? Let us rather enjoy the bird in beautiful and free nature; let us enjoy the bird of prey for his glorious flight, for his proud air, let us enjoy the song-bird for his heart-refreshing song, for his gay plumage; in short, let us spare the bird for his own sake."[47]

Notes

1. Hans von Berlepsch, *Der gesamte Vogelschutz, seine Begründung und Ausführung* (Gera, 1899), 1–2.
2. Ernst Hartert, *Einige Worte der Wahrheit über den Vogelschutz* (Neudamm, 1900), 20–21.
3. Curt Floericke, *Vogelbuch. Gemeinverständliche Naturgeschichte der mitteleuropäischen Vogelwelt*, 2nd ed. (Stuttgart: Köhler, 1922), 69.
4. See Orvar Löfgren, "Our Friends in Nature. Class and Animal Symbolism," *Ethnos* 50 (1985): 184–213.
5. A survey of the contemporary literature may be found in Karl Russ, *Zum Vogelschutz. Eine Darstellung der Vogelschutzfrage in ihrer geschichtlichen Entwicklung bis zur Gegenwart* (Leipzig: Vogt, 1882).
6. See "Chronik der Waldverheerungen durch Raupenfraß," in *Der besorgte Forstmann*, no. I (Weimar, 1798), 19–24.
7. Johann Matthäus Bechstein, *Forstinsectologie oder Naturgeschichte der für den Wald schädlichen und nützlichen Insecten nebst Einleitung in die Insectenkunde überhaupt, für angehende und ausübende Forstmänner und Cameralisten* (Gotha: Hennings, 1818), 51–52.
8. Johann Friedrich Naumann, *Johann Andreas Naumann's Naturgeschichte der Vögel Deutschlands, nach eigenen Erfahrungen entworfen*, pt. 4 (Leipzig: Fleischer, 1824), 74.
9. Johann Matthäus Bechstein, *Kurze aber gründliche Musterung aller bisher mit Recht oder Unrecht von dem Jäger als schädlich geachteten und getödeten Thiere nebst Aufzählung einiger wirklich schädlichen, die er, diesem Berufe nach, nicht dafür erkennt* (Gotha: Ettinger, 1792), 2.
10. Ibid., 5.
11. Casimir Graf von Wodzicki, "Über den Einfluss der Vögel auf die Feld- und Waldwirthschaft im Allgemeinen, wie besondere auf die waldschädlichen Insecten," *Journal für Ornithologie* 1 (1853): 294.
12. See Alfred E. Brehm, "Unsre Bodenwirthschaft und die Vögel. Vortrag, gehalten am 19. September 1873 im internationalen Congresse der Land- und Forstwirthe zu Wien," *Journal für Ornithologie* 22 (1874): 26–39.
13. Berlepsch, *Der gesamte Vogelschutz*, 73–74.

14. Berlepsch, *Der gesamte Vogelschutz*, 80.
15. Erwin Stresemann, "Beiträge zu einer Geschichte der deutschen Vogelkunde," *Journal für Ornithologie* 73 (1925): 596.
16. Johann Friedrich Naumann, "Beleuchtung der Klage über Verminderung der Vögel in der Mitte von Deutschland," *Zeitschrift für die gesammte Ornithologie* 2 (1849): 132–134.
17. Ibid., 140.
18. Ibid., 140.
19. Ferdinand von Droste-Hülshoff, *Die Vogelschutzfrage. Ein Referat* (Münster: E. C. Brunn, 1872), 37.
20. Karl Theodor Liebe and Georg v. Wangelin, "Referat über den Vogelschutz," in *Karl Theodor Liebe: Ornithologische Schriften*, ed. Carl R. Hennicke (Leipzig: Malende, 1893), 132.
21. Karl Theodor Liebe, "Zum Vogelschutz," in *Ornithologische Schriften*, 72–73.
22. Wilhelm Heinrich Riehl, *Land und Leute*, vol. 1 of *Die Naturgeschichte des Volkes als Grundlage einer deutschen Social-Politik* (Stuttgart: J. G. Cotta, 1854), 35.
23. Wilhelm Heinrich Riehl, "Die Volkskunde als Wissenschaft. Ein Vortrag," in *Culturstudien aus drei Jahrhunderten*, ed. Wilhelm Heinrich Riehl (Stuttgart: J. G. Cotta, 1859), 205–229.
24. Riehl, *Land und Leute*, 30.
25. Karl Theodor Liebe, "Ornithologische Rundschau in Ostthüringen 1877–1879," in *Ornithologische Schriften*, 430.
26. Otto Schnurre, *Die Vögel der deutschen Kulturlandschaft* (Marburg: Ellwert, 1921), 8.
27. Ibid., 44.
28. "Entwurf eines Gesetzes, betreffend den Schutz von Vögeln," in *Sammlung sämtlicher Drucksachen des Reichstages*, 7. Legislatur-Periode, II. Session 1887/1888, Nr. 90 (Berlin, 1888), 6.
29. Ibid.
30. Also see the stock of files of the Geheimes Staatsarchiv Preußischer Kulturbesitz Berlin, I.HA., Rep.87B, Landwirtschaftsministerium "Das Vogelschutzgesetz vom 22. März 1888 und 30. Mai 1908 und die hierzu gestellten Änderungsanträge."
31. For a more extensive discussion of this issue, see Friedemann Schmoll, "Vogelleichen auf Frauenköpfen. Ein Streitfall aus der Geschichte des Vogelschutzes," *Rheinisch-westfälische Zeitschrift für Volkskunde* 44 (1999): 155–169.
32. Carl Georg Schillings, "Vogelausrottung für Frauenputz," *Flugschrift des Dürerbundes*, no. 88 (Munich, 1922): 8–9.
33. Ludwig Klages, "Mensch und Erde," in *Mensch und Erde. Sieben Abhandlungen*, 3rd ed. (Jena: E. Diederichs, 1929), 20.
34. Carl R. Hennicke, *Vogelschutzbuch* (Stuttgart: Strecker und Schröder, 1911), 6–9.
35. Ludwig Reinhardt, *Kulturgeschichte der Nutztiere* (Munich: E. Reinhardt, 1912), 696.
36. Schilling, "Vogelausrottung für Frauenputz," 9.
37. See Reinhardt Johler, "Vogelmord und Vogelliebe. Zur Ethnographie konträrer Leidenschaften," *Historische Anthropologie* 5 (1997): 1–35; Friedemann Schmoll, "Kulinarische Moral, Vogelliebe und Naturbewahrung. Zur kulturellen Organisation von Naturbeziehungen in der Moderne," in *Natur—Kultur. Volkskundliche Perspektiven auf Mensch und Umwelt*, ed. Rolf Wilhelm Brednich, Annette Schneider, and Ute Werner (Münster: Waxmann, 2001), 213–227.
38. Also see the file "Umfang und Werth des Krammetsvogelfangs in Preußen," in Geheimes Staatsarchiv Preussischer Kulturbesitz, Berlin, I. HA, Rep. 87 B, Landwirtschaftsministerium, Nr. 19980.
39. Johann Andreas Naumann, *Der Vogelsteller oder die Kunst allerley Arten von Vögeln*

sowohl ohne als auch auf dem Vogelherd bequem und in Menge zu fangen (Leipzig: Schwickert, 1789), 3.
40. Christian Ludwig Brehm, *Vollständiger Vogelfang* (Weimar: Voigt, 1855), 1.
41. Reinhardt, *Kulturgeschichte der Nutztiere*, 656.
42. For such descriptions, see Heinrich Gätke, *Die Vogelwarte*, ed. Rudolf Blasius (Braunschweig, 1900), 379; Brehm, *Vollständiger Vogelfang* (1855), 49, and *Das Leben der Vögel, dargestellt für Haus und Familie* (Glogau: Flemming, 1861), 419.
43. Gottfried Eigner, *Naturpflege in Bayern* (Munich: Lindauer, 1908), 66.
44. Such rhetoric can be found in the following publications: Ernst Rudorff, "Heimatschutz," *Grenzboten* 56, no. 2 (1897): 401–402; Paul Schultze-Naumburg, *Kulturarbeiten*, vol. 9: *Die Gestaltung der Landschaft durch den Menschen* (Munich: Callwey, 1917), 315; Ernst Rudorff, "Über das Verhältnis des modernen Lebens zur Natur," *Preußische Jahrbücher* 45 (1880): 270.
45. On the international significance of birds and bird protection in agriculture, see Michael Shrubb, *Birds, Scythes and Combines: A History of Birds and Agricultural Change* (Cambridge: Cambridge University Press, 2003). On bird protection in North America, see American Ornithologists' Union, *Fifty Years' Progress of American Ornithology, 1883–1933* (Lancaster, Penn.: American Ornithologists' Union, 1933); Peter J. Schmitt, *Back to Nature. The Arcadian Myth in Urban America* (New York: Oxford University Press, 1969); Mark V. Barrow: *A Passion for Birds: American Ornithology after Audubon* (Princeton: Princeton University Press, 1998).
46. This transcontinental interest in bird protection helps to explain the enormous impact of Rachel Carson's *Silent Spring* (New York: Houghton Mifflin, 1962) in Europe. See Raymond Dominick, *The Environmental Movement in Germany: Prophets and Pioneers, 1871–1971* (Bloomington: Indiana University Press, 1992), 123.
47. Eberhard v. Riesenthal, "Über Vogelschutzbestrebungen," *Mitteilungen über die Vogelwelt* 12 (1912): 144.

Chapter 8

Protecting Nature between Democracy and Dictatorship

The Changing Ideology of the Bourgeois Conservationist Movement, 1925–1935

JOHN ALEXANDER WILLIAMS

R<small>ETURNING TO NATURE</small> became a common aim of organized popular-cultural organizations in early twentieth-century Germany. Conservationist, "life reform" (*Lebensreform*), hiking, and youth movements gained a substantial degree of popular support between 1900 and 1914 by endeavoring to improve both the German individual and the German nation through exposure to the rural environment. These naturist projects were by no means peculiar to Germany, but they became arguably most popular there in response to the unusually rapid industrialization and urbanization of the country. When total war, economic disaster, and political unrest in the years 1914–1923 further destabilized the country, back-to-nature movements redefined their mission, pledging to help the nation overcome its catastrophe.

The apparent ease with which the Nazis took up back-to-nature ideas has prompted historians to search for ideological continuities. Until the 1980s the dominant view was that pre-1933 naturism helped sow the seeds of fascism in German culture. This argument stemmed from the broader thesis that modern German intellectuals and artists followed a special path (*Sonderweg*), turning to a neo-Romantic, reactionary, antiurban, and völkisch ideology. This allegedly predisposed many Germans to reject liberal democracy and embrace authoritarian political systems. Expressing this thesis in his 1970 study on antiurbanism, historian Klaus Bergmann approvingly quoted Kurt Tucholsky to the effect that, "It started out green and ended bloody red."[1]

Since the 1980s, however, the "special path" thesis has been undermined by historians who have questioned the capitalist modernization theory and teleological reasoning at its core.² Nor is it any longer acceptable to draw a direct line of continuity between antimodern thought and Nazism, since historians have shown that even the Third Reich was shot through with elements of technological, economic, and cultural modernism.³ More generally, recent research in environmental history has demonstrated that attitudes and practices toward nature in Germany have always been intertwined with changing social relations, politics, and economic factors.⁴

Recent studies of German conservationism have moved beyond the approach that focuses on Germany's special path, and they paint a more complex portrait of their subject. Raymond Dominick illuminates some of the most important continuities, but also *discontinuities*, between Weimar and Nazi conservationism.⁵ William Rollins argues that the Wilhelmine preservationist movement embodied the cultural and political reformism of the educated middle class.⁶ Karl Ditt provides useful comparisons with British and American conservationism.⁷ Thomas Lekan shows that the movement's changing representations of the landscape were complicated by a strong tendency toward regionalism.⁸ My own 1996 article concerns the ambiguity of the concepts of nature and Heimat, showing how this influenced conservationist rhetoric and practice.⁹

This essay focuses on a period that has not yet attracted much attention, namely the middle years of the Weimar Republic through the *Gleichschaltung*, or synchronization, that the Nazis enforced. In this era the leading spokesmen of the conservationist movement redefined their project as a battle against a "homeland emergency" (*Heimatnot*). This sense of crisis derived from fears, typical of the Weimar bourgeoisie, of a powerful working class and an unruly young generation. Rising to the self-imposed challenge of overcoming the emergency, late Weimar conservationists formulated a new ideal narrative of returning to nature that conflated nature, the nation, and the regional "homeland," or "Heimat." That narrative offered an alternative to the uncertainties and divisions of Weimar—a socially unified, racially homogeneous natural/national Heimat. Conservationists hoped to pacify workers and young people by communicating this narrative of progress through propaganda and the teaching of homeland studies.

This ideological change greatly influenced the conservationists' relationship to urban-industrial modernity. While late Weimar conservationism retained a certain nostalgia for the rural-agrarian past, the movement largely jettisoned its long-standing critique of industrialization, dismissing it as outmoded and too "romantic." Instead they redefined late nineteenth-century *Landschaftspflege*, or landscape planning, as a self-consciously modernist project of landscape cultivation that aimed to integrate industrial technology into the landscape in rational, yet aesthetically pleasing and environmentally acceptable ways.[10]

Far from rejecting modernity, then, leading conservationists envisioned a forward-looking, socially homogeneous nation in which nature itself would be well ordered and in harmony with the work of human beings. The greatest commonality between late Weimar conservationism and Nazism was that both harnessed concepts of an orderly nature and an undivided society to their agenda for restoring the nation. Yet while many leading conservationists welcomed the overthrow of the Weimar Republic, the transition to Nazism was not seamless. For the Nazis injected conservationism with their own strain of brutality, exploiting the concept of a homogeneous natural Heimat to help justify their all-encompassing project of imperialist war and racial "cleansing."

Stabilizing the Homeland

Organized *Naturschutz*, or nature conservation, had its roots in the late nineteenth-century response among the educated bourgeoisie to the rapid growth of cities and the accompanying damage to rural landscapes. It became an organized movement during the fin de siècle, an era of popular self-mobilization around a wide range of mass reform initiatives.[11] An array of state-affiliated and private nature conservation associations were founded in the first decade of the twentieth century. Conservationism was an integral part of a broader Heimat protection movement that aimed to protect regional cultures and landscapes as diverse components of the German nation. The centrality of a pastoral homeland ideal led conservationists to focus less on wilderness, of which little remained in Germany, than on the lived-in rural landscape in which human beings and nature coexisted.[12]

The vast majority of the movement's leaders and rank-and-file were members of the educated middle class—schoolteachers, professors, lawyers, doctors, forestry experts, and writers—with a smaller proportion of aristocrats.[13] Their social and political conservatism and rural focus limited the movement's appeal to the working class. They also alienated industrial capitalists and property owners, who successfully fought efforts to impose compulsory conservation measures. By 1914 many had settled for lobbying the state to protect unique "natural monuments" and to put pressure on local industrial polluters.[14] Prominent conservationists joined semiofficial advisory commissions. But the state governments were hardly motivated to accept many of the commissions' recommendations. The Bavarian government, for instance, declared, "By no means will economic interests be sacrificed for these endeavors. On the contrary, [the state conservation commission] will see to it that these idealist interests will be recognized without their interfering with economic goals."[15]

Thus German conservationists had reason to take heart in the postwar republic's apparent interest in their project. Article 150 of the Weimar constitution placed "monuments of art, history, and nature, as well as the landscape" under state protection. This pledge encouraged the movement as it underwent a period of refounding in the early 1920s. Regional Heimat protection organizations retained their importance, but in 1922 a new national umbrella organization, the People's League for Nature Conservation (Volksbund für Naturschutz), was founded in Berlin. Alongside the well-established German League for Heimat Protection (Deutscher Bund für Heimatschutz, founded 1904), the Volksbund came to enjoy close contact with and some financial support from the Prussian and national governments.[16]

In the process of refounding, conservationists delineated the new problems to be overcome. Germany's key internal weakness, they argued, was rebelliousness among both the industrial working class and the generation of adolescents and young people. That middle-class conservationists took this view is hardly surprising. Fear of the equalizing implications of labor movements and urban mass society had long been an integral part of the educated bourgeoisie's sense of class identity. The experience of a revolution in which young working men and women were at the forefront greatly increased this anxiety. Germany had suddenly become a mass democracy in which Marxist parties and unions had unprecedented power. Moreover, the bourgeoisie underwent a severe economic decline in the early 1920s that culminated in the disastrous hyperinflation of 1923. As Bernd Weisbrod has shown, when the economic uncertainty of the working class came to be shared by the Weimar middle classes, the sense of security—"the very essence of 'bourgeois' existence"—was lost. Far from encouraging solidarity with workers, this decline seems only to have made the bourgeoisie insist more vigorously on its moral and cultural superiority.[17]

Concerns about the young generation grew among the educated bourgeoisie during the early Weimar years. Adults had been greatly disturbed by the appearance of mass socialist and bourgeois youth movements since before the war, and some had launched projects of youth welfare and "cultivation" intended to regain control over the socialization of adolescents. Now a rising mass culture offered young people even more opportunities for autonomy during their leisure hours. Chronic youth unemployment and the economically uncertain future of their children also fueled bourgeois adults' worries about the next generation.[18]

Early Weimar conservationist rhetoric expressed these social anxieties through images of nature suffering under the onslaught of the urban masses. For example, the Bavarian State Advisory Committee for the Care of Nature warned in 1919 that the "social aftereffects" of the Munich Soviet Republic—code language for the revolutionary working class—were contaminating Lake Königssee, "irrevocably destroying the idyllic peace of the lake and the surrounding mountains." Furthermore, Bavarian game had come under threat from unrestrained poaching since the revolution: "Animals are suffering under the greed, ignorance, and superstition of a segment of the populace and under the brutality and thoughtlessness of youth."[19] Although this kind of rhetoric became less panic-stricken after 1920, there remained a tendency to castigate the unruly masses for tramping loudly through the rural environment, littering the landscape, and mistreating wild animals and plants.

Two intertwined goals emerged from this social interpretation of Germany's problems. First, conservationists hoped to pressure state and national governments into privileging nature protection over private property rights. Singular natural monuments such as ancient trees, caves, and rock formations would benefit from state oversight; and larger protected natural areas would be established as well.[20]

The second and broader goal was to strengthen the commitment of workers and young people to nature as the way to restore national health and stability. But how would this appreciation for nature be taught? The conservationists did not develop a coherent answer to that question in the early 1920s. Instead they revived the ideal back-to-nature narrative that had been prevalent in the Wilhelmine movement. In the 1919 reprint of Konrad Guenther's 1912 book *Nature Conservation*, for instance, the author described a worker walking home from the factory through woods and fields. Guenther wrote hopefully that the worker would naturally experience a sense of contentment and freedom in the countryside, "For only in nature is there no difference between poor and rich, high and low; only there does it cost nothing to gain a wealth of knowledge and happiness."[21] The appreciation of nature served a pacifying, class-transcendent function in this narrative. Yet early Weimar conservationists remained vague as to how that mentality could be induced among workers to whom it might not come naturally.

This began to change in the mid-1920s. By 1924 it was evident that the conservationist project had stalled at the state level. State governments had generally refused to develop laws that would implement Article 150 because, as the Bavarian Interior Minister bluntly explained, "an effective conservationist policy would have to provide for the compensation of private property owners, and the means to do so are not available."[22] Nor had the central government proved willing to introduce a nationwide conservation law. Indeed, the governments were rejecting more proposals for nature preserves than they supported.[23]

Lamenting this absence of state support, leading conservationists became convinced that, as one leader announced, "Conservationism only has a future if it becomes popular."[24] But how were socially elitist conservationists to appeal effectively to a mass public that they had long bypassed? They quickly arrived at a consensus that popular pedagogy was the answer. Thus in the mid-1920s, conservationists turned their attention to a project of teaching the young generation to love and respect the rural landscape of the homeland.

This shift in the conservationists' goals led to a stronger rhetorical emphasis on the "Heimat" concept. The language of the Heimat movement must have seemed well suited to the new aim of popularization. There appeared to be no limit to the readership for such celebrants of the region as Hermann Löns and the Plattdeutsch writers.[25] The many Heimat protection organizations bent on preserving local and regional traditions enjoyed the patronage of state interior and cultural ministries, primarily for nationalist reasons. The Reich Interior Minister lent his support to the Rhineland Heimat League, for instance, because its purpose was "to put the existing clubs and interest groups in the service of German cultural propaganda."[26] Moreover, educators were advocating new curricula in "homeland studies" (*Heimatkunde*) in primary and secondary schools as a way of helping young people transcend social divisions. They assisted the governments of Bavaria, Baden, Prussia, Württemberg, Thuringia, and Saxony in developing guidelines for the teaching of homeland studies.[27]

The conservationists thus began in various ways to advocate popular homeland studies in the hopes of strengthening their influence. Various state advisory commissions fabricated their own homeland studies courses. The Volksbund für Naturschutz began to lobby the state for homeland studies in the schools, at the same time launching an appeal to the public to take part in organized Sunday hikes, field trips to nature preserves, lectures, study courses, public exhibitions, and museum tours. And new organizations that combined conservation with homeland studies sprang up throughout Germany.[28]

Yet when Weimar conservationists shifted their focus to homeland studies, they unwittingly created a conceptual problem for themselves. For *Heimat* was an extraordinarily slippery and unstable concept containing an abundance of potentially conflicting meanings. There were several reasons for this. First, the concept had strong connotations of regional particularism that were potentially at odds with a project of national unity. In a country as culturally diverse as Germany, and in a federalist political system in which the national government suffered from an ongoing crisis of legitimacy, the individual's strongest emotional affiliation tended to be with a locale or region.[29] Indeed, Heimat language often boosted centrifugal political forces. Some local governments in Bavaria, for instance, cultivated regional identity as a means of resisting what they saw as "exaggerated centralism in the state and in the Reich."[30] Such rhetoric had an influence in many regional preservationist movements as well.[31] Even though many Heimat protection activists viewed regional identity as a component of national identity, there was no guarantee that the love of the region would commit a young person to greater national unity.

Second, each individual's relationship to the Heimat was uniquely shaped by personal experience and memory. As pedagogic experts pointed out, teachers of homeland studies had to take into account that every person's local environment was "individually appropriated through experience, shot through with spirituality, and thoroughly colored by personal perspective."[32] This subjective, unique reception of the homeland by each individual therefore also held the potential to undermine a collective commitment to national unity.

Finally, the *Heimat* concept was ideologically vague enough to be appropriated by groups with divergent political aims. The Social Democratic hiking organization "Friends of Nature" was only one of many examples. The Friends of Nature developed an alternative concept of Heimat that was both regionalist and socialist. They advocated "social hiking" by industrial workers as a way of raising political consciousness. The working inhabitants of both town and country, they asserted, suffered from economic backwardness and social inequality. Gaining knowledge about one's homeland thus meant comprehending the depredations of capitalism on the common people.[33] Such political appropriations of the *Heimat* concept were clearly at odds with the bourgeois conservationists' goal of social pacification.

Given this multiplicity of possible homelands, *Heimat* was something of a conceptual Trojan horse for the conservationist movement. It would not be a simple task to instrumentalize homeland studies in order to marshal support for conservation, pacify the masses, and inculcate the desire for national unity. Leading conservationists apparently began to realize this in the late 1920s, for they undertook a rhetorical effort to clarify and standardize the *Heimat* concept. As the fragile political and economic stability of the mid 1920s broke down, a growing sense of dread further galvanized that effort. By 1930 the middle class faced a political resurgence of the Marxist Left, an economic depression of unprecedented magnitude, an upsurge of social conflict, and political polarization. In standardizing the Heimat concept, the bourgeois conservationists were thus attempting both to simplify their chief conceptual category and to justify the movement anew in a period of national crisis.

They made two important standardizing changes in the representation of Heimat. First, they conflated the homeland with nature itself. Nature was recast as the stable material foundation of the social order. In this fully anthropocentric conception of the "natural homeland," nature's significance derived almost exclusively from its meaning for society.[34] Second, they identified this "natural homeland" with the totality of the German nation. Conservationists now depicted the national crisis through the image of an endangered natural/national Heimat. As their solution to this "homeland emergency," they developed a new back-to-nature narrative in which conservation would help save the nation. The result would be a strengthened natural/national homeland protected from both social conflict and environmental destruction.

The late Weimar movement offered this narrative over and over again to the public. In a 1929 speech, for instance, Heinrich Hassinger of the Rhineland Heimat League warned of worsening social conflict. During the French occupation of the Rhineland, he explained, Rhinelanders had united behind the homeland, and their appreciation for the nation had grown. Unfortunately, Germans as a whole were not yet a united people, because "the antipathy of one against another has hindered the final, great, national unity that would bring us together under one flag of common ideas and convictions." This was "one of the great emergencies of our time, ranking with unemployment, the housing shortage, and the disintegration of the family under the burden of poverty." Hassinger acknowledged that *Heimat* meant many things to people, but he urged that the concept be brought more clearly into line with the nation. Teaching about and celebrating the regional homeland must bind young Germans to the entire nation:

> The world is complicated and the homeland is always simple. The longing for home that is alive within us is the most natural, the most deep-seated, and the most elementary of longings. . . . He who has found a *Heimat* in his people has dropped an anchor; never again will he be tossed about by the wind and waves of life. Secure in his locality, love will grow in him and energy will be set free so that he can serve the nation.[35]

Concluding with a metaphor of homogeneity, Hassinger announced that the hearts of those who share a feeling for the homeland beat in the same rhythm: "We feel ourselves understood, and we understand."[36]

Probably the most representative text of late Weimar conservationism, however, was an anthology entitled *Der deutsche Heimatschutz* published in 1930 by the newly founded Society of the Friends of German Heimat Protection. Comprised of an impressive roster of political, cultural, and economic elites, the society aimed to communicate the vision of a "truly national culture" to the people "so that it will become self-evident for everyone."[37] The ultimate goal was a national community (*Volksgemeinschaft*) that would transcend regional, class, and political divisions.[38]

It would be a mistake to interpret this organization's ideology as evidence of peculiar antimodern and antirationalist sentiments. The society claimed that its goals were superior to the critique of industrial modernity that had characterized the prewar movement:

> Protection of the homeland is not a form of romantic longing for the "good old days," but the vigorous pursuit of good *new* days. We desire not things that are similar in form to the past, but things that are equivalent in value. The Heimat protection movement stands in the middle of contemporary life; it takes the view that a merely material civilization must be reshaped into a culture based on the soul and on morality.[39]

Furthermore, *Der deutsche Heimatschutz* rejected class elitism in favor of communitarian racial nationalism. The authors blamed the problem of social division not on industrial workers, but rather on "cosmopolitanism" and "foreign elements." For example, Karl Wagenfeld, leader of the Heimat protection movement in Westphalia, wrote:

> The lack of connection with *Heimat* and with traditional world-views weakens the sense of morality and trivializes feeling; and it is reflected in action that is unmoored, impulsive, and ruthless. And so the proletariat has grown . . . racially, culturally, and socially inferior, into cultureless and dissatisfied masses who complain that, "We have become strangers, and the machines are idols of cold iron upon which gray priests sacrifice our youth and spill our blood."[40]

There is a telling shift in this passage away from the threatening image of an inferior German proletariat to the "gray priests" who have enslaved workers. Judging by some of his other writings, it seems clear that Wagenfeld's gray priests stood for Jewish capitalists.[41]

In blaming the homeland emergency on fundamentally strange and different outside influences, the authors of this anthology deployed the demonizing language of racism. The society further claimed "a deeper *scientific* knowledge of the foundations of the German national character,"[42] rhetorically subordinating both the individual and the region to a racially homogeneous natural/national homeland. The author Karl Giannoni, for instance, played down regional diversity by defining the regional homeland as a mere building block of *national* identity:

> *Heimat* does not signify simple exceptionalism, but rather it is the partial expression of a larger national geographic space (*Volksraum*). Furthermore, the individual in the homeland is the partial expression of a singular national character (*Volkstum*). Both the national space and the national character are truly to be experienced and understood on the level of the homeland. We receive our national character from the hand of the *Heimat*.[43]

In another article Anton Helbok directly attacked regional pride, using Bavaria as his negative example. The peculiarities of Bavarian physiognomy, psychology, and culture, he wrote, had only originated since the tribal settlement of that area. Therefore they were determined not by tribal Bavarian blood but by geography and climate. *German* blood was primary. "Culturally blinkered" celebrants of particular regions fostered "a philistine local patriotism, . . . degrading the figure of the inward-looking German into a dwarf who looks ridiculous next to the modern, active individual."[44]

The anti-Semitism and obsessive harping about "German blood" in this anthology signaled leading conservationists' appropriation of the kind of racial nationalism that had rapidly become dominant on the moderate and far Right since the war.[45] While it is possible to find examples of such language in preservationist sources even before the war, this biologistic strain of nationalism had hitherto been in the minority, subordinated to a more culturally based understanding of national identity and to strong currents of regional pride. But now, even a prominent "racial hygienist" such as Dr. Eugen Fischer could contribute to the discourse of the society. In his article "Homeland and Family," Fischer wrote that a German's feeling of connectedness to the national homeland was the product of thousands of years of heredity: "As the people itself says, it is in the blood." A healthy people could absorb a modicum of blood from another race, so long as that influence was kept within reasonable bounds. But the family—"the location at which two [generational] links in the hereditary chain meet and intertwine to bring into existence a new chain reaching into the future"—was being destroyed by the pernicious influence of the metropolis with all its "cosmopolitan" influences.[46]

Der deutsche Heimatschutz was a seminal text in the formulation of a new ideal back-to-nature narrative, setting the tone of conservationist rhetoric into the mid-1930s. The narrative began with the problem of internal and external dangers that together signaled a crisis of the racial nation. Conservation was then presented as the solution. By popularizing conservation through "homeland studies" and organized activities such as hiking, the movement would be doing its part to help the German people overcome each specific threat to its well-being. The result would be an ideal natural/national Heimat

Table 8-1
Structure of the Late Weimar Back-to-Nature Narrative

Endangered Heimat >>>	*Conservationism* >>>	*Strengthened* Heimat
rebellious workers and youths		contented workers and youths
masses ruin the environment		masses respect the environment
class and generational conflict		social harmony
regional particularism		national pride
foreign influences		racial purity
lack of rootedness in the nation		commitment to the nation
romanticism		modernity
national weakness		national power
decline		progress

that was healthy, unified, and strong. The structure of this late Weimar back-to-nature narrative, with all its normative oppositions, can be outlined as in table 8.1.

Planning Landscapes

What effects did this new narrative have on the conservationists' view of nature itself? Two important changes occurred. First, there was a tendency to cast natural diversity in a negative light and to view a homogeneous natural order as preferable. Soon a "purer" and simpler nature took precedence over visual and biological diversity. Many late Weimar conservationists seem to have transferred their fear of social complexity and the hopeful vision of a united national community onto the metaphorical field of nature.[47]

This trend can be illustrated by juxtaposing three texts from 1912, 1926, and 1932 by the prominent conservationist Konrad Guenther. Guenther wrote in 1912 that although the "friend of nature" had nothing against the "moderate and rational exploitation of nature," the "inexhaustible diversity" of the earth was rapidly fading and there was a dire need to preserve what remained of the natural habitat.[48] In a 1926 article, he reformulated this proto-ecological viewpoint using the mechanistic metaphor of a clock. Nature was an "organic work of art" whose plant and animal species hung together by invisible threads. If one species were exterminated, the damage would become apparent at some other place in the system.[49] Yet by 1932 Guenther saw nature solely in terms of the natural/national Heimat. The forest was the "primal homeland" of the Germans, "a living landscape for the Germans' health and spirit."[50] Because the flora and fauna on German territory had coexisted with Germans for thousands of years, they were now strictly German animals and plants.[51] Guenther took a further step away from natural diversity in the following passage:

> The forest promotes individual life. Its animals are mostly loners, forming herds much more rarely than the animals of the steppes. The Germanic tribes lived like this as well, affirming only to a limited degree the equality of their blood. The former influence of the separating trees and impassable wilderness is still alive in the [Germans'] addiction to segregating themselves into parties and associations.[52]

Guenther thus transformed a description of natural diversity into a metaphor for the chaotically divided political culture of the late Weimar Republic.

This inclination to cast diversity in negative terms was intertwined with, and reinforced by, a second change in the conservationists' view of nature. As the movement's leaders took up their forward-looking goals, they harnessed nature to a modernist vision of national economic progress. In the early 1930s they began to consider their earlier reverence for the traditional agrarian landscape to be outdated and overly "romantic." Many conservationists now argued that the ideal German landscape must embody a synthesis between the old and the new. Indeed, human beings had a right to impose order on nature. This shift to a modernist ideology of progress increasingly led many prominent conservationists to accept the encroachment of industrial capitalism into the rural landscape.

These tendencies in the conservationists' conception of nature had practical consequences, for they promoted the acceptance of a policy of *Landschaftspflege*.[53] The word can be translated as either "landscape care" or "landscape planning." The conceptual difference between the two translations parallels the shift from a traditionalist, agrarian-romantic view of the landscape to the more rationalist, modernist view found in late Weimar conservationism. Wilhelmine conservationists had aimed for "landscape care" in their efforts to preserve the rural landscape from industry and mass society. A good example is the 1909 book *The Disfigurement of Our Land*, in which preservationist Paul Schultze-Naumburg railed against "the majority of people [who] seem to believe that our country will become ever more beautiful and magnificent through an industrious program of construction and increasing economic prosperity." They see "the ugliness of modern times as a necessity."[54] This attack on the modernist ideology of industrial-technological progress fit squarely within the "landscape care" tradition.

But "landscape planning," which entails the active shaping of nature, expresses better the trend in the late Weimar movement's discussion of conservation policy. The leading Prussian conservationist Walther Schoenichen, for instance, wrote that the German landscape was "a genuinely life-or-death matter for the entire people" that could only be solved by the conservationists' "active influence in the configuration (*Ausgestaltung*) of the landscape."⁵⁵ The new aim of many conservationists extended beyond the protection of singular natural monuments and limited areas to a wider-ranging method of landscape planning that included afforestation and the regulation of agricultural land-use.⁵⁶ Most significantly, conservationists began to call for the orderly and harmonious integration of new industrial and transportation facilities into the landscape. This synthesis of nature and industrial technology was to be achieved by a coalition of industrialists, scientists, engineers, architects, and conservationists—what one conservationist called a "spontaneous alliance between technicians and all who were hitherto opposed to technology."⁵⁷

Proponents of landscape planning took pains to convince the public and the state that they were optimistic and forward-looking in their view of technological change. It should be absolutely clear, wrote one, that when conservationists battled gas stations and billboards it was not because they were opposed to economic progress, but rather "because of the harsh ugliness with which these things disfigure the landscape."⁵⁸ Conservationists also distanced themselves from the "romantic" prewar movement. Writing in *Der deutsche Heimatschutz*, Friedrich Haßler dismissed Wilhelmine conservationism as outdated:

> In order to wage battle against the abuses of the time, the [earlier] movement had to take a position directly opposed to technology. In many things, it thought in more romantic terms than are appropriate today. Perhaps the early movement did not recognize that technology had already embarked on the path toward self-discipline.⁵⁹

Haßler advised Weimar conservationists not to stand in the way of the economic and technological developments, which were "absolutely necessary in a land as economically damaged as Germany" after the Treaty of Versailles. "Without falling back into false romanticism," he concluded, technicians, industrial planners, and engineers would see to it that the "character of the landscape" was harmed as little as possible.⁶⁰ The conservationists' self-appointed task within the landscape planning project was to act as midwives in the blending of nature and technology. According to Carl Fuchs, conservationists would henceforth try to influence industrialists and the state to (1) "destroy as few of the beauties of the homeland as possible or as is absolutely necessary"; (2) guarantee that technical facilities in the rural landscape served the national economy, not just private enterprise; and (3) see to it that "these new enterprises . . . create new beauties where they destroy old ones."⁶¹

It is important to recognize that this policy shift was closely related to the conservationists' pedagogic project of indoctrinating the masses in a commitment to the natural/national homeland. Given the acceptance of industrial workers into the new ideal narrative of progress toward the harmonious national community, there was both a practical and a symbolic logic to the incorporation of the modern nation's industrial workplaces into the rural landscape. By 1933, conservationist ideology was neither simply antimodern nor antirationalist. Its primary component was not "the fateful fetish of agrarian romanticism,"[62] but the utopian vision of a peaceful, racially homogeneous national community, a vision that rhetorically brought together an allegedly more harmonious past and a bright new technological future in order to help the German people overcome its crisis in the present. Thus the bourgeois conservationists' response to the crises of late Weimar brought them into close ideological proximity to another project of overcoming the "homeland emergency"—National Socialism.

From Weimar to Nazi Conservation

Following Hitler's accession to power in 1933, the *Gleichschaltung* or "synchronization" of organized conservationism proceeded rapidly and for the most part voluntarily. The German League for Heimat Protection voted unanimously to incorporate itself into the new Reich League for Nationhood and Homeland (Reichsbund für Volkstum und Heimat). Many smaller regional and local Heimat groups did so as well, although synchronization was slower at this level, and some regional organizations maintained a nominal independence.[63] Yet many state officials pointed to the continued existence of local Heimat organizations as a threat to the centralization of the state's power. Some regional Heimat protection organizations even used the centralizing impetus against their competitors, denouncing as traitors any individuals or groups that insisted on the continued celebration of the regional homeland.[64] Leading conservationists such as Walther Schoenichen and Hans Klose moved easily into high offices in the Reichsbund.[65] Ideological self-coordination took place in conservationist journals, which busily relocated the love of nature within "a new National Socialist cultural tradition, the expression of a new era":

> The unique German character originated out of the German forest, and out of that forest flow new energies into our blood. . . . Lakes, swamps, waterfalls, characteristic rock formations and other geological curiosities—all of these belong to the face of the German landscape. Honor and love that face as the face of our great mother! Protect and preserve it![66]

Why did bourgeois conservationists accept Nazism so quickly? After all, by 1931 there were around five hundred nature preserves in Germany.[67] Yet there is little evidence that conservationists had actually succeeded in popularizing their project, which helps explain why they would support a new, extremely populist regime with potentially unlimited power to carry out conservationist measures. Moreover, the movement's spokesmen had never overcome their disappointment with the Weimar government's failure to pass a national conservation law. As Schoenichen announced in his *Conservation in the Third Reich* (1934), the movement expected the Nazis to fulfill "many significant demands and wishes of our people for which sympathy has heretofore been lacking."[68] Conservationists also hoped that the Nazis would force through a more comprehensive policy of teaching homeland studies and conservation in the schools.[69] These hopes for dictatorial conservation measures were in keeping with the more general sense among bourgeois elites that only authoritarian methods could save the country.

There were also several ideological affinities between late Weimar conservationism and National Socialism. In each case, however, the Nazis took these affinities to extremes—a pattern that conservationists willingly accepted. First, the definition of the national emergency as a problem of social division was common to both. But the Nazis went much further, specifically attacking socialism, the democratic system of Weimar, and "racial enemies." Second, the natural/national Heimat was the conceptual foundation for a homogeneous, unified national community in both Weimar conservationist and Nazi rhetoric. But the Nazis infused this utopian ideal more thoroughly with racial nationalism and anti-Semitism. Many prominent conservationists now followed their lead by recasting their arguments in overtly racist terms. Konrad Guenther became one of the loudest exponents of the notion that people with blond hair and blue eyes had a much stronger feeling for nature than others, and that the conservation of nature would aid in the "racial cleansing" of Germany.[70] In an article entitled "'The German People Must be Cleansed'—And the German Landscape?," Walter Schoenichen declared that the hereditary makeup of a nation directly determined its relationship to nature. "In the past the German soul always drew strength from German nature and the German landscape when it became necessary to overcome foreign influences and help Germanness break through again." Therefore the nation had to "keep the nature of our homeland free of all influences that are antithetical to Germanness." He then reverted to his tried-and-true attack on mass culture as manifested in billboards, kitsch, and "foreign ways of building." But now the blame lay with the "merchant's spirit," Schoenichen's code term for the influence of "Jewish" capitalism.[71]

The Nazis used the notion of a racially homogeneous natural/national Heimat as a rationale for persecuting minorities. The direct ideological connection between a homogeneous nature and a "cleansed" national community was particularly obvious in their discussions of the relationship of Jews to the landscape. For instance, some Nazis and their followers on the local level attempted to eradicate "Jewish influence" from nature itself. In the spring of 1933, complaints from the small Franconian town of Streitberg began to arrive at the Bavarian State Chancellery. Some townsfolk were annoyed by the Jewish ownership of local caves called the Bing-Höhle. According to the editor of the *Nürnberger Zeitung*, the local people, who were "the oldest and most dependable National Socialists in the area," were tired of being deprived of this source of tourist income. As a result, their anger toward the Jewish owners of the caves was "growing markedly."[72] Meanwhile the Nazis were destroying the family Benario, who owned the caves. They removed the head of the family from his professorship at a Nuremberg business school, and they imprisoned the son in the new concentration camp at Dachau. He was shot "while trying to escape."

Even though this persecution signified the de facto "Aryanization" of the caves, more letters arrived from the Streitberg townspeople complaining that this natural monument was still "disfigured simply by having to bear a Jewish name." As one anonymous writer declared, "our Lord God did not create these caves in order that Jews could make money off them; rather he placed them on German soil so that the German people could enjoy them."[73] In February 1934, the Bavarian State Chancellory directed the district authorities to turn the caves over to the town. The caves were henceforth to have a different name, and the cave's rightful owners were to be notified that "any further exploitation by non-Aryans of this nature monument for private profit is under the present circumstances undesirable."[74] This incident exemplifies the way in which the rhetoric of a homogeneous natural/national Heimat could be used to justify the "Aryanization" of the landscape.

A third ideological affinity between late Weimar conservationism and Nazism was the concept of "landscape planning." The "synchronized" conservationists continued to see themselves as working within the modern industrial system to protect the landscape from the worst depredations. But they adapted their rhetoric in an effort to turn racial nationalism to the purpose of influencing the new regime. Schoenichen, for instance, sensed the dangers of the Nazi public works program of waterway regulation and land reclamation. In his 1933 *Appeal of the German Landscape to the State Labor Service* he acknowledged that, "The German landscape will have to make sacrifices if the treasures and forces dormant within it are to be placed in the service of the great process of healing." But conservationists could only tolerate these sacrifices when encroachments were truly unavoidable and did not destroy "irrevocable values" of the homeland. "Just as much emphasis should be placed on the loving care of the natural scenery," Schoenichen concluded, "as on a given project's practical economic significance." The Labor Service should preserve the landscape by ensuring that "that which is new integrates itself into what already exists and harmoniously adapts itself to the whole."[75]

Conservationists enthusiastically welcomed the construction of the Autobahn because of the government's evident commitment to the principle of careful landscape planning. As Dietmar Klenke has shown, their acceptance of the regime's aim to motorize Germany was facilitated by state propaganda glorifying the Autobahn as a symbol of "national unification through struggle." The architects of the Autobahn generally endeavored to conform the road to the natural features of the landscape, a practice that conservationists deemed superior to other countries' coldly utilitarian "rape" of the landscape through the use of straight lines. The landscape planning ideology of the Autobahn combined conservationist, modernist, and nationalist ideas, and this apparent harmonizing of technology and nature became one of the chief reasons why many conservationists supported the regime.[76]

But the moment of truth for conservation in the Third Reich came with the promulgation of the Reich Conservation Law in 1935. The new government in its centralizing mode placed conservation under the jurisdiction of Hermann Göring. This hunter and self-proclaimed nature lover established a central state agency for conservation and personally drove through the passage of the national law. The law placed under state protection rare plants, nongame animals, natural monuments, and established nature preserves. It also pledged to preserve previously unprotected areas "on account of rarity, beauty, distinctiveness, or because their scientific, cultural, forestry, or hunting significance lies in the general interest."[77] Conservationists throughout Germany praised the law for finally privileging conservation over private property rights, and for "offering a means of maintaining the entire landscape in a *heimatlichen* condition, so that Germans with their German sensibilities can feel themselves permanently at home there."[78]

Yet the law was a Pyrrhic victory. Its purpose, in my view, was less to protect nature than to set limits on conservation as the Nazi regime began to prepare for war.[79] The economic and military interests of the state clearly predominated in the law, which stated that conservation policy was "not allowed to set restrictions on areas in use by the army, important public roads, shipping, or essential economic endeavors."[80] Subsequent state conservation decrees in the Nazi era contained similar caveats, protecting the interests of hunters and farmers as well. The regime's policy was only to uphold the rights of nature if they did not interfere with its plans for economic autarky and military buildup.

Another sign that Nazi conservation was a sham was the regime's increasingly negative stance toward leading conservationists after 1935. Even though conservationists continued to pay homage to "the need of a fully new National Socialist order to fashion the national community,"[81] the Reichsbund für Volkstum und Heimat was disbanded in 1935, and henceforth any state-affiliated conservationists who spoke up for conserving nature against the "general interest" were sacked. General Forestry Minister von Keudell, for instance, who was dedicated to forest conservation, lost his job in 1937 due to the state's need to harvest more timber. The Forestry Ministry subsequently declared that, "The purpose and goal of forest development is not a natural forest, but a naturally *economic* forest."[82] Even Walther Schoenichen was abruptly forced into retirement from the Prussian state's conservation advisory commission in 1938.[83] Most of the Weimar conservationists were not around to protest when concerns about protecting nature became fully subordinate to war after 1939.[84]

During the Weimar era, bourgeois conservationists attempted to conceptualize new relationships between the individual, nature, and the nation. Their purpose was to find ways of helping the German people overcome the trauma of war and revolution, as well as to restore the educated bourgeoisie's cultural control over urban workers and young people. These goals were stymied by the rapid changes and chronic instability of the Weimar era. Leading conservationists looked with growing urgency after 1925 for ways in which their project might help restore order. In an attempt to inspire popular support for nature protection, they developed a pedagogic approach. Their ideal back-to-nature narrative envisioned teaching young people and workers to appreciate their "natural Heimat," thereby cultivating a morally healthy, socially peaceful, and patriotic populace.

However, the existing discourse of Heimat held the potential to undermine the conservationist goal of popularization and pacification, because it contained elements of regional particularism, political ambiguity, and individual subjectivity. In the movement's published rhetoric after 1926, we find a process of redefining and standardizing the Heimat concept. The reformulated back-to-nature narrative that had become dominant by 1930 represented the homeland as national rather than regional, socially harmonious rather than divisive, and modern rather than backward-looking. This vision of the homeland also came to include a strong current of racial nationalism that rhetorically bound all people of German blood to each other and to the landscape.

This racist language categorized "foreign influences" as the key to Germany's crisis, implying that they had to be overcome in order to reach the goal of the national community.

As late Weimar conservationists recast their activity as a "contemporary method of healing society,"[85] their ideas of how best to protect rural nature changed markedly. Conceiving of natural diversity as a negative metaphor for the contemporary homeland emergency, many conservationists arrived at an overtly modernist position according to which the imposition of order on the landscape was beneficial to the nation. The movement increasingly advocated a rationalistic policy of "landscape planning" in order to bring advanced technological facilities into harmony with the rural landscape. Thus although they continued to criticize certain aspects of urban-industrial society, the conservationists did not choose the traditionalist, antimodern, antirationalist path of "cultural despair." Instead they partook of the modernist faith in industrial progress, envisioning the melding of nature and technology as a step forward into a better future. In this respect they were similar to other "progressivist" conservation movements of the time; but they differed fundamentally from the environmentalist movements that took shape in Western Europe and elsewhere after the Second World War and that generally rejected the ideology of industrial progress.[86]

The racial-nationalist and modernist instrumentalization of nature in late Weimar conservationism found a positive reception among the National Socialists. They too offered a utopian alternative to the uncertainties and divisions of Weimar—a socially unified, racially homogeneous national community. This ideological commonality, as well as the conservationists' desire for a restoration of social order and a strong state commitment to preserving nature, helps explain the smooth incorporation of the conservationist movement into the Third Reich. However, although the Nazis appeared to support both nature preservation and the harmonious integration of modern technology into the landscape, their national conservation law in fact limited conservation to only those measures that would not interfere with the regime's political goals of economic autarky and imperialism. In practice the Nazis exploited nature in thoroughly modern ways, imposing "order" on the landscape, extracting the material resources needed for war, and constructing "natural laws" that they then used to justify mass murder.

Notes

1. Klaus Bergmann, *Agrarromantik und Großstadtfeindschaft* (Meisenheim am Glan: Hain, 1970), 361. This interpretation figures prominently in, among others, George L. Mosse, *The Nationalization of the Masses: Political Symbolism and Mass*

Movements in Germany from the Napoleonic Wars through the Third Reich (New York: Fertig, 1975); Ulrich Linse, Ökopax und Anarchie: Eine Geschichte der ökologischen Bewegungen in Deutschland (Munich: Deutscher Taschenbuch Verlag, 1986); Anna Bramwell, Ecology in the Twentieth Century: A History (New Haven: Yale University Press, 1989); Joachim Wolschke-Bulmahn, Auf der Suche nach Arkadien: Zu Landschafstidealen und Formen der Naturaneignung in der Jugendbewegung und ihrer Bedeutung für die Landespflege (Munich: Minerva, 1990).
2. Beginning with David Blackbourn and Geoff Eley, The Peculiarities of German History: Bourgeois Society and Politics in Nineteenth-Century Germany (Oxford: Oxford University Press, 1984). For critiques of the "special path" thesis as it pertains to naturist projects, see John A. Williams, "Giving Nature a Higher Purpose: Back-to-Nature Movements in Weimar Germany, 1918–1933" (Ph.D. diss., University of Michigan, 1996); William Rollins, A Greener Vision of Home: Cultural Politics and Environmental Reform in the German Heimatschutz Movement, 1904–1918 (Ann Arbor: University of Michigan Press, 1997); Thomas Rohkrämer, Eine andere Moderne? Zivilisationskritik, Natur und Technik in Deutschland 1800–1933 (Paderborn: Schöningh, 1999).
3. Detlev J. K. Peukert, Inside Nazi Germany (New Haven: Yale University Press, 1987); Joachim Radkau, Technik in Deutschland: Vom 18. Jahrhundert bis zur Gegenwart (Frankfurt am Main: Suhrkamp, 1989); Ian Kershaw, The Nazi Dictatorship, 4th ed. (Oxford: Oxford University Press, 2000).
4. See, for instance, Franz-Joseph Brüggemeier and Thomas Rommelspacher, eds., Besiegte Natur: Geschichte der Umwelt im 19. und 20. Jahrhundert (Munich: C. H. Beck, 1987), and Blauer Himmel über der Ruhr: Geschichte der Umwelt im Ruhrgebiet, 1840–1990 (Essen: Klartext, 1992); William Cronon, ed., Uncommon Ground: Toward Reinventing Nature (New York: W. W. Norton, 1995).
5. Raymond Dominick, The Environmental Movement in Germany: Prophets and Pioneers, 1871–1971 (Bloomington: Indiana University Press, 1991), 84–115. By contrast, Gert Gröning and Joachim Wolschke-Bulmahn, Die Liebe zur Landschaft: Teil III: Der Drang nach Osten (Munich: Minerva, 1987), and Michael Wettengel, "Staat und Naturschutz 1906–1945: Zur Geschichte der Staatlichen Stelle für Naturschutz in Preußen und der Reichsstelle für Naturschutz," Historische Zeitschrift 256, no. 4 (1993): 355–399, both draw a rather straight line of ideological continuity from the earliest Wilhelmine conservationists to the Nazis.
6. Rollins, Greener Vision of Home.
7. Karl Ditt, "Naturschutz zwischen Zivilisationskritik, Tourismusförderung und Umweltschutz: USA, England, und Deutschland, 1860–1970," in Politische Zäsuren und gesellschaftlicher Wandel im 20. Jahrhundert, ed. Matthias Frese and Michael Prinz (Paderborn: Schöningh, 1996), 499–533.
8. Thomas Lekan, "Regionalism and the Politics of Landscape Preservation in the Third Reich," Environmental History 4, no. 4 (1999): 384–404.
9. John A. Williams, "'The Chords of the German Soul are Tuned to Nature': The Movement to Preserve the Natural Heimat from the Kaiserreich to the Third Reich," Central European History 29, no. 3 (1996): 339–384. Joachim Radkau and Frank Uekötter, eds., Naturschutz und Nationalsozialismus (Frankfurt am Main: Campus, 2003), which was not available at the time of writing, presumably benefits from the recent scholarship on nature protection.
10. Rolf-Peter Sieferle, Fortschrittsfeinde? Opposition gegen Technik und Industrie von der Romantik bis zur Gegenwart (Munich: C. H. Beck, 1984), acknowledges the growth of strong technocratic tendencies among Weimar conservationists. This argument is closely related to that of Jeffrey Herf, Reactionary Modernism: Technology, Culture, and Politics in Weimar and the Third Reich (Cambridge: Cambridge

University Press, 1984), who points to an ideology of technological progress among certain reactionary intellectuals. Neither author, however, moves beyond the reductive claim based on Germany's special path that irrationality was the key continuity between right-wing modernism and Nazism.
11. Jürgen Reulecke, *Geschichte der Urbanisierung in Deutschland* (Frankfurt am Main: Suhrkamp, 1985); Geoff Eley, ed., *Society, Culture, and the State in Germany, 1870–1930* (Ann Arbor: University of Michigan Press, 1996); Diethart Kerbs and Jürgen Reulecke, eds., *Handbuch der deutschen Reformbewegungen 1880–1933* (Wuppertal: Hammer, 1999); Kevin Repp, *Reformers, Critics, and the Path of German Modernity: Anti-Politics and the Search for Alternatives, 1890–1914* (Cambridge, Mass.: Harvard University Press, 2001); Friedrich Lenger, ed., *Towards An Urban Nation: Germany Since 1780* (Oxford: Berg, 2002).
12. Celia Applegate, *A Nation of Provincials: The German Idea of Heimat* (Berkeley: University of California Press, 1990); Edeltraud Klueting, ed., *Antimodernismus und Reform: Zur Geschichte der deutschen Heimatbewegung* (Darmstadt: Wissenschaftliche Buchgesellschaft, 1991); Alon Confino, *The Nation as a Local Metaphor: Württemberg, Imperial Germany, and National Memory, 1871–1918* (Chapel Hill: University of North Carolina Press, 1997); Elizabeth Boa and Rachel Palfreyman, *Heimat—A German Dream: Regional Loyalties and National Identity in German Culture 1890–1990* (Oxford: Oxford University Press, 2000); Rollins, *Greener Vision of Home.*
13. For the early organizational history see Walther Schoenichen, *Naturschutz, Heimatschutz: Ihre Begründung durch Ernst Rudorff, Hugo Conwetz und ihre Vorläufer* (Stuttgart: Wissenschaftliche Verlagsgesellschaft, 1954); Andreas Knaut, *Zurück zur Natur! Die Wurzeln der Ökologiebewegung* (Greven: Gilda, 1993); Dominick, *Environmental Movement in Germany*, 3–80.
14. Dr. Haupt, "Die Reinhaltung der Gewässer, eine Aufgabe des Heimatschutzes," *Mitteilungen des Landesvereins Sächsischer Heimatschutz* 2 (1911/1912): 455–459.
15. Cited in Arne Andersen and Reinhard Falter, "'Lebensreform' und 'Heimatschutz'" in Friedrich Prinz and Marita Kraus, eds., *München, Museumsstadt mit Hinterhöfen: Die Prinzregentenzeit 1886–1912* (Munich: C. H. Beck, 1988), 299.
16. Anonymous, "Fünf Jahre Volksbund Naturschutz!," *Naturschutz* (1927): 230–235.
17. Bernd Weisbrod, "The Crisis of Bourgeois Society in Interwar Germany," in *Fascist Italy and Nazi Germany*, ed. Richard Bessel (Cambridge: Cambridge University Press, 1996), 23–39, here 29.
18. Detlev J. K. Peukert, *Grenzen der Sozialdisziplinierung: Aufstieg und Krise der deutschen Jugendfürsorge, 1878 bis 1932* (Cologne: Bund, 1986); Mark Roseman, *Generations in Conflict: Youth Revolt and Generation Formation in Germany, 1770–1968* (Cambridge: Cambridge University Press, 1995); John A. Williams, "Steeling the Young Body: Official Attempts to Control Youth Hiking in Germany, 1913–1938," *Occasional Papers in German Studies* 12 (July 1997), and "Ecstasies of the Young: Sexuality, the Youth Movement, and Moral Panic in Germany on the Eve of the First World War," *Central European History* 34, no. 2 (2001): 162–189.
19. Bayerisches Hauptstaatsarchiv Munich (hereafter BHStAM), MK 40501: *Bayerischer Landesausschuss für Naturpflege, Jahresbericht XI–XVIII (1916–1923).*
20. B. Wolf, *Das Recht der Naturdenkmalpflege in Preußen* (Berlin: Bornträger, 1920), 1.
21. Konrad Guenther, *Der Naturschutz* (Freiburg, 1912; rpt. Stuttgart: Franckh, 1919), iii–iv, 13.
22. BHStAM, MK 14475: Bayerisches Ministerium des Innern to Ministerium für Unterricht und Kultus (24 September 1921).
23. For instance, Staatsarchiv München (hereafter STAM), LRA München 19708: Isartalverein and Bund Naturschutz in Bayern to Bezirksamt München-Land (16

December 1924), requesting woodland protection in the Isar Valley; rejected in a letter of 12 May 1926.
24. Hans Klose, "Naturschutzring Berlin-Brandenburg," *Naturschutz* 8 (1927): 8.
25. Wettengel, "Naturschutz," 375.
26. BHStAM, MK 51152: Reichminister des Innern to Ministerialrat Sperr (26 January 1925).
27. Eduard Spranger, "Der Bildungswert der Heimatkunde," in *Handbuch der Heimaterziehung II* (Berlin, 1923), 20–21; Margarete Götz, *Die Heimatkunde im Spiegel der Lehrpläne der Weimarer Republik* (Frankfurt am Main: Peter Lang, 1989); Applegate, *Nation of Provincials*, 153–155.
28. Anonymous, "Fünf Jahre Volksbund Naturschutz!," 231; BHStAM, MK 51166: "Kurze Darlegung der Entstehung des Nordbayerischen Verbandes für Heimatpflege und Heimatforschung" (ca. 1925).
29. See Applegate, *Nation of Provincials*.
30. BHStAM, MK 51152: Kreistag des Kreises Schwaben und Neuburg to Landesamt für Denkmalpflege München (16 November 1929).
31. See Lekan, "Regionalism."
32. Spranger, "Bildungswert," 7.
33. Theo Müller, "Ein 'Berg frei' dem neuen Jahr!," *Die Naturfreunde: Mitteilungsblatt für den Gau Rheinland* 5 (1925): 2.
34. This tendency was by no means peculiar to modern Germany. See, for instance, Thomas Dunlap, *Nature and the English Diaspora: Environment and History in the United States, Canada, Australia, and New Zealand* (Cambridge: Cambridge University Press, 1999).
35. Heinrich Hassinger, "Heimat und Volk," *Wandern und Schauen* 5 (November 1929): 3.
36. Ibid., 4.
37. BHStAM, MK 51147: Richard Weinmann to Bayerisches Ministerium für Unterricht und Kultus (29 September 1930). In addition to the influential conservationists Werner Lindner, Paul Schultze-Naumburg, and Walter Schoenichen, the society counted among its members high-ranking governmental ministers and legislative figures; several high clergymen of both the Protestant and Catholic churches; members of the judiciary; industrialists; aristocrats; museum directors; lawyers; professors; military officers; authors; journalists; and even Konrad Adenauer, then mayor of Cologne. Gesellschaft der Freunde des Deutschen Heimatschutzes, ed., *Der deutsche Heimatschutz: Ein Rückblick und Ausblick* (Munich: Kastner und Callwey, 1930), 6–8.
38. Karl Giannoni, "Heimat und Volkserziehung," in *Heimatschutz*, 62.
39. Ibid., 60.
40. Karl Wagenfeld, "Industrie und Volkstum," in *Heimatschutz*, 73.
41. His anti-Semitism is overt in Karl Wagenfeld, "Heimatschutz Volkssache," *Heimatschutz* 1 (Winter 1925/1926): 1.
42. Karl Hahm, "Heimatschutz und Heimatpflege," in *Heimatschutz*, 90.
43. Giannoni, "Volkserziehung," in *Heimatschutz*, 57.
44. Helbok, "Mensch und Volk," in *Heimatschutz*, 17.
45. See Paul Weindling, *Health, Race and German Politics Between National Unification and Nazism, 1870–1945* (Cambridge: Cambridge University Press, 1989).
46. Eugen Fischer, "Mensch und Familie," in *Heimatschutz*, 35.
47. There were, however, a few rare examples of conservationists who offered relatively sophisticated ecological analyses in favor of natural diversity, i.e., Heinrich Kraft, "Natur und Mensch," in *Heimatschutz*, 9–11.
48. Guenther, *Naturschutz*, 3, 111, 275.

49. Konrad Guenther, "10 Leitsätze für den Deutschen und seine Heimatnatur," *Naturschutz* 7 (1926): 319–320.
50. Konrad Guenther, *Die Heimatlehre vom Deutschtum und seiner Natur* (Leipzig: Voigtländer, 1932), 8.
51. Ibid., 16.
52. Ibid., 8.
53. It is also evident in various writings concerning the rational exploitation of natural resources, i.e., Alfons Diener von Schönberg, "Des deutschen Waldes Not," *Mitteilungen des Landesvereins sächsischer Heimatschutz* (1932): no page numbers.
54. Paul Schultze-Naumburg, *Die Entstellung unseres Landes*, 3rd ed. (Meiningen: Bund Heimatschutz, 1909), 70–71.
55. Schoenichen, "Entwicklung," 226.
56. Wettengel, "Naturschutz," 374.
57. Richard Weinmann, *Von der Kulturmission der Heimatarbeit* (Windsheim: Delp, 1932), 15. Lekan, "Regionalism," implies that regional Heimat protection organizations retained the earlier meaning of "landscape care" even into the Nazi era and used it to resist the state's destruction of the landscape. The point is well taken; but there is also evidence that at least some regional conservationists followed national leaders in advocating "cultivation." One Bavarian, for example, announced that humanity has a "natural right to dig the signs of his existence, culture, and civilization into the landscape." Georg Wolf, "Industriebauten in der Landschaft," *Blätter für Naturschutz und Naturpflege* 11 (1929): 38.
58. Hans Schwenkel, "Gegner des Heimatschutzes," *Naturschutz* (1930/1931): 27.
59. Friedrich Haßler, "Heimatschutz und Technik," in *Heimatschutz*, 184.
60. Ibid., 186.
61. Carl Fuchs, "Heimatschutz und Volkswirtschaft," in *Heimatschutz*, 152.
62. Dominick, *Environmental Movement in Germany*, 114.
63. Lekan, "Regionalism," 391–399.
64. See, for example, the comments of the president of Lower Saxony on local Heimat organizations in Niedersächsisches Hauptstaatsarchiv Hannover, Hann. 174 Fallingbostel Nr. 2/3: Regierungspräsident Niedersachsens to the Oberpräsidealen in Hannover (2 December 1934). On the denunciation of regional celebrations, see the documentation of Richard Weinmann's attack against the chief administrator of the Franconian League for the latter's "one-sided stress on the tribal idea" in BHStAM, MK 51166, file entitled "Heimatpflege und Heimatforschung. Nordbayerischer Verband in Nürnberg, 1925–1939."
65. Dominick, *Movement*, 96–102; Wettengel, "Naturschutz," 379–381.
66. STAM, LRA Wolfratshausen 41918: Dr. R. Wiesend, "Die Kulturaufgaben des Landbürgermeisters im neuen Reiche," reprint from *Blätter für Naturschutz und Naturpflege* (November 1933).
67. Ludwig Sick, *Das Recht des Naturschutzes* (Bonn: Röhrscheid, 1935), 21.
68. Walther Schoenichen, *Naturschutz im Dritten Reich* (Berlin: Bernmühler, 1934), i.
69. BHStAM, MK 40501: Bund Naturschutz in Bayern to Bayerisches Staatsministerium für Unterricht und Kultus (4 May 1933).
70. Konrad Guenther, *Deutsches Naturerleben* (Stuttgart: Steinkopf, 1935).
71. Walther Schoenichen, "'Das deutsche Volk muß gereinigt werden': Und die deutsche Landschaft?," *Naturschutz* (1932/1933): 205.
72. BHStAM, MWi 2675: Karl A. Stauder to Bayerisches Staatsministerium (30 July 1933).
73. BHStAM, MWi 2675: Anonymous, untitled statement (page numbered 24103).
74. BHStAM, MWi 2675: Staatskanzlei des Freistaates Bayern to Bezirksamt Ebermannstadt (2 February 1934).

75. Walther Schoenichen, *Appell der deutschen Landschaft an den Arbeitsdienst* (Berlin: Neumann, 1933), 3–8. On the Nazi state's regulation of waterways, see Wolf Schmidt, ed., *Von "Abwasser" bis "Wandern": Ein Wegweiser zur Umweltgeschichte* (Hamburg, 1986), 115–116.
76. Dietmar Klenke, "Autobahnbau und Naturschutz in Deutschland: Eine Liaison von Nationalpolitik, Landschaftspflege und Motorisierungsvision bis zur ökologischen Wende der siebziger Jahre," in *Politische Zäsuren*, ed. Frese and Prinz, 465–498. For a differing interpretation, see Thomas Zeller, "'The Landscape's Crown': Landscape, Perceptions, and Modernizing Effects of the German Autobahn System, 1939–1941," *Technologies of Landscape: Reaping to Recyling*, ed. David E. Nye (Amherst: University of Massachusetts Press, 1999), 218–238.
77. BHStAM, MK 51183: "Reichsnaturschutzgesetz vom 26. Juni 1935." On the history of the law and Nazi conservationist practice, see Dominick, *Environmental Movement in Germany*, 104–111; Wettengel, "Naturschutz," 382–397.
78. Werner Weber and Walther Schoenichen, *Das Reichsnaturschutzgesetz vom 26. Juni 1935* (Berlin: Neumann, 1936), 31.
79. See also Wettengel, "Naturschutz," 385–386.
80. Cited in Weber and Schoenichen, *Reichsnaturschutzgesetz*, 31.
81. Werner Weber, "Naturschutz im Rahmen der völkischen Gestaltungsaufgaben," in *Der Schutz der Landschaft nach dem Reichsnaturschutzgesetz: Vorträge auf der Ersten Reichstagung für Naturschutz in Berlin am 14. November 1936*, ed. Hans Klose, Hans Schwenkel, and Werner Weber (Berlin: Neumann, 1937), 42.
82. Quoted in Wettengel, "Naturschutz," 386. See also the paper by Imort in this volume and Heinrich Rubner, *Deutsche Forstgeschichte 1933–1945: Forstwirtschaft, Jagd und Umwelt im NS-Staat* (St. Katharinen: Scripta Mercaturae, 1985).
83. Wettengel, "Naturschutz," 386.
84. On the influence of the *Landschaftspflege* concept in the regime's imperialist "Generalplan Ost," cf. Gröning and Wolschke-Bulmahn, *Liebe zur Landschaft*.
85. Walther Schoenichen, "Aus der Entwicklung der Naturdenkmalpflege," in *Heimatschutz*, 226.
86. Samuel Hays, *Conservation and the Gospel of Efficiency: The Progressive Conservation Movement, 1890–1920* (Cambridge, Mass.: Harvard University Press, 1959); Kirkpatrick Sale, *The Green Revolution* (New York: Hill and Wang, 1993); Ramachandra Guha, *Environmentalism: A Global History* (London: Longman, 2000).

Chapter 9

Protecting Nature in a Divided Nation

Conservation in the Two Germanys, 1945–1972

SANDRA CHANEY

THE TWO GERMANYS inherited a common tradition in conservation that provided foundations for postwar efforts to protect nature. In the quarter century between the end of World War II and the emergence of global environmentalism, groups and individuals concerned about the natural world broadened the scope of nature conservation (*Naturschutz*) as they confronted unprecedented threats to nature and human health that accompanied the "economic miracles" in East and West Germany. In an era when both governments and the general public prioritized economic growth and tolerated the exploitation of natural resources as one of the costs of prosperity, conservation appeared to stand in the way of progress. Convinced that their cause was more relevant than ever, however, conservationists in both countries developed justifications for their work that conveyed the expanded responsibilities of nature conservation. In the process, they also modified a shared tradition in conservation to reflect the priorities of two distinct political and economic systems. The consequences of following diverging paths in conservation were evident before, but especially by, the early 1970s when the imperiled human environment became a focus of international discussions and government policy.

Several studies have shed light on the different roads the two Germanys followed since the early 1970s, when environmental conditions in the Federal Republic began to improve, but those in the GDR worsened. East Germany responded to the worldwide energy crisis by cutting expensive fuel imports from the Soviet Union, and by depending almost solely on the brown coal (with sulfur impurities) it had in abundance. The regime also continued to prioritize economic growth by developing highly polluting heavy industry. West Germany, by contrast, could better afford to import raw materials and goods produced in heavy industry, to scale back coal production and import oil, and to develop and use expensive technologies to reduce industrial pollution. The Federal Republic's democratic institutions also enabled political leaders, a free press, and citizens to make informed decisions and take action on behalf of the environment.[1] These comparisons help to explain why two environments emerged in the FRG and GDR between 1970 and 1990, but what about the years prior to the 1970s, when, as historian Raymond Dominick explains, neither the capitalist Federal Republic nor the communist GDR grappled with environmental problems very effectively?[2]

Shifting attention to the other end of the history of a divided Germany reminds us that there was not one, but many distinct reform traditions that gradually coalesced into a multidimensional program of reform to protect the environment (*Umweltschutz*) by 1970. Of these traditions, the urban public health movement, including pollution control, has received the most attention from scholars. This essay examines less widely discussed changes and continuities in the tradition of conservation in East and West Germany, analyzing debates about the "proper" treatment of nature and some of the policies and practices that shaped how it was used. Although ideology did not determine the fate of nature (neither socialism nor capitalism can boast of a stellar record), it did shape laws, institutions, and mentalities, which together influenced conservationists' response to the question of how, why, and for whom nature ought to be protected. The affluence that developed under capitalism, and the free expression fostered by democratic institutions, enabled West German conservationists to implement a more flexible response to this central question. Initially they emphasized the material benefits of their work for economic recovery and prosperity, but by the latter 1950s, they also underscored the less tangible advantages of conservation in terms of health, regional planning, and basic human rights. Despite promises that socialism would create optimum conditions for conservation by eliminating capitalism and private property and by introducing centralized planning and communal land ownership, East German conservationists had less flexibility in their work. To convince party officials that protecting nature was "productive" labor that contributed to economic growth, they emphasized rational use of natural resources, while also pursuing the less politically charged work of protecting small nature reserves.

By the early 1970s, when the international community debated the "limits to growth" on a planet with finite resources, both Germanys had begun to implement impressive environmental reforms. But the GDR postponed costly cleanup until after the "real" worldwide struggle against capitalist imperialism had been won. By contrast, the Federal Republic used its affluence to begin paying the high cost of cleaning up the environment, aware of the nonmaterial returns that such an investment would bring.

From Nature Conservation to Environmental Protection: Protecting Nature in the Federal Republic, 1945–1972

In the immediate postwar period, conservationists throughout occupied Germany struggled against popular assumptions that nature conservation was primarily a cultural affair involving the protection of natural monuments and scenic parts of the countryside—a luxury concern suited for a future time of stability. As the western zones absorbed over twelve million refugees from the east, confronted shortages of food, fuel, and housing, and coped with the loss of a quarter of its forests and farmland, leading conservationist Gert Kragh warned in language tinged with völkisch phraseology that "the loss of living space, the lack of numerous branches of production and the cessation of participation in world trade and commerce will force us to exhaust the . . . local landscape." The very future of the German people, he wrote, depended on keeping the land it had fertile and productive to support intensified use in the immediate and more distant future.[3]

By necessity, traditional nature conservation was broadened after the war to include resource conservation and preserving, restoring, and increasing the health and fertility of entire landscapes. This task, referred to as *Landschaftspflege* (landscape planning), evolved out of the Enlightenment tradition of garden and landscape design, which envisioned shaping nature to improve it and accommodate development. By the 1930s, landscape care had begun to be a component of land-use and regional planning, an activity of growing importance for the state in many industrialized nations. In Germany, landscape planning acquired negative associations in World War II when influential landscape architects tried to implement grandiose plans to "Germanize" land in the Eastern occupied territories.[4] Rather than break with this tradition that had been tainted by those trying to implement a racialized version of their craft, conservationists after the war coupled nature conservation with landscape planning in rhetoric and in practice to respond to the unprecedented threats to nature that came with recovery and rapid reconstruction. Although racial definitions of "landscape" disappeared, nationalistic understandings persisted in some circles well into the 1950s. Furthermore, even in the early 1960s, some conservationists held firmly to the belief that the physical and spiritual condition of a people was reflected in and dependent upon the aesthetic appearance and biological health of the land. In general, conservationists hoped to preserve the sound practices in nature preservation and landscape care that had developed before and during the Nazi regime, and pledged to promote them in conformity with democracy.[5] As a minority voice trying to influence the majority, conservationists adapted their discourse to larger public debates of the postwar period, promising to aid in economic revival and democratic renewal.

To resume their work, conservationists built on the legal and administrative foundations for protecting nature that had been laid through the Reich Conservation Law (Reichsnaturschutzgesetz, RNG) of 1935. Passed largely because of personal intervention by Reich Forest Master Hermann Göring, the RNG had marked the culmination of efforts since the 1920s to draft national conservation legislation. Impressed by the breadth of the law and related ordinances, occupation authorities allowed the RNG to remain in effect after deleting sections such as those permitting the confiscation of property without compensation and creating an office for expropriation and resettlement. Although the RNG provided the means to preserve plants, animals, natural monuments, reserves, and scenic parts of the countryside, the law offered little help for protecting vast sections of land, or for reclaiming areas damaged by war or threatened by postwar emergency measures.[6]

Under the Federal Republic, the states acquired primary jurisdiction over conservation and administered the RNG through lower (*Kreis*, or county level), superior (*Bezirk*, or district level), and supreme conservation offices now in state ministries of agriculture, culture, or the interior.[7] To aid officials who often lacked time and expertise to address conservation adequately, the RNG continued a practice introduced in some German states around the turn of the century of having independent agencies at each level of government headed by honorary commissioners.[8]

Scholars have examined the continued influence in the Federal Republic of landscape architects who had been prominent during the Third Reich such as Konrad Meyer, Heinrich Wiepking, and Alwin Seifert. Only recently has more attention been given to the fate of conservation officials and rank-and-file commissioners.[9] Hans Klose, director of the Reich Conservation Agency since 1939, remarked in 1946 that weeding out volunteer commissioners with ties to National Socialism had made the situation of conservation in most districts "so critical that one can view the future only with considerable mistrust."[10] Despite such concerns, conservation officials and commissioners emerged relatively unscathed by the war and resumed their work with a high degree of continuity that was typical, particularly among civil servants. Klose himself did not escape accusations during the occupation that he was a Nazi sympathizer, and worse, an unpurged Nazi. Though prominent in conservation during the Third Reich, Klose had never joined the party; he also ended up in the British zone with its more lenient denazification procedures. According to Klose's own description, he "was a good democrat," just like the monarchists in England. The bronze bust of William II that adorned his temporary office in the Lüneburg Heath attested to his nationalist, conservative political views.[11]

Commissioners who survived the war and passed political inspection resumed their work, but with minimal financial support, and in some cases, without maps and files documenting protected areas or copies of the RNG and ordinances that had been lost during the war. Conditions improved slightly by 1947 when twenty-four individuals formed the Working Association of German Commissioners for Conservation (Arbeitsgemeinschaft Deutscher Beauftragter für Naturschutz und Landschaftspflege, ABN). Established initially to preserve uniformity in conservation among the states, the ABN evolved into a professional organization that charted the course of official conservation through the 1960s as commissioners' work regularly intersected with urban development and regional planning.[12]

The ABN cooperated closely with the Federal Institute for Conservation (Bundesanstalt für Naturschutz und Landschaftspflege, BANL), an agency that originated in Prussia in 1906 but became the national conservation institution of the Reich in 1935, lodged in the Ministry of Agriculture under Göring. After the war, Klose devoted his years before retirement to preserving the BANL.[13] In addition to conducting research, the BANL resumed publication of the journal, *Natur und Landschaft* (*Nature and Landscape*), worked with the ABN to organize workshops for commissioners, and helped sponsor an annual German Conservation Day, attended also by public officials and private groups, that focused on issues ranging from regional planning to the impact of tourism and agriculture on nature.

To handle the additional responsibilities that came to be associated with official nature conservation after the war required transforming West Germany's 575 honorary commissioners into full-time, uniformly trained professionals who helped guide decisions about economic development through land-use planning. But shedding their image as well-intentioned members of a beautification club (according to one commissioner) who merely responded to damage after the fact proved difficult. Until the late 1960s, some commissioners at the state level, and most at the district and county levels, conducted their work as volunteers with little financial support and in addition to their occupation (often as a teacher, forester, landscape architect, administrative official, or engineer). As honorary commissioners with some expertise but little authority, they could only advise the nonspecialist officials responsible for conservation![14]

Despite such weaknesses, commissioners scored modest successes in the late 1940s and 1950s, primarily through traditional nature preservation in the countryside and public awareness activities that involved delivering lectures or radio programs, preparing exhibits, and writing publications. During the occupation they urged state governments to include conservation in their constitutions. In the British zone, in particular, they cooperated with citizens and political leaders to protest allied "dismantling" of Germany's forests, warning that overcutting exacerbated erosion, disrupted the water supply, and ultimately hindered the economic recovery of Europe.[15] In the context of agricultural reforms that consolidated fields to boost production in the 1950s, commissioners advocated soil conservation by preserving hedges and wetlands and planting shelterbelts.[16] With mixed results they offered testimony in public hearings over construction projects that might harm nature or the water supply, and cooperated with private organizations, the press, and parliamentarians in southern Germany to block plans to dam rivers for hydroelectric power and build more chair lifts on scenic mountains.[17]

Continuing a long tradition of collaboration between private organizations and official conservation, commissioners worked with long-standing groups and numerous new "working" alliances and "protective associations" established in the late 1940s and early 1950s to conserve forests, preserve wildlife, fight water and noise pollution, and draft conservation legislation.[18] Among the largest of the new organizations was the German Conservation Ring (Deutscher Naturschutzring, DNR), established in 1950 largely through the efforts of Klose, who said he wanted to prevent the state from having a monopoly on conservation as it had during the Third Reich. In the mid-1950s this unwieldy umbrella organization united sixty-one organizations—nature users and nature protectors—and claimed to represent 760,000 West Germans. By 1970 the DNR boasted that the one hundred groups registered as members represented the will of two million people.[19] In the 1950s the DNR and West Germany's other, largely middle-class associations, several of them under the patronage of prominent political leaders, typically relied on executive bodies to define their goals in collaboration with state-sponsored conservation, and eschewed conflict with government officials in favor of compromise. While some old-guard conservationists feared that too many private groups might fragment efforts to protect nature when a united front was desperately needed, more organizations strengthened the base of support and diversified justifications for pursuing the ever-widening agenda of postwar conservation. The existence of so many different groups with a specific focus, but with a common top-heavy structure and overarching message that preached restraint in a time of rapid economic recovery, supports the view that West German democracy established strong foundations in the paternalistic culture of the 1950s.[20] By the 1960s the DNR and other groups were more willing to use confrontational tactics, pressing political parties to take a stance on conservation in election years and resorting to lawsuits.[21]

Even in the conservative climate of the early postwar years, when most people sought a return to political stability, the cause of defending nature—and the meanings associated with it—compelled some West Germans to resort to confrontation. In the context of opposition against the allied occupation, political leaders initiated a "top-down" protest movement against foreign authorities for overcutting timber in Bizonia. In an act of patriotism aimed at preventing the continued "dismantling" of forests, they established the Protective Association of the German Forest (Schutzgemeinschaft Deutscher Wald, or SDW) in 1947. Their reference to the many distinct regional forests as one unified German forest expressed a desire for national unity that was rooted in nature and linked with positive cultural traditions inspired by woods. Chapters of the SDW enlisted youth to plant trees and educated the public about the ecological and economic importance of well-tended forests for the future of Germany and Europe. The organization also urged people to remember the cultural significance of the country's woodlands, asserting in the words of national SDW president and noted political figure, Robert Lehr (CDU), that the German character was "deeply anchored in the forest, in the Heimat of fairy tales and sagas."[22]

Relying on arguments that similarly linked nature to a sense of national identity, several hundred nature enthusiasts responded to the DNR's appeal in 1956 to write Chancellor Konrad Adenauer, urging him to reject France's demand to canalize the West German stretch of the scenic Moselle River. Despite their warnings that the "sacrifice" of this "jewel of the German Heimat" would transform the Moselle into a "chain of drainage basins for French industrial effluents," Adenauer had little choice but to build the canal; it was a small price to pay for the return of the Saar and progress toward European integration.

Protecting the Heimat also served as the rallying cry for Baden-Württemberg's Black Forest Society in forming a coalition of "Heimat-loving loyal citizens" to stop a utility company from damming the Wutach River, "the lifeblood" of the region's geological wonder, the Wutach Gorge. In a controversy that dragged on throughout the 1950s leaders of the coalition (including a district and county conservation commissioner) tested options available in a democracy, organizing public lectures, writing pamphlets and press releases, collecting over 180,000 signatures on a petition, and working with parliamentarians to prepare a major question in the state legislature. In this particular case, conservationists prevailed; the gorge was saved.[23]

Although some scholars have maintained that Heimat protection lacked persuasive power after the war because of its antimodern emphasis and association with National Socialism, these examples of protest in the 1940s and 1950s indicate that some West Germans concerned with protecting nature continued to draw inspiration from that middle and upper middle-class tradition.[24] A German concept that became the force behind a cultural movement in the late nineteenth century, *Heimatschutz* (literally, the defense or protection [*Schutz*] of one's homeland [*Heimat*]) conveyed a somewhat nostalgic impulse to preserve the nature and culture of an area as an aesthetic whole, while accommodating economic modernization.[25] In the first postwar decade, protecting nature in one's Heimat sometimes occurred in the context of attempts to recover from the humiliation of defeat and to reconstruct a positive national identity. By defending forests, rivers, and landscapes that gave local, regional, or even national homelands their distinct charm, West Germans were able to reaffirm their unique natural and cultural heritage. In addition, at a time when many West Germans avoided involvement in politics, the call to preserve the Heimat united people in their attachment to a particular place, inspiring them to become engaged in seemingly nonpolitical causes while becoming politicized in the process.

As highways, canals, railroads, and power lines transformed rural and urban areas, leaving no area untouched, appeals to preserve "pristine" or "primeval" landscapes of the Heimat expressed an unrealistic understanding of nature, but also a desperate attempt to save at least some parts of the country from being irrevocably changed by rapid postwar economic recovery. Acknowledging that they sounded overly sentimental and alarmist, conservationists echoed their predecessors in insisting that they did not oppose economic "progress." But in an era seemingly obsessed with materialism, they argued, assuming a presumably selfless, moralizing stance, progress sometimes meant restraint and an appreciation for the intangible value to present and future generations of natural places that were becoming ever more rare.

Overlapping with this traditional rhetoric that conveyed aesthetic sensibility, moral responsibility, and an emotional attachment to a particular place were scientific justifications for conservation that acknowledged overriding economic concerns. Particularly since the mid-1930s, when the Nazi regime had passed the RNG but had simultaneously compromised nature conservation in the battle for economic production, and later, in the war for living space, conservationists understood that defending nature against repeated attacks was inadequate. Only by assuming an active role in economic planning would they be able to steer development in directions less destructive of nature. During the lean years of the occupation, conservationists continued to emphasize that fertile land, well-managed forests, a reliable supply of clean water, and "green spaces" in urban areas were vital to economic growth and material prosperity. The public, however, continued to associate nature conservation and landscape planning with measures to protect nature in rural areas *from* people, not *for* them.

As cities spilled over into the countryside, blurring the distinction between rural and urban, conservationists in the late 1950s promoted their cause in ways more likely to resonate with a growing urban population enjoying prosperity and more free time. By 1960, nearly a third of West Germany's 56 million people (up from 50.8 million in 1950) lived in cities with over 100,000 inhabitants. The country's rapid urbanization brought a higher standard of living, but also worsening air, water, and noise pollution, and the steady replacement of scenic, fertile land with urban, industrial sprawl.[26] According to doctors and psychologists who relied in part on North American research, these changes contributed to higher instances of fatigue, anxiety, heart disease, and cancer—symptoms associated with "diseases of civilization" or "manager sickness." To reduce the health risks of working long hours at a hectic pace in the seemingly unnatural urban environment that humans had built, some medical practitioners and conservationists recommended rest and relaxation in tranquil natural settings.[27] Worries in the 1950s that city life threatened public health contained echoes of an earlier critique of urban, industrial society and ambivalence toward modernity, but also reflected higher expectations for health and well-being that accompanied an improved standard of living.

In this context Alfred Toepfer, a wealthy Hamburg businessman, philanthropist, and chairman of the Nature Park Society (Verein Naturschutzpark, VNP), publicly advocated the creation of spacious, quiet, nature parks in some of West Germany's most popular areas, from the North Sea to the Alps. Extensive media coverage and promotional films in 1956 popularized Toepfer's vision to make scenic landscapes publicly accessible while promoting health and continued prosperity: after spending time in nature (quietly and orderly), West Germans would return to their jobs, fully restored, and ready to work.[28] Despite initial objections of property owners who feared restrictions on the use of their land if it were included in the parks, this private initiative with a decisively social orientation developed into a publicly funded program that required the expert guidance of regional planners, a professional group of growing importance. Rather than single out scenic, relatively untouched landscapes for preservation (because so few remained), planners sought to revitalize less prosperous areas of the country that were being hit hard by the shrinking agricultural sector and by population loss as people migrated to cities in search of better-paying jobs. Planners envisioned transforming these rural areas into "model landscapes" that would address the growing demand for recreation areas, revitalize the local economy through tourism, and demonstrate how to "order" the country's limited space more efficiently—ultimately to distribute the fruits of prosperity more equitably.[29]

Between 1957 and 1963, twenty-six parks covering over 3.75 million acres were established across the country, linked to urban centers by an expanding network of highways. By the early 1970s, fifty-three parks had been erected on an estimated 15 percent of the total area of the Federal Republic.[30] The program helped satisfy growing consumer demand for nonmaterial goods in the form of recreational spaces, but tolerated the pollution produced by an affluent consumer society rather than combating it directly. In addition, the existence of numerous "nature parks" gave the false impression that steps were being taken to protect nature. To the regret of some conservationists, the majority of funds for these parks went toward recreational facilities; very little of the land included in the parks was stringently protected.[31]

Despite shortcomings, the nature park program occupied a central position in conservationists' efforts to help guide development by participating in land-use and spatial planning. As researchers with the BANL emphasized, in the early 1960s West Germany's many nature parks, 750 nature reserves, 3,800 protected landscapes, and 38,000 natural monuments, had done little to halt haphazard development, the "overconsumption" of fertile land, or worsening pollution. In numerous publications, conservationists insisted on the need to restore ecological "order to the landscape," particularly in and near conurbations, through effective area planning. As protecting nature became inseparable from guiding the growth of urban centers, conservationists had to address more directly than previously the problems that plagued cities, including pollution.[32] In the late 1950s and early 1960s heightened media coverage of air, water, and noise pollution, including Willy Brandt's much-derided promise to bring blue skies back to the Ruhr in his failed campaign for chancellor in 1961, as well as more data quantifying the extent of damage to forests, soil, water, and human health caused by pollution, compelled conservationists to warn that West Germans' entire "living space" was threatened and in need of protection.[33] As DNR chairman, Hans Krieg, insisted in 1961 in an appeal to a Bundestag deputy, West Germany needed a ministry that would be responsible for public health, climate, the quality of water and air, the protection of plants, animals, reserves and nature parks, and the control of pesticides![34] Krieg was calling for a ministry of the environment, but would have to settle for a new Ministry for Health Affairs, created that same year, to address air, noise and water pollution, and public health.[35]

It is curious that Krieg used the term *Lebensraum* prominently in his appeal for administrative reform. Undeniably tainted because of its use by leading Nazis to express expansionist aims that culminated in genocide, the concept gradually became normalized in conservationist discourse by the early 1960s. In its most basic definition *Lebensraum* means "habitat," but also implies an understanding of the reciprocal relationship between a species and its surroundings—in this case the human species. To a greater degree than concepts such as *Naturschutz* or *Landschaftpflege*, *Lebensraum* expressed the growing concern to protect the quality of life of people living under increasingly unhealthy environmental conditions. For a time in the 1960s, before the term "environment" became popular, the concept "living space" provided a way to define a new arena of political concern: the "denatured" environment in city and country that compromised public health and the quality of life.

Even the German Horticulture Society's often-cited Green Charter of Mainau (1961) relied on the concept *Lebensraum* to voice alarm over the extent to which humans had created surroundings harmful to their health and well-being. "The basic foundations of our life," the charter declared, "have become endangered because vital elements of nature are being dirtied, poisoned and destroyed." According to the charter, "the worth of human beings is threatened where the natural environment is damaged." Grounded in the Basic Law's protection of human dignity, liberty, and right of inheritance, the statement assumed that "individual and ... political freedom can develop only in a living space that has healthy conditions for existence." The charter communicated an understanding that the cumulative effects of pollution, pesticides, haphazard development, and the loss of green spaces threatened not only nature, but the quality of life and liberty of people wherever they lived—in the city or the country. Though the progressive aims of the charter aroused little notice at the time because of their association with an elite circle of "wise men," among them Duke Lennart Bernadotte, conservationists applauded the charter's encompassing vision and its demand for effective regional planning to correct disorderly development fostered in part by liberal economic policies of the "miracle years."[36]

Reflecting a confidence in technology often associated with the 1960s, noted conservationist Konrad Buchwald, a self-described "precursor of ecology from the right," argued that if humans had built their unhealthy "ersatz" environment of industrialized farms, managed forests, and urban industrial sprawl, they could restore a more healthy natural order to it as well. With this mind-set, and with the awareness that their work could not be separated from urban development, conservationists sought to transform the spaces where West Germans conducted their daily lives into greener, healthier surroundings.[37] By the latter 1960s, conservation received attention from an expanding corps of professionals who had completed rigorous programs at technical colleges and universities. This new generation emphasized the importance of scientific evidence, technical expertise, and ecology more so than their predecessors, many of whom had been shaped by a humanist curriculum. Some of these new experts found employment as full-time commissioners employed by the state, but far more worked in institutes or with independent agencies outside of the bureaucracy. As professionals they prepared studies and official reports that estimated the productive capacity of landscapes, and that measured the ability of ecosystems to tolerate stresses from agriculture, industry, housing, and increased consumption. Participation in international meetings, such as the 1968 UNESCO-sponsored Man and the Biosphere conference, made clear, however, that researchers the world over lacked data to assess the cumulative effects of multiple "stresses" on human health.[38]

But with ample evidence of the many pressures that contributed to a "denatured" environment, and with no unaltered nature left in the country, some conservationists worked to ensure that at least parts of the environment remained natural. In 1966 DNR President and noted television figure, Bernhard Grzimek and a small group of conservationists followed an international trend in advocating the establishment of the country's first national park in the Bavarian Forest where stringent protection would be balanced with tourism and research. Plans for national parks had been debated in Germany since the turn of the century, but even most proponents concluded that their country lacked the space to preserve large tracts of land where economic uses were limited, if not prohibited. In the 1950s, promising efforts to erect a park in the Bavarian Alps had failed because of conflict over hunting rights. Grzimek's initiative a decade later, however, succeeded primarily because it promised help for a rural region in economic trouble. Located along the Iron Curtain, the area around the Bavarian Forest had the highest rate of unemployment and the lowest income per capita in the country.

After regional planners had failed to resuscitate the economy by attracting new industry, the idea to develop tourism through a national park gained support. Grzimek initially envisioned returning state-owned forested land to a supposedly "more natural" state that approximated its condition in the early Middle Ages, complete with beaver, lynx, and buffalo. When the Bavarian government officially opened the thirty-thousand-acre national park, however, a less controversial and less drastic "remodeling" of this managed forest began. Officially opened in 1970 as the "crowning point" of European Conservation Year (ECY) in the Federal Republic, the park's historic significance was all but lost in news coverage that spotlighted the most obvious concession to tourism, the reintroduction of native and nonnative wildlife. Nonetheless, erecting a national park in a remote corner of the country where traditional economic use was phased out almost entirely, marked a small victory for conservationists who had preached for years about restraint, respect for nature, and responsibility toward future generations. The national park also reflected a desire to restore a "more natural" order to parts of the environment, a desire nurtured by affluence but made seemingly necessary by environmental problems that West Germans could not afford to ignore.[39]

By the mid-1960s West Germany had passed federal guideline laws to manage the water supply (1957), reduce air pollution (1959), and practice conservation in conjunction with regional planning (1965), and also had established a Ministry of Health Affairs (1961) to oversee water, noise, and air pollution control. But as international conferences and media coverage of devastating oil spills, dying rivers, dangerous pesticides, and garbage "avalanches" indicated, too little had been done to prevent "poisoning" of the "environment," a concept popularized only in 1970.[40] During European Conservation Year (1970), a year-long public awareness campaign planned by the Council of Europe and organized in West Germany largely by the DNR, ABN, and BANL, conservationists explained that "Grandpa's conservation is dead." Nature conservation now meant nothing short of protecting humans in their environment.[41]

But it was the SPD/FDP coalition government under Brandt that captured media headlines and convinced the public that "the environment" urgently needed protection. The social-liberal coalition's environmental program of the early 1970s was developed in response to rising concern about pollution and urban sprawl at home and abroad, particularly in the United States, and in an effort to win public confidence after hefty criticism from conservatives over *Ostpolitik*. The new interior minister, Hans-Dietrich Genscher, also hoped to increase his influence and that of his party (FDP) by assuming responsibility for the complex of problems that he dubbed "environmental protection."[42] Prompted by these considerations, the government unveiled an action program on the environment in September 1970 that emphasized technical measures to reduce pollution, insisted that polluters must pay cleanup costs, and urged support for environmental reforms at all levels of society. The program called for strengthening federal authority over air, noise and water pollution and conservation, and urged passage of several laws, many of them already under consideration, to control emissions, reduce noise pollution, regulate waste disposal, ban DDT, and update the water maintenance law.

Conservationists applauded the government's attempt to centralize environmental protection (which was unsuccessful in the case of conservation and water management). Yet they criticized the program for emphasizing quick-fix technical measures to reduce pollution at the expense of long-term "ecological" solutions that involved protecting habitats, restoring ecological "health" to areas compromised by development, and drafting ecologically sound, legally binding land use plans. "Biological *Umweltschutz*," which they had pursued for years, they claimed, was the necessary complement to "technical environmental protection."[43] Conservationists, however, had no intention of resisting the absorption of nature conservation by environmental protection. A long tradition had not died, but had been modernized and transformed into a small, but integral, part of the daunting challenge of protecting the local, national, and global environment. From across the border, however, East Germans claimed that the Federal Republic had not yet tackled the greatest threat to the environment: capitalism.

From Nature Conservation to Socialist *Landeskultur* (Resource Planning): *Protecting Nature in the German Democratic Republic, 1945–1972*

In the Eastern zone of occupation Soviets military officials and new German political leaders introduced land reform in the autumn of 1945, expropriating estates over 100 hectares from Nazi and Junker "plunderers of the people" for redistribution among poorer farmers, agricultural laborers, and newly arriving refugees. But land reform exacerbated the problem of food shortages as new owners, many of them inexperienced and lacking seed and basic farm equipment (because Soviets confiscated machinery as reparations), had to share limited food with Russian commanders intent on feeding their soldiers. In response to shortages of food and natural resources East Germans drained wetlands for cultivation and hunted rare animals, nearly killing off the Elbe River's beaver population. Like people elsewhere in Europe, they coped with fuel shortages by felling trees and extracting peat from moors, even in protected areas.[44]

Conservationists in the eastern zone, like those in the western sectors, insisted that their work was more critical than ever to cope with increased pressures on limited natural resources. As had been the case since 1935, official conservation remained highly centralized in the GDR, promising desirable uniformity, but typically involving bureaucratic delays for local initiatives.

The RNG remained the legal basis for official conservation temporarily, and by 1947 each of the five states had appointed honorary commissioners who recruited volunteers in counties and towns to begin recovering maps of protected areas and inventorying natural monuments, reserves, and protected species.[45] In January 1951 the Ministry of Agriculture assumed control over conservation, and aside from a three-year interruption in the mid-1950s, remained responsible for it until 1975. After administrative reforms in 1952, which replaced the old states (*Länder*) with fifteen districts too weak to challenge central authority, officials in charge of agriculture and forestry oversaw conservation at the district and county levels.[46]

The East German counterpart of the BANL was established in 1953, when the Ministry of Agriculture approved the creation of a research institute for conservation within the German Academy of Agricultural Sciences (Deutsche Akademie der Landwirtschaftswissenschaften, DAL). The Institute for Land Research and Conservation (Institut für Landesforschung und Naturschutz, ILN), based in Halle, together with its four branch offices and learning center on Lake Müritz, conducted research, trained commissioners, and in the absence of independent agencies, advised district and county conservation officials.[47] Under the leadership of Hermann Meusel, director from 1953 to 1963, the ILN played a decisive role in drafting the new conservation law of 1954 and in ensuring that the GDR kept pace with international trends in expanding the scope of conservation beyond narrow preservation. The ILN also preserved ties with West Germans (with much less success after the early 1960s) while building strong relations with conservationists in socialist countries.[48]

The ILN also supervised the work of commissioners who performed many of the same tasks as their West German colleagues. Commissioners in the GDR promoted public awareness of conservation through lectures, publications, and exhibits. They helped to safeguard the nesting grounds of East Germany's unique population of cranes, eagles, heron and other rare birds, and intervened to protect endangered beaver when industrial effluents compromised their habitats. Commissioners aided the DAL in instructing farmers on agricultural production cooperatives (Landwirtschaftliche Produktionsgenossenschaften, LPGs) to practice contour plowing and protect hedges around fields, and worked with the officially sanctioned Nature and Heimat Friends in planting trees and shelterbelts to control erosion. They took action with mixed results when citizens used nature reserves for garbage dumps, camped in unauthorized areas, or when officials permitted cooperatives to build vacation homes on protected waterfronts.[49]

In general, however, commissioners faced tremendous obstacles in achieving even the most basic goals. They sounded little different from their West German colleagues when they justifiably complained about inadequate funding and expressed frustration that the few officials they advised on matters related to conservation were swamped with other responsibilities. At the highest level, in the central Ministry of Agriculture, a "one-man collective" (in the words of a frustrated commissioner) oversaw conservation throughout the country.[50] At the district level, officials responsible for conservation preferred to allocate more funds and personnel to "productive" measures that were integral to the regime's economic plans, such as melioration and reforestation.[51]

An additional handicap in the GDR was that the ILN and commissioners had limited freedom to influence the regime's questionable policies. In 1957, for example, Meusel failed to convince higher-ups to build the country's first nuclear power plant in a location that did not overlap with the nature reserve Stechlin Lake, the "pearl of the Mark" near Berlin.[52] When the ILN followed West Germany's lead in advocating "quiet oases" for public health and recreation, organizing the 1958 conservation week around this theme, officials in the Ministry of Agriculture hampered the initiative to broaden conservation. They censored an essay collection that included Meusel's proposal for setting aside 10 percent of the country as recreational landscapes, as well as an article by a Berlin physician that outlined health risks in the workplace and promoted relaxation in nature as a form of preventative medicine.[53] Even with respect to more mundane matters conservationists encountered resistance from officials. When an industrious commissioner distributed his own newsletter in 1958 without official approval, he was reprimanded, especially because his communication criticized the "mass extermination of crows as a means of controlling pests."[54] In 1962 the Ministry of Agriculture scolded "conservation aides" (*Naturschutzhelfer*) in some districts and counties who had formed their own "nature watch" to oversee reserves, informing them that there was "absolutely no legal basis" for such an independent organization.[55]

Though far more numerous in the GDR than in the FRG, commissioners never evolved into a professional group with specialized education that would enable them to influence planning decisions. By the mid-1960s there were six thousand commissioners; all but about 230 of them, however, were probably volunteer conservation aides at the *Kreis* (county) level or lower who monitored and posted signs near reserves and publicly promoted conservation. Involving citizens broadened the base of support for protecting nature and freed commissioners to deal with administrative tasks, but the arrangement also reinforced the assumption that nature conservation meant narrow preservation and could be handled by volunteers in their spare time. Some volunteers, however, took their work quite seriously. Why else would officials in the central Ministry of Agriculture complain about troublesome aides and commissioners who dared to "raise complaints against the methods of the state apparatus that did not correspond to the facts"?[56]

To an even greater degree than in West Germany commissioners in the GDR worked hand-in-glove with citizens to promote conservation. In January 1949, long-standing conservation and Heimat protection organizations were subsumed under the League of Culture for the Democratic Renewal of Germany (Kulturbund zur demokratischen Erneuerung Deutschlands, as of 1958, Deutscher Kulturbund, DKB), one of several mass organizations the state and party created to achieve ideological conformity. The DKB united approximately 200,000 East Germans from the intelligentsia who previously had belonged to groups as diverse as the Goethe Society and the Heimat League. Within the DKB, the Central Commission of "Nature and Heimat Friends" relied on district and Kreis commissions to supervise citizens engaged in activities often only remotely related to conservation. Though exact numbers are not known, most conservation commissioners belonged to the NHF, serving as leaders who organized activities. While some Nature and Heimat Friends protected historic monuments, others established nature trails, studied photography, geology, area botany, or ornithology, or revamped Heimat museums to reflect socialist traditions. During the 1950s, membership in the NHF doubled, from 20,500 members in 1951 to over 45,000 in 1958. But the upsurge did not bring about a corresponding increase in people who participated in conservation-related tasks.[57]

Since the inception of the NHF, its leaders popularized what was supposedly a uniquely socialist approach to conservation, one that intentionally parted ways with Germany's middle class dominated tradition of protecting nature. As the dedicated communist Karl Kneschke explained to fellow NHF members at their first annual meeting in 1950, being a friend of nature required severing ties with groups they had belonged to in the past, even the historically working class socialist "Friends of Nature" founded at the turn of the century.[58] Further widening the ideological divide between present and past, the NHF adopted "guiding principles" in 1954 that urged members to combat "misanthropic and mystical views" and other "remnants of conservative organizations," including conservation groups. Despite decades of capitalist greed and a fascist war that inflicted deep wounds in nature, under socialism, leaders promised, the landscape would "be tended and designed according to the progressive achievements of science and the tenets of our planned economy."[59]

In the GDR, even more so than in West Germany, the concept of Heimat protection remained central to conservation rhetoric. But, publications emphasized, old middle-class understandings of *Heimat* that had been "criminally misused" under fascism were no longer operative. They had been replaced by a socialist vision of the Heimat in which the natural resources and places of production, the forests and fields, serve those who work. This socialist Heimat no longer was "the private property of the large land owner and the capitalist," but a place that was designed by and belonged to all workers equally.[60] DKB publications made it clear that protecting nature was regarded as but one aspect of the more important task of building and defending a peace-loving Heimat of patriotic socialists. According to prominent conservationist Hugo Weinitschke, the Heimat was the realm where nature and culture intersected. In the land where socialist relations prevailed in production and guided human interaction, people developed a socialist consciousness, acquiring nonexploitative ways of thinking and acting with each other and with nature. As the new socialist Heimat shaped the individual, the socialist individual in turn shaped the Heimat, including nature.[61]

Translated into practice, designing the socialist Heimat involved members in activities not unlike those that engaged West German conservationists. For example, the NHF in Pirna in Saxony appealed to the NHF Central Commission in Berlin in 1951 to use its influence to halt extensive tree cutting in the Bastei nature reserve in the tourism area Sächsische Schweiz, while another local group informed the district government of New Brandenburg in 1952 that Waren city officials were filling Herren Lake with trash.[62] Together with commissioners, the NHF organized local nature walks and sponsored outdoor activities for the Ernst Thälman Pioneers and the Free German Youth.[63] Similar to West Germany's SDW, NHF members organized tree-planting actions, enlisting teachers and school children to revitalize abandoned sandpits, quarries, and slag heaps. In 1956 the NHF took part in limited reclamation in the recently opened "Schwarze Pumpe" brown coal mining area near Hoyerswerda.[64]

That same year the organization introduced an annual competition for "the beautiful village," calling on citizens to plant trees, tidy up their towns, and tend city parks, gardens, and cemeteries. Cash prizes were awarded to the groups with the highest productivity in terms of hours worked. When some participants complained that collectivization was compromising the appearance of villages and surrounding countryside in 1958, they were urged to develop a new aesthetic that found beauty in the separation of living quarters and place of work and that appreciated the unique qualities of LPGs, machine tractor stations, and large complexes for animal husbandry. Beginning in 1957 and continuing through the 1970s, the NHF sponsored an annual nationwide conservation week in cooperation with the ILN, informing people about preservation, public health and recreation, pollution control, and careless development, but did so with limited effect and little governmental support.[65]

According to NHF members, their most significant accomplishment in the 1950s was helping to draft East Germany's new conservation law of 1954. While those involved in the fairly public debate about this legislation agreed that the country's economic growth had expanded the responsibilities associated with conservation, necessitating legal reform, the law that was passed by the East German parliament in August 1954 contained little that was new. Its preamble did express a more ambitious and scientific approach to conservation in promoting study of the "entire household of the landscape" to "preserve and increase the fertility of the land." It also firmly established "protected landscape area" (*Landschaftsschutzgebiet*) as a legal category, strengthening the legal basis for preserving larger areas of land for recreation.[66] Conservationists expressed relief in passing new legislation that reminded a largely ignorant population of the importance of protecting nature, and probably enjoyed updating the RNG, if only to a limited extent, before their West German colleagues who were engaged in a similar (unsuccessful) process. But some regretted that the law sanctioned only the placement of honorary commissioners, not the restoration of independent conservation agencies.[67]

While the primary architects of the law, most notably ILN director Hermann Meusel, defended the legislation, a small group of experts led by Professor Georg Pniower, landscape architect and Director of the Institute of Horticulture and Resource Planning at Humboldt University, dismissed it as a slightly revised version of the RNG that offered little means for using nature for productive purposes. Pniower believed that only more encompassing legislation could serve as the basis for *Landeskultur*, a desirable planning and exploitation of natural resources. This was a concept that referred to the use of technology to harness nature's productive potential and maximize the economic productivity of water, soil, forests, and other resources—ultimately to serve human progress. Two years after passage of the new law Pniower continued to attack his colleagues at the ILN in public and in writing for being romantic conservationists of the old school who had failed to read the "signs of the times."[68]

The ongoing debate over the impact of the law revealed a tension within the circle of conservationists in the 1950s and 1960s, and in East German society as a whole, between those who openly embraced socialism and those who did not or who were critical of the obvious gap between official ideology and actual practice. The latter, including Meusel [a member of the Socialist Unity Party (SED)], was primarily concerned about keeping conservation scientifically up-to-date, while the former, such as Pniower, wanted socialist ideology to transform conservation into resource planning. Indeed, by equating the protection of nature with the rational use of natural resources, resource planning justified conservation in ways that were compatible with official ideology. According to socialism, nature was essential to economic production and continued social reproduction, but had no value until it was transformed by labor into something of quantifiable worth. Thus, according to socialism's labor theory of value, preserving nature was "unproductive" labor that detracted from national wealth by using workers in activities that did not produce goods of measurable value. By contrast, rationally using natural resources constituted "productive" labor that yielded goods to enhance the country's wealth.[69]

Pressure to explain the relevance of conservation increased in the late 1950s in the context of disputes within the SED. In 1958 the SED concluded that the process of building the foundations of socialism had been realized. After purging the party of opponents, Walter Ulbricht announced at the Fifth Party Congress in July the regime's intention to finish the process of building socialism by developing a "superstructure" that would cultivate the "new man of the socialist epoch."[70] Within the context of the SED's offensive, the Ministry of Agriculture and the leadership of the DKB worked frenetically to put conservation on the proper ideological course. Tipped off by Kurt Kretschmann, an active socialist, veteran commissioner, and director of the learning center on Lake Müritz who was frustrated by the poor performance of official conservation and its lack of committed socialists,[71] the Central Committee of the SED ordered Minister of Agriculture, Hans Reichelt, to develop "a clear political and technical conception" for conservation.[72] In drafting "Principles of Conservation in the GDR," a process completed only in January 1960, officials in the Ministry of Agriculture faulted Germany's middle-class conservation tradition that excluded the working classes and succumbed to Göring's "degenerate" form of nature conservation. Reflecting heightened cold war tensions, early drafts accused the "Adenauer State" of selling out nature in pursuit of NATO membership and rearmament (e.g., canalizing the Moselle, letting British troops train in the Lüneburg Heath nature reserve). Drafts denounced the ILN and commissioners for retaining unofficial ties with West Germany and with the predominantly "capitalist" International Union for the Conservation of Nature (IUCN), which still had not taken a stance against nuclear armaments. Ministry officials, probably relieved to strike back at those who had criticized them for years, accused commissioners of espousing "petty bourgeois" attitudes toward nature and for lacking devotion to the party (of roughly two hundred county commissioners, only eleven belonged to the SED).[73] Before the guidelines could be finalized and published, however, they had to be approved by the organizations essential to their implementation, the ILN and the DKB. Their involvement ensured that ministry officials at least acknowledged the expanded dimensions of conservation and pressured state and party bureaucrats to support conservation—now presumably "socialist"—more than previously.[74]

The final version announced that conservation was integral to building socialism and fulfilling the regime's seven-year plan (1959–1965) that projected further increases in production, consumption, and worker productivity. With predictable phraseology, the guidelines claimed that only under socialism—where all of nature's resources were owned in common and where a centrally planned economy enforced conservation—was it possible to protect nature, promote scientific research, manage natural resources, and reduce pollution. But the final version of principles also repeated appeals to set aside landscapes for public health and recreation, and subtly faulted East Germany's industry for failing to reduce water and air pollution and its mining enterprises for ravaging land without reclaiming it for later productive uses.[75] After the Soviet Union passed a new conservation law in 1960, East Germans followed their lead, returning to the task of modifying their guidelines. The Ministry of Agriculture issued updated "Principles of Socialist Resource Planning" in 1964, making them a binding extension to the conservation law and formalizing the concept that prioritized economic use of nature.[76]

For its part in completing the socialist transformation of society the DKB scrutinized members, conceding that the diverse special committees of Nature and Heimat Friends were united by a common interest in a particular field, but not a shared commitment to socialism.[77] Not long after the construction of the Berlin Wall in August 1961, the leadership lamented that the hearts of Nature and Heimat Friends "still have not become enflamed" with political conviction.[78] Some NHF members, however, believed that socialism provided favorable conditions for effective conservation, if only officials, cooperatives, and citizens would end their exploitative treatment of nature inherited from the capitalist past.[79] Writing in 1961, editor of the NHF journal, Reimar Gilsenbach satirized the "good socialist" director of a chemical enterprise who emptied effluents into a nearby river, regretting only that the factory was not located on a larger waterway to better absorb pollutants. Equally frustrating was the labor union chairman who received permission to build a youth hostel on the protected shore of a lake, claiming that socialism had resolved the contradiction between preservation and development.

Exasperated by the persistent gap between socialist theory and practice Gilsenbach asked, "What good is the small nature reserve that pleases a dozen experts when simultaneously a million people are forced to breathe air that is full of soot and stinking of waste gas." Not narrow preservation but "total mastery of nature," he wrote, would enable modern industrial society to protect nature more effectively in order to use it more intensively.[80]

With little land in private hands East Germany boasted that it had better conditions than the Federal Republic for protecting nature for its people. The GDR, however, fulfilled this promise only to a limited extent. Early efforts to establish a national park in the Sächsische Schweiz (1954–1955), for example, did not get beyond the planning stage. This was not because of bureaucratic indifference, but because of inadequate water and sewer treatment facilities and conservationists' concern that the polluted Elbe "contradicted the character of a national park."[81] As noted earlier, the ILN, together with the DKB and the official tourist service, also developed plans in the late 1950s for a system of large landscape preserves for public recreation on 10 percent (later 14 percent) of the territory of the GDR. To justify the creation of these "quiet oases" they explained that a majority of the country's three hundred protected landscapes were limited in size and unsuited for weekend outings, causing the Baltic Coast, the Sächsische Schweiz, and the country's few other large protected landscapes to be overrun by tourists. Initially, however, officials blocked the campaign, most likely because conservationists' arguments acknowledged labor-related illnesses, and seemed merely to parrot West German claims. Most significantly, the plan necessitated conservationists' involvement in (and criticism of) area planning, an important source of the regime's power.

Conservationists refused to give up and had modest success in the 1960s. In the absence of a public arena and free press in which an entrepreneur-philanthropist like Toepfer could advance such a vision, they appealed to district governments to take action. Unfortunately, they had production quotas to meet and were slow to approve plans for using land for such "unproductive" purposes, a view that prevailed despite assertions about a correlation between outdoor recreation and increased worker productivity.[82] After prodding by the NHF and the ILN and awareness of the success of West Germany's nature park program, district governments endorsed proposals for several recreation areas beginning in the early 1960s, Lake Müritz among the first. Even in 1966, however, little had been done to transform this area into a public park. More promising was Kohrener Land near Leipzig, a recreational landscape planned with considerable involvement of the district's DKB.[83]

The regime turned its attention to recreation in the mid-1960s, when it introduced the five-day-work-week every two weeks and relaxed travel restrictions to Eastern Bloc countries. According to reports by the government's Committee for Tourism and Hiking, of primary concern were complaints from foreigners and East Germans about the poor quality of accommodations in popular vacation areas, evidence that existing facilities were not being used efficiently, and fears of revenue loss—all problems that were to be addressed with better state planning of the tourist industry.[84] In the context of these government efforts, the protection of scenic landscapes for public recreation acquired greater legitimacy. At the end of the 1960s, East Germany had protected landscapes on an estimated 18 percent of its territory, but few were adequately protected. Most were much smaller than West Germany's nature parks and were not easily accessed by East Germans living in the most heavily industrialized regions. This latter concern was partially addressed in the early 1970s with the reclamation of strip mining areas near Leipzig, Halle, and in the district of Cottbus.[85]

These ongoing efforts to protect nature for public recreation or scientific research were obstructed also because some land that might have been set aside ended up serving as hunting grounds for a privileged few. In the mid-1950s, for example, the president of the DAL, Hans Stubbe, urged the central Ministry of Agriculture to reverse a decision allowing Soviet troops to hunt on "valuable property of the people"—a prized nature reserve in Rostock District.[86] An even greater problem was the Interior Ministry's chain of "special hunting areas," some that included nature reserves, for use by party elites or given outright to loyal officials, Stubbe among them.[87] Such arrangements decided upon at the highest political level give added meaning to the frustrated comments of one researcher with the ILN who remarked that "in West Germany resignation is directed against the pervasive power of industry; with us [it is directed against] the omnipotence of the bureaucracy."[88]

The greatest obstacle blocking effective conservation in the 1960s was the regime's seven-year plan that forced through the final phase of collectivization and exacerbated already visible pollution problems. Working through approved channels, NHF leaders and ILN scientists took advantage of the regime's mild thaw, applying steady pressure on government and party officials to realize that "the rapid increase in production requires equally planned, farsighted care of [the] . . . natural resources of our socialist society" if costly damage was to be avoided.[89] Of immediate concern to commissioners, ILN researchers, and NHF members was the rapid transformation of farmland and forests into industrial landscapes. Open-pit mining of brown coal in the southern part of the country gobbled up thousands of hectares of land annually, leaving crater landscapes in its wake. Despite complaints by citizens and officials since the 1950s, reclamation followed at a lumbering pace.[90] Like their West German colleagues, ILN researchers reported in 1967 that sulfur dioxide and other airborne pollutants released through burning lignite coal threatened public health and contributed to significant economic losses in forestry and farming.[91]

The introduction of industrialized farming in 1967—about a decade after the FRG—created more problems. Conservationists in both Germanys reported a loss of species and declining soil fertility because of land consolidation, hedge removal, draining wetlands, and the application of chemical fertilizers and pesticides. But the GDR's practice of separating crop production from animal husbandry caused added harm. Toxic runoff containing nitrates from heavily fertilized plant monocultures and concentrated waste from farms with thousands of animals jeopardized already limited supplies of ground and surface water.[92] Using politically expedient rhetoric, conservationists continued to blame water pollution on the far-reaching effects of Germany's capitalist past, but they also subtly faulted the "egoism" of East German enterprises for their role in causing 87 percent of the country's waterways to be classified as heavily or very heavily polluted. But natural factors contributed to the problem: East Germany's water cycle provided less than half the water annually per person than the FRG, and its rivers had a lower self-cleaning capacity, owing to a lower rate of flow.[93] Although conservationists hoped to play a key role in guiding economic development they wielded little influence. They were most successful with classical preservation. Because of their efforts, the GDR had 361 nature reserves in 1961, up from 148 in the mid-1950s; by 1967 there were 637 reserves scattered around the country.[94]

At the end of the 1960s the GDR had laws and ordinances requiring reclamation of strip mining areas (1951), protecting hedgerows (1953), safeguarding drinking water and preventing flooding (1963), and empowering local governments to require industries to reduce or remove pollutants (1967). In addition, the constitution of 1968 acknowledged conservation in its many dimensions.[95] Despite such measures, leading NHF member and ILN researcher Hugo Weinitschke complained that most officials lacked a commitment to conservation and some continued to think that the main focus of nature conservation was preserving "the beauties of nature."[96] Following years of prompting by the NHF, the ILN, and others, as well as international attention to environmental protection, the Volkskammer passed the "Law for Socialist Landeskultur" in May 1970. This legislation and subsequent ordinances addressed air, water, and noise pollution, pesticide control, and solid waste disposal, but enforcement by district governments, especially vague standards for water and air quality, was uneven.[97]

As indicated by the name of the law, conservationists did not embrace the increasingly popular concept of environmental protection immediately, arguing that it referred only to defensive technical measures to control pollution. By contrast, they believed, the more encompassing phrase, "socialist resource planning," involved "designing the relationship between man and the environment" in accordance with socialism to support "further improvement of peoples' material and cultural standard of living." Only after the GDR was accepted as a member of the United Nations in 1973 and participated fully in international discussions did public discourse adopt the western importation, environmental protection (*Umweltschutz*).[98] Irrespective of terminology, the East German government was determined to gain international recognition and win the support of its own people by demonstrating a commitment to environmental reforms. In 1971, the same year that the West German government presented its Program on the Environment, the new leader of the SED, Eric Honecker, unveiled a five-year plan at the Eighth Party Congress calling for pollution control, recycling, rational land-use, and accelerated reclamation. The government created a new environment ministry, supported more environmental research, and encouraged the NHF to lead controlled public discussions of these recent achievements.

In the hopeful climate generated by promising reforms and a relatively open debate, a few daring economists asserted that nature and environmental protection was "productive" labor that prevented expensive cleanup measures and yielded immeasurable benefits, some of which would become evident only in the future.[99] The regime also encouraged citizens to file petitions for help with environmental problems, but was unprepared for the flood of complaints that signaled mounting environmental concern.[100] When the cost of clean up proved to be too high, party officials explained that change could come gradually, after the battle against capitalism had been won.

When the UN Conference on the Human Environment opened in Stockholm, in June 1972, the GDR and other eastern bloc countries except Romania were conspicuously absent. They had boycotted the meeting because the GDR was given voice, but not the same right to vote as the Federal Republic. Smarting from their absence at the gathering where they hoped to present their country as a model of environmental reform, East Germans accused western nations of issuing alarmist predictions to achieve political ends.[101] The slogan "only one earth," Weinitschke asserted in reference to the publication that served as a catalyst for conference deliberations,[102] was an attempt by capitalist countries to convince the world that the "principle enemy of humanity is environmental pollution." Such propaganda, he claimed, only blinded people from realizing that the real global struggle was against "imperialist nations."[103]

West German political leaders questioned exaggerated predictions about "limits to growth" that were associated with the Club of Rome and published on the eve of Stockholm, but East Germans dismissed them outright as evidence of a "crisis in capitalism," a "coming-home-to roost" for imperialist nations that had exploited the Third World to achieve prosperity. But acknowledging limits to growth would have contradicted socialism's faith that rational use of resources enabled economic growth to continue without harm to the environment. Furthermore, cutting production would have threatened the regime whose legitimacy was based in part on its ability to provide citizens with consumer goods.[104]

Despite ideological differences that the GDR emphasized, the two Germanys drew closer diplomatically the year after Stockholm, during the high point of Ostpolitik. Negotiations by the Inner-German Frontier Commission in November 1973 sought to develop plans for addressing shared environmental problems, such as protecting the water supply and cleaning up rivers that linked the two countries—especially the Werra and Weser, rivers highly salinized by potash industries, and the Elbe. Because these and most other connecting rivers flowed out of East Germany into the Federal Republic, the GDR was at a distinct disadvantage in terms of cleanup. Indicative of the mind-set that stalled environmental reforms until the collapse of the regime, the Politburo reported just prior to the talks that the GDR would secure "no economic advantages" from paying the high cost of purifying these rivers. Within a limited amount of time and distance, the water would flow out of the country without affecting East Germany's use! Because of the expense involved, officials planned to reduce pollution over a long period and with financial support from West German users who stood to benefit most.[105] Talks broke down in 1974 when West Germany opened its Federal Environmental Office in West Berlin, though minor environmental agreements were concluded in the 1970s. Negotiations resumed in 1980.[106]

AFTER WORLD WAR II, conservationists in both Germanys used the RNG as a point of departure for continuing the tradition of classical preservation while addressing the ever-widening dimensions of conservation in advanced industrial societies. During the 1950s, in neither country did conservationists challenge the prevailing faith in economic progress, arguing instead that conservation ensured development and prosperity. But many conservationists in the FRG continued to sound like old-fashioned cultural pessimists, faulting the hubris of technology, excessive materialism, and society's alienation from nature for causing increased harm to the natural world. Their East German counterparts preached a more hopeful (and overly optimistic) message rooted in socialism's traditional faith in technology, insisting that total mastery of nature would enable humankind to preserve parts of nature while freeing up resources for continued economic productivity.

Though West German conservationists were sometimes ignored, they used the freedoms available in their newly restored democracy, advancing their cause through private associations, the media, and political institutions. As conservationists responded to the consequences of affluence in the late 1950s and 1960s, they asserted the relevance of their work to public health, regional planning, and civil liberties—arguments with a decisive social orientation. In the process, they helped implement a nature park program, establish a national park, and participated in land-use planning—actions that shaped the contours of the land in the Federal Republic.

Conservation in West Germany benefited from, and was shaped by, the country's democratic institutions. To some extent, conservation helped to transform those same institutions. By the early 1970s the tradition inherited from the past was forced to adopt more democratic strategies as thousands of West Germans formed grassroots single-issue initiatives, resisting absorption by traditional conservation groups. In 1972, fifteen citizens' initiatives joined together to work for environmental issues nationwide by forming the Federal Association of Citizens' Initiatives for Environmental Protection (Bundesverband Bürgerinitiativen Umweltschutz, BBU). Initially the BBU looked for support from the FDP-led Federal Ministry of the Interior, but when the government sought to reduce West Germany's dependence on oil in 1973 by relying more on nuclear power, citizens' initiatives concentrated on the antinuclear campaign, often working outside of established political channels.[107] But citizens' initiatives shared with traditional groups the expectation that they be given a voice in deciding how the state would protect citizens' rights to a healthy environment.

In East Germany, conservationists operating within the restricting parameters of state socialism continued the long tradition in classical preservation. They were able to do so partly because it was comparatively low in cost, had limited effects on industry, could be pursued at the local level, and was not overtly political. But the state that promised more effective conservation did little to protect nature for its own people or to enable conservationists to restore and improve the health of the land to support multiple uses over the long term. East Germans managed to set aside several recreational landscapes, and emphasized that they did so without having to battle property owners, but neither did they have the freedom of association and expression to challenge the regime when it used "valuable property of the people" to create farms that were vast monocultures or to erect hunting reserves for an elite few. While conservationists continually reminded party officials that increased use of natural resources necessitated greater planning and foresight, they went largely unheeded. Forcing nature conservation to conform to socialism by emphasizing resource planning, a practice that prioritized economic uses of nature, ultimately obstructed conservationists in pursing the many other dimensions of their work. With no public sphere in which a free press, private organizations, or individual citizens could effectively challenge the state and party, the regime had little incentive to alter its emphasis on production, despite costs to nature and the environment.

After the initial euphoria in the early 1970s, the pace of environmental reform slowed in both Germanys because of the challenge and expense of implementing policies in the midst of a global energy crisis. Frustrated by the government's inattention to environmental decline, and encouraged by electoral success of the West German Greens, NHF members pressed for a new organization at the end of the 1970s. In agreeing to replace the NHF with the Society for Nature and the Environment (Gesellschaft für Natur und Umwelt, GNU) within the DKB in 1980, the regime hoped to contain criticisms of its policies while engaging in limited dialogue. By seeking greater leverage in environmental discussions, however, the GNU came to be viewed as part of the state apparatus, not the opposition. This perception was reinforced in the late 1980s by the GNU's distrust and criticism of ecological opposition groups such as ARCHE whose young members questioned the regime's environmental policies and "misused" the sanctuary provided by the Protestant church.[108] When the regime collapsed, so too, did the influence of the GNU, an organization that had helped to modernize (in politically acceptable ways) a long tradition in conservation.[109]

Notes

The West German section of this essay is revised from "For Nation and Prosperity, Health and a Green Environment in West Germany, 1945–1970," in *Nature and German History*, ed. Christoph Mauch (Berghahn Books, 2004), 93–118.

1. Maria Haendcke-Hoppe and Konrad Merkl, *Umweltschutz in beiden Teilen Deutschlands* (Berlin: Duncker und Humblot, 1986); Raymond Dominick, "Capitalism, Communism, and Environmental Protection. Lessons from the German Experience," *Environmental History* 3, no. 3 (July 1998): 311–332.
2. Dominick, "Capitalism, Communism, and Environmental Protection," 320.
3. Gert Kragh, "Gesunde Landschaft bedingt die Zukunft des Volkes," 13 December 1945, Bundesarchiv, Koblenz, Bundesamt für Naturschutz (hereafter BAK B 245/137).
4. Gert Gröning and Joachim Wolschke-Bulmahn, *Die Liebe zur Landschaft*, pt. 3, *Der Drang nach Osten. Zur Entwicklung der Landespflege im Nationalsozialismus in den "eingegliederten Ostgebieten" während des Zweiten Weltkriegs* (Munich: Minerva, 1987).

5. Hans Huth, "Report on the Present Situation of Nature Protection in the American, British and French Occupied Zones of Germany" (June 1948), 6, BAK B 245/220.
6. *Reichsgesetzblatt* 1, no. 68 (1935): 821–826. See also Michael Wettengel, "Staat und Naturschutz 1906–1945. Zur Geschichte der Staatlichen Stelle für Naturdenkmalpflege in Preußen und der Reichsstelle für Naturschutz," *Historische Zeitschrift* 6, no. 4 (1993): 382–387; Hans Klose to Karl Duve, 27 May 1947, BAK B 245/137.
7. Walter Mrass, *Die Organisation des staatlichen Naturschutzes und der Landschaftspflege* (Stuttgart: Eugen Ulmer, 1970), tables 10–17.
8. Andreas Knaut, "Die Anfänge des staatlichen Naturschutzes. Die frühe regierungsamtliche Organisation des Natur- und Landschaftsschutzes in Preußen, Bayern und Württemberg," in *Umweltgeschichte: Umweltverträgliches Wirtschaften in historischer Perspektive*, ed. Werner Abelshauser (Göttingen: Vandenhoeck und Ruprecht, 1994), 143–162.
9. On landscape architects after 1945, see esp. Gert Gröning and Joachim Wolschke-Bulmahn, *Die Liebe zur Landschaft*, pt. 1, *Natur in Bewegung. Zur Bedeutung natur- und freiraumorientierter Bewegungen der ersten Hälfte des 20. Jahrhunderts für die Entwicklung der Freiraumplanung* (Munich: Minerva, 1986).
10. Klose to Richard Lohrmann, 4 July 1946, BAK B 245/253. Max Bromme, "Natur- und Landschaftsschutz im Regierungsbezirk Wiesbaden," *Verhandlungen Deutscher Beauftragter für Naturschutz und Landschaftspflege* (hereafter *Verhandlungen*) 2 (1948): 70, notes that a majority of officials and commissioners in Hesse had to leave "for political reasons."
11. On the postwar continuities in personnel, see Willi Oberkrome, *Deutsche Heimat: Nationale Konzeption und Regionale Praxis von Naturschutz, Landschaftsgestaltung und Kulturpolitik in Westfalen-Lippe und Thüringen (1900–1960)* (Paderborn: Schöningh, 2004), 396–404. On Klose and the Nazi party see Dominick, *The Environmental Movement in Germany: Prophets and Pioneers, 1871–1971* (Bloomington: Indiana University Press, 1992), 119. On his support for the Nazi regime's colonial schemes see Wettengel, "Staat und Naturschutz," 396. For allegations against him and his political leanings see Klose to Richard Lohrmann, 2 May 1947, BAK B 245/253; and Klose to Ilse Waldenburg, 17 November [1947?], BAK B 245/255.
12. Klose to Federal Ministry of Agriculture, 27 June 1950, BAK B 245/247; Kragh, "Der Beitrag der Naturschutzstellen bei Planungsarbeiten im Städtebau und in der Landschaft," 24 October 1950, BAK B 245/153; Konrad Buchwald, "Unsere Zukunftsaufgaben im Naturschutz und in der Landschaftspflege," *Verhandlungen* 11 (1957): 30–35.
13. Klose, *Fünfzig Jahre staatlicher Naturschutz* (Giessen: Brühlscher Verlag, 1957), 42–50.
14. Klose, "Über die Lage der Landes- und Bezirksstellen," *Verhandlungen* 2 (1948): 5; Wilhelm Lienenkämper, "Gedanken zur Tätigkeit der Naturschutzbeauftragten," *Verhandlungen* 3 (1949): 32–36; "Haushaltsmittel für Naturschutz und Landschaftspflege," *Verhandlungen* 10 (1956): 120–121, 126–128; Mrass, *Organisation*, 40–41, and tables 10–17, 23, 25.
15. Dominick, *Environmental Movement*, 120; Kurt Borchers, *Der Wald als deutsches Volksgut* (Lüneburg: Im Kinau, 1948).
16. Wilhelm Lienenkämper, *Grüne Welt zu treuen Händen. Naturschutz und Landschaftspflege im Industriezeitalter* (Stuttgart: Franckh, 1963), 89–91.
17. Monika Bergmeier, *Umweltgeschichte der Boomjahre 1949–1973: das Beispiel Bayern* (Münster: Waxmann, 2002), chap. 3; Sandra Chaney, "Visions and Revisions of

Nature: From the Protection of Nature to the Invention of the Environment in the Federal Republic of Germany, 1945–1975" (Ph.D. diss, University of North Carolina, Chapel Hill, 1996), chap. 3; Dominick, *Environmental Movement*, 125–134.

18. Such alliances include the Protective Association of the German Forest (Schutzgemeinschaft Deutscher Wald), the Alliance for Protection of Germany's Waters (Vereinigung Deutscher Gewässerschutz), and the German Working Group for Fighting Noise (Deutscher Arbeitsring für Lärmbekämpfung). The Interparliamentary Working Association for a Sustainable Economy (Interparlamentarische Arbeitsgemeinschaft für eine naturgemäße Wirtschaftsweise) led in drafting conservation and environmental legislation.

19. Klose, "Aufruf zur Bildung des Deutschen Naturschutzrings," 20 May 1950, BAK B 245/236; "Zur Begründung des Deutschen Naturschutzrings," *Verhandlungen* 4 (1950): 126; Deutscher Naturschutzring, *25 Jahre Deutscher Naturschutzring* (Siegburg: Buch- und Offsetdruckerei Daemisch-Mohr, 1975).

20. Diethelm Prowe, "The 'Miracle' of the Political-Culture Shift. Democratization between Americanization and Conservative Reintegration," in *The Miracle Years. A Cultural History of West Germany, 1949–1968*, ed. Hanna Schissler (Princeton: Princeton University Press, 2001), 451–458.

21. DNR, "Aktuelle Informationen: Der Deutsche Naturschutzring fragt die Parteien," *Informationsbrief* 19/20 (September 1965): 59–64; Hubert Weinzierl, *Die große Wende im Naturschutz* (Munich: BLV Verlagsgesellschaft, 1970), 12; Dominick, *Environmental Movement*, 211.

22. Josef Fochepoth, "German Reaction to Defeat and Occupation," in Robert G. Moeller, ed., *West Germany under Construction: Politics, Society, and Culture in the Adenauer Era* (Ann Arbor: University of Michigan Press, 1997), 73–89. On the SDW see Chaney, "Visions and Revisions of Nature," chap. 2. Lehr is quoted in "Westdeutschland feiert den 'Tag des Baumes,'" *Nachrichtenblatt für Naturschutz* 23, nos. 5/6 (May/June 1952): 13.

23. Chaney, "Water for Wine and Scenery, Coal and European Unity: Canalization of the Mosel River, 1950–1964," in *Water, Culture, and Politics in Germany and the American West*, ed. Susan Anderson and Bruce Tabb (New York: Peter Lang, 2001), 236–238; Chaney, "A 'Democratic Movement of the People' Saves the Wutach Gorge: A Case Study in Early West German Environmental Activism, 1949–1960," in *Shades of Green*, ed. Nathan Stolzfuss, Christof Mauch, and Douglas Weiner (Rowman and Littlefield, forthcoming).

24. On postwar weaknesses of Heimat protection, see Arne Andersen, "Heimatschutz. Die bürgerliche Naturschutzbewegung," in *Besiegte Natur: Geschichte der Umwelt im 19. und 20 Jahrhundert*, ed. Franz-Josef Brüggemeier and Thomas Rommelspacher (Munich: C. H. Beck, 1989), 156–157; and Rolf Peter Sieferle, *Fortschrittsfeinde? Opposition gegen Technik und Industrie von der Romantik bis zur Gegenwart* (Munich: C. H. Beck, 1984), 227. Compare these accounts with Oberkrome, *Deutsche Heimat*.

25. William H. Rollins, *A Greener Vision of Home. Cultural Politics and Environmental Reform in the German Heimatschutz Movement, 1904–1918* (Ann Arbor: University of Michigan Press, 1997); Andreas Knaut, *Zurück zur Natur! Die Wurzeln der Ökologiebewegung* (Greven: Kilda-Verlag, 1993).

26. Christoph Kleßmann, *Zwei Staaten, eine Nation. Deutsche Geschichte 1955–1970* (Bonn: Bundeszentrale für politische Bildung, 1988), 30–32; Arne Andersen, *Der Traum vom guten Leben. Alltags- und Konsumgeschichte vom Wirtschaftswunder bis heute* (Frankfurt am Main: Campus, 1997), 127–135; Axel Schildt, "From Reconstruction to 'Leisure Society': Free Time, Recreational Behaviour and the Dis-

course on Leisure Time in the West German Recovery Society of the 1950s," *Contemporary European History* 5, no. 2 (1996): 191–197.
27. Joachim Bodamer, *Gesundheit und technische Welt* (Stuttgart: Ernst Klett, 1960); Alfred Marchionini, "Gesundheit, Freizeit und Naturpark," *Naturschutzparke*, no. 17 (February 1960): 3–7.
28. Alfred Toepfer, "Naturschutzparke—eine Forderung unserer Zeit," *Naturschutzparke*, no. 7 (Autumn 1957): 173; "Schatzkammer der Natur," and "Naturschutzparke—Kraftquellen unseres Volkes," films produced by Eugen Schuhmacher, 1956, Verband Deutsche Naturparke [VDN] Archive, Niederhaverbeck, Lüneburg Heath. For newspaper clippings of Toepfer's public announcement see VDN Archive, binder "Neue Naturparke. Die einzelnen Gebiete und Presse, 1956."
29. Dittrich to Toepfer, 10 September 1959, VDN Archive, binder no. 45, "Bundesbehörde und Anstalten"; Erich Dittrich, "Der Ordnungsgedanke der Landschaft und die Wirklichkeit," *Verhandlungen* 12 (1959): 127–134.
30. Hans-Dietmar Koeppel and Walter Mrass, "Natur- und Nationalparke," in *Natur- und Umweltschutz in der BRD*, ed. Gerhard Olschowy (Hamburg and Berlin: Paul Parey, 1978), 803–804; *Natur und Landschaft* [N & L] 50, no. 10 (1975): 266–273. Parks ranged in size from 38 square kilometers [9,500 acres] (Harburger Berge, Hamburg) to 2,908 square kilometers [727,000 acres] (Altmühltal, Bavaria) and included privately owned land.
31. Udo Hanstein, *Entwicklung, Stand und Möglichkeiten des Naturparkprogramms in der Bundesrepublik Deutschland*, Beiheft 7, *Landschaft + Stadt* (Stuttgart: Eugen Ulmer, 1972), 34–35, 41–42.
32. Kragh, "Ordnung der Landschaft—Ordnung des Raumes. Bericht über den Deutschen Naturschutztag Bayreuth, vom 22. bis 27 Juni 1959, *N & L* 34, no. 8 (1959): 113–116.
33. Heinrich Lohmeyer, "Unser Lebensraum ist in Gefahr!," *N & L* 36, no. 3 (1961): 33–36; "Die Bedrohung unseres Lebensraumes und Maßnahmen zu seiner Pflege und Erhaltung im Zahlenspiegel," *N & L* 39, no. 8 (1964): 124–133.
34. Hans Krieg to unidentified Bundestag deputy, 24 October 1961, BAK B 245/235.
35. Dominick, *Environmental Movement*, 200.
36. "Grüne Charta von der Mainau," *N & L* 36, no. 8 (1961): 151. The charter led to the creation of the German Council for Landscape Planning (Deutscher Rat für Landespflege), a presumably independent body comprised of academics, landscape architects, and prominent individuals who commissioned studies and advised the federal president and national and state ministries on issues ranging from canalizing the Moselle to managing solid waste. On the charter and the DRL see Jens Ivo Engels, "Ideenwelt und politische Verhaltensstile von Naturschutz und Umweltbewegung in der Bundesrepublik 1950–1980" (Habilitationsschrift, University of Freiburg, 2004), 116–141.
37. Konrad Buchwald, *Die Zukunft des Menschen in der industriellen Gesellschaft und die Landschaft* (Braunschweig: Hans August-Stolle Verlag, 1965). See also Oberkrome, *Deutsche Heimat*, 398–399.
38. UNESCO, *Use and Conservation of the Biosphere. Proceedings of the Intergovernmental Conference of Experts on the Scientific Basis for Rational Use and Conservation of the Resources of the Biosphere, Paris, 4–13 September 1968* (Paris: UNESCO, 1970); Olschowy, "Zur Belastung der Biosphäre," *N & L* 44, no. 1 (1969): 3–6; Günter Küppers, Peter Lundgreen, and Peter Weingart, *Umweltforschung—die gesteuerte Wissenschaft? Eine empirische Studie zum Verhältnis von Wissenschaftentwicklung und Wissenschaftspolitik* (Frankfurt am Main: Suhrkamp, 1978), 104–105; Kai F. Hünemörder, "Vom Expertennetzwerk zur Umweltpolitik:

Frühe Umweltkonferenzen und die Ausweitung der öffentlichen Aufmerksamkeit für Umweltfragen in Europa (1959–1972)," *Archiv für Sozialgeschichte* 43 (2003): 275–296,
39. Chaney, "Visions and Revisions of Nature," chap. 7.
40. Cover story, "Vergiftete Umwelt," *Der Spiegel*, 5 October 1970, 74–96.
41. Weinzierl, *Die große Wende im Naturschutz*, 81.
42. Wolfgang Erz, "Europäisches Naturschutzjahr 1970—und was wurde erreicht?," *N & L* 45, no. 12 (1970): 409–411; Horst Bieber, "Langsam stirbt der Umweltschutz," *Die Zeit*, 20 October 1978, 8; Klaus-Georg Wey, *Umweltpolitik in Deutschland. Kurze Geschichte des Umweltschutzes in Deutschland seit 1900* (Opladen: Westdeutscher Verlag, 1982), 201–205; Franz-Josef Brüggemeier, *Tschernobyl, 26 April 1986. Die ökologische Herausforderung* (Munich: Deutscher Taschenbuch Verlag, 1998), 208–213; Hans-Dietrich Genscher, *Erinnerungen* (Berlin: Siedler Verlag, 1995), 125–132; Karl Ditt, "Die Anfänge der Umweltpolitik in der Bundesrepublik Deutschland während der 1960er und frühen 1970er Jahre," in Matthias Frese, Julia Paulus, and Karl Teppe, eds., *Demokratisierung und gesellschaftlicher Aufbruch. Die sechziger Jahre als Wendezeit der Bundesrepublik* (Paderborn: Schöningh, 2003), 305–347.
43. Küppers et al., *Umweltforschung*, 127–132; Bernd Guggenberger, "Umweltpolitik und Ökologiebewegung," in Wolfgang Benz, ed., *Die Geschichte der BRD*, vol. 3, *Economics* (Frankfurt am Main: Fischer Taschenbuch Verlag, 1989), 414; Olschowy, "Stellungnahme zum Sofortprogramm für Umweltschutz der Bundesregierung," *N & L* 46, no. 4 (1971): 103–105.
44. Norman Naimark, *The Russians in Germany: A History of the Soviet Zone of Occupation, 1945–1949* (Cambridge, Mass.: Belknap Press, 1997), 141–166; Markus Rösler, Elisabeth Schwab, and Markus Lambrecht, eds., *Naturschutz in der DDR* (Bonn: Economica-Verlag, 1990), 13–14.
45. Kurt Kretschman, "Auferstanden aus Ruinen. Aus Kurt Kretschmanns persönlichen Erinnerungen an die Naturschutzentwicklung in der DDR," *Naturschutz heute* 1 (1994): 48–49, available at http://www.nabu.de/nabu/history/ddr.htm (accessed 10 August 2000); Rösler et al., 20.
46. Minister of Agriculture, GDR, to Minister for Volksbildung (Wandel), GDR, 5 January 1951, microfilm, Bundesarchiv, Berlin-Lichterfelde, Ministry of Agriculture, file no. 3752 (hereafter BArch DK 1/3752), ff. 24–26. See also Hermann Behrens, Ulrike Benkert, Jürgen Hopfmann, and Uwe Maechler, *Wurzeln der Umweltbewegung. Die "Gesellschaft für Natur und Umwelt" (GNU) im Kulturbund der DDR. Ein Beitrag zur Geschichte der ökologischen Bewegung in den neuen Bundesländern*, Forum Wissenschaft: Studien 18 (Marburg: Verlag des Bundes demokratischer Wissenschaftlerinnen und Wissenschaftler, 1993), 36–37.
47. Herbert Ant, "Das Institut für Landesforschung und Naturschutz in Halle/Salle," *N & L* 42, no. 5 (May 1967): 105–107; Rösler et al., *Naturschutz in der DDR*, 109–117.
48. See Hermann Meusel to Ministry of Agriculture, GDR, 10 March 1958, BArch DK 107/77/30.
49. On the activities of commissioners see esp. Kurt Kretschmann [commissioner, Oberbarnim District, Brandenburg], activity reports in BArch DK 1/3752, ff. 80, 128, 133, 196, 205, 222; and Walter Gotsmann, "Aus der Tätigkeit eines Kreisbeauftragten für Naturschutz," in Reimar Gilsenbach, *Reichtum und Not der Natur* (Dresden: Sachsenverlag, 1955), 16–24. On protecting endangered beaver and birds see Kretschmann, "Vernichtung der Adler und des Bibers im Schorfheidegebiet," 19 June 1952, BArch DK 1/3752; and Alfred Weber, "Die

Aufgaben des Naturschutzes an der Ostsee," in Gilsenbach, *Reichtum und Not der Natur*, 13–14.
50. Kretschmann to Central Committee, SED, Abteilung Landwirtschaft (Mitbach), 1 December 1957, microfilm, BArch DK 1/ 1039, f. 16.
51. Meusel, "Denkschrift der Deutschen Akademie der Landwirtschaftswissenschaften zu Berlin über die Verbesserung der Arbeit auf dem Gebiet des Naturschutzes," 21 April 1955, microfilm, BArch DK 107/25/13; Kretschmann to Central Committee, SED, 1 December 1957, microfilm, BArch DK 1/1039, ff. 14–17.
52. Meusel to Otto Grotewohl, Chairman, Council of Ministers, GDR, 15 February 1957, BArch DK 107/ 77/30.
53. Dr. Schwarz, Managing Director, Sektion Landeskultur und Naturschutz, DAL, "Aktenvermerk, betr. Naturschutzbrochüre der Naturschutzwoche 1958," 27 March 1958, microfilm, BArch DK 107/77/30.
54. The newsletter is discussed in Ministry of Agriculture, GDR, "Anlage 1" [to accompany "Richtlinien des Naturschutzes"], 28 September 1958, microfilm, BArch DK1/1039, f. 20.
55. ILN and Ministry of Agriculture, "Eine wichtige Richtigstellung: Naturwacht—Naturschutzhelfer," *Aus der Arbeit der Natur- und Heimatfreunde im Kulturbund zur Demokratischen Erneuerung Deutschlands* (hereafter AdA) 9, nos. 7/8 (1962): 218.
56. Behrens et al., *Wurzeln der Umweltbewegung*, 43; Rösler et al., *Naturschutz in der DDR*, 98; quote from Ministry of Agriculture, GDR, "Anlage 1," 28 September 1958, microfilm, ff. 20–21.
57. Mary Fulbrook, *Anatomy of a Dictatorship. Inside the GDR, 1949–1989* (New York: Oxford University Press, 1995), 60–61; "Entwurf! Geschäftsordnung der Natur- und Heimatfreunde im Kulturbund für demokratischen Erneuerung Deutschlands," n.d. [1950?], Stiftung Archiv der Parteien und Massenorganisation der DDR im Bundesarchiv, DY 27 2848 (hereafter SAPMO-BArch DY 27/2848); Behrens et al., *Wurzeln der Umweltbewegung*, 30–36, 44.
58. Karl Kneschke, "Aus dem Schlussreferat anläßlich der 1. Zentralkonferenz der Natur- und Heimatfreunde," 12 November 1950, SAPMO-BArch DY 27/378.
59. "Leitsätze der Natur- und Heimatfreunde. Vorschlag für die Konferenz am 3. und 4. Juli 1954 in Weimar," SAPMO-BArch DY27/917, 271–275.
60. Reimar Gilsenbach, "Heimat und Vaterland," *AdA*, no. 3 (1958): 61, 62. For detailed coverage of the convergence of socialism, conservation, and Heimat preservation, see Oberkrome, *Deutsche Heimat*, 289–379.
61. Comments by Hugo Weinitschke, in "Stenografisches Protokoll der Sitzung des Präsidialrates [of the DKB] am 21. November 1975," SAPMO-BArch DY27/958, 207–230.
62. NHF, Stadt Wehlen, to Central Commission, NHF, 15 November 1951, microfilm, BArch DK 1/3752, f. 88; District Council, Neubrandenburg, Abteilung Landwirtschaft (Marschall) to Ministry of Agriculture, GDR, 10 December 1952, microfilm, BArch DK 1/3753, f. 47.
63. "Pfingsttreffen der Natur- und Heimatfreunde in Bad Freienwalde, am 12. und 13. Mai 1951," microfilm, flyer in BArch DK 1/3752, f. 221; Behrens et al., *Wurzeln der Umweltbewegung*, 34–35, 158–161.
64. "Protokoll über die von der Landesleitung des Kulturbundes zur demokratischen Erneuerung Deutschlands . . . einberufene Sitzung zur Pflanzaktion am 16.1.51," BArch DK 1/ 3752, f. 60; Dr. Knorr, Report on NHF agenda of 1956 in "Stenographisches Protokoll der Sitzung des Präsidialrates [of DKB]," 4 November 1955, SAPMO-BArch DY27/919, 261; Behrens et al., *Wurzeln der*

Umweltbewegung, 46; Gerhard Würth, *Umweltschutz und Umweltzerstörung in der DDR* (Frankfurt am Main: Peter Lang, 1985), 112.
65. Gisela Werner-Stockmann, "121 Gemeinden veränderten ihr Gesicht. Erfahrungen beim Wettbewerb 'Das schöne Dorf im Bezirk Halle,'" *AdA*, no. 2 (1957): 33–35; Harald Rüssel, "Die Wandlung des Begriffes 'Schönes Dorf' bei der sozialistischen Umgestaltung der Landwirtschaft," *AdA*, no. 7 (1958): 165–166; Behrens et al., *Wurzeln der Umweltbewegung*, 46–47; Würth, *Umweltschutz und Umweltzerstörung in der DDR*, 106.
66. Nikola Knoth, "Die Naturschutzgesetzgebung der DDR von 1954," *Zeitschrift für Geschichtswissenschaft* 39, no. 2 (1991): 163–172.
67. Hermann Meusel, "Unsere Arbeit volksverbindend zwischen Ost und West," *Verhandlungen* 8 (1954): 68–69.
68. Georg Pniower, "Naturschutz im Spiegel der Landeskultur," *Natur und Heimat* (hereafter *N & H*) 1 (1952): 4–7; continued in vol. 2 (1952): 4–9; Pniower's comments in "Aus der Diskussion," *AdA*, nos. 8–9 (1956): 207–210; and Pniower, "Bemerkungen zur Problematik der Landeskultur," to DAL and the Central Committee of the SED, March 1956, BArch DK 107/25/13. See also Oberkrome, *Deutsche Heimat*, 284–289.
69. Rösler et al, *Naturschutz in der DDR*, 73–74; Werner Gruhn, "Aktuelle Aspekte der DDR-Umweltpolitik," *Deutsche Studien* 19 (December 1981): 432; Joan DeBardeleben, *The Environment and Marxism-Leninism. The Soviet and East German Experience* (Boulder: Westview Press, 1985), 269.
70. Monika Gibas, "Ideologie und Propaganda," in *Die SED. Geschichte, Organisation, Politik. Ein Handbuch*, ed. Andreas Herbst, Gerd-Rüdiger Stephan, and Jürgen Winkler (Berlin: Dietz Verlag, 1997), 253–254; Kleßmann, *Zwei Staaten, eine Nation*, 309.
71. Kretschmann to Walter Ulbricht (General Secretary, SED), 24 September 1957, microfilm, BArch DK 1/1039, ff. 8–12; Kretschmann to SED Central Committee, Abteilung Landwirtschaft (Mitbach), 1 December 1957, microfilm, BArch DK 1/1039, ff. 14–17.
72. Franz Mellentin [Head of Agriculture, SED Central Committee] to Minister of Agriculture, Hans Reichelt, 7 January 1958, microfilm, BArch DK 1/1039, f. 7. Reichelt, later environment minister, belonged to the Democratic Farmers' Party of Germany (Demokratische Bauernpartei Deutschlands).
73. N.a., "Der Weg des Naturschutzes in Deutschland," "4. Entwurf," 20 June 1958, microfilm, BArch DK 1/3754, ff. 88–98; Ministry of Agriculture, "Anlage 1," 28 September 1958, ff. 18–25.
74. Weinitschke's comments in "Stenographische Aufzeichnungen der Sitzung der AG Naturschutz am 9. Dezember 1959 in der DAL," 10 December 1959, BArch DK 107/77/30.
75. Heinrich [Representative of Minister of Agriculture, GDR], "Grundsätze des Naturschutzes in der Deutschen Demokratischen Republik," 13 November 1959, microfilm, BArch DK 1/3754, ff. 23–29; Kleßmann, *Zwei Staaten, eine Nation*, 308–313. The second five-year plan, introduced in 1956, was halted in 1959 when the SED followed the lead of the Soviet Union and announced the GDR's own seven-year economic program (1959–1965).
76. Würth, *Umweltschutz und Umweltzerstörung in der DDR*, 95–96; drafts in BArch DK 107/A9/35.
77. DKB, Bezirksleitung Erfurt, "Bemerkungen zur Situation der Natur- und Heimatfreunde und Vorschläge für deren Veränderungen," n.d. [April 1959?], SAPMO-BArch DY27/3307.

78. N.a. [NHF Central Commission], "Bericht über die bisherige Erfüllung des Arbeitsplanes 1961 der Kommission Natur- und Heimatfreunde des Präsidialrates," n.d. [August 1961?], SAPMO-BArch DY 27/2842.
79. Herbert Bauer [Chairman, NHF Central Commission] and Karl-Heinz Schulmeister to SED Central Committee, "Zu einigen Problemen des Naturschutzes und der Landschaftspflege," n.d. [sent 29 March 1961], SAPMO-BArch DY 27/2947.
80. Gilsenbach, "Wohin gehst du, Naturschutz?," *N & H* 10, no. 7 (1961): 350–351.
81. Meusel, "Landschaftsschutzgebiete als Erholungszentren, *N & H*, no. 5 (1959): 26. On planning for a national park see BArch DK 107/77/47; Annegret Nickels, "Wird die Sächsische Schweiz Deutschlands erster Nationalpark?," *N & H*, no. 8 (1954): 252–253; and Kurt Wiedemann, "Landschaftsschutz für die Sächsische Schweiz," *N & H*, no. 5 (1958): 152–155, no. 6 (1958): 176–178.
82. Bauer and Schulmeister to SED Central Committee [sent 29 March 1961], SAPMO-BArch DY27/2947.
83. "Protokoll der Arbeitsausschußsitzung vom 15.3.67 in Waren" [Neubrandenburg District], SAPMO-DY27/3969; "Konzeption. Vorschlag zur Entwicklung des Naherholungsgebietes 'Kohrener Land,'" n.d. [1966?], SAPMO-BArch DY 27/3969.
84. Komitee für Touristik und Wandern, DDR, "Stellungnahme zum Stand und der weiteren Entwicklung des Erholungswesens in der DDR," n.d. [1964?], SAPMO-DY 27/3693.
85. Hans Stubbe, "Erklärung des Präsidenten der Deutschen Akademie der Landwirtschaftswissenschaften zu Berlin zur 8. Naturschutzwoche anläßlich einer Pressekonferenz am 6. May 1964," SAPMO-BArch DY27/3506; Ludwig Bauer and Hugo Weinitschke, *Landschaftspflege und Naturschutz. Eine Einführung in ihre Grundlagen und Aufgaben* (Jena: VEB Gustav Fischer Verlag, 1967), 188–191; Gruhn, "Umweltschutz in der DDR," *Deutschland Archiv* 5, no. 10 (1972): 1048; Ilka Nohara-Schnabel, "Zur Entwicklung der Umweltpolitik in der DDR," *Deutschland Archiv. Zeitschrift für Fragen der DDR und der Deutschland Politik* 9, no. 8 (August 1976): 818; Würth, *Umweltschutz und Umweltzerstörung in der DDR*, 106–107.
86. DAL President to Ministry of Agriculture, GDR, 8 November 1954, microfilm, BArch DK 1/6694, ff. 109–110.
87. Ministry of the Interior, GDR, "Sonderjagdgebiete des Ministeriums des Innern," 27 May 1955, microfilm, BArch DK1/6944, ff. 116–130; Ministry of Agriculture, GDR, "Vorlage an das Politbüro des ZK über Maßnahmen zur Verbesserung des Jagdwesens in der Deutschen Demokratischen Republik," July 1956, and accompanying "Anlage. Aufstellung der zur Zeit bestätigten Sonderjagdberechtigten (ausser den Mitgliedern des Politbüros)," BArch DK 1/VA/1617.
88. Sektion Landeskultur und Naturschutz, "Bericht über die 1. Sitzung der Arbeitsgemeinschaft 'Naturschutz' am 3 März 1955 in Berlin," 11 March 1955, BArch DK 107/25/13.
89. Bauer and Schulmeister to SED Central Committee [sent 29 March 1961].
90. Winrich, Ministry of Agriculture, GDR, "Dienstreisebericht über den Perspektivplan für Devastierung und Rekultivierung in Leipzig am Montag, dem 20.5. und im Halle am Dienstag, dem 21.5.1957," microfilm, BArch DK 1/3694, f. 49.
91. Bauer and Weinitschke, *Landschaftspflege und Naturschutz*, 165; Stubbe, "Erklärung des Präsidenten der Deutschen Akademie der Landwirtschaftswissenschaften zu Berlin zur 8. Naturschutzwoche anläßlich einer Pressekonferenz am 6. May 1964."

92. Reichelt, "Die Landwirtschaft in der ehemaligen DDR—Probleme, Erkenntnisse, Entwicklungen," *Berichte über Landwirtschaft* 70 (1992): 117–136; Andreas Kurjo, "Landwirtschaft und Umwelt in der DDR. Ökologische, rechtliche und institutionelle Aspekte der sozialistischen Agrarpolitik," in *Umweltprobleme und Umweltbewußtsein in der DDR*, ed. Redaktion Deutschland Archiv (Cologne: Verlag Wissenschaft und Politik, 1985), 39–78.
93. Bauer and Weinitschke, *Landschaftspflege und Naturschutz*, 157–158; Gilsenbach, "Wohin gehst du, Naturschutz," *N & H* 10, no. 6 (1961): 309; Rösler et al., *Naturschutz in der DDR*, 37; Dominick, "Capitalism, Communism, and Environmental Protection," 324.
94. Rösler et al., *Naturschutz in der DDR*, 133–135; Bauer and Weinitschke, *Landschaftspflege und Naturschutz*, 232–233.
95. Würth, *Umweltschutz und Umweltzerstörung in der DDR*, 26–31; Rösler et al., *Naturschutz in der DDR*, 24.
96. Weinitschke, "Stellungnahme zu den Thesen für den Naturschutz" [drafted by the State Committee for Forestry, Agricultural Council, GDR], n.d. [June 1967?], SAPMO-BArch DY 27/2685.
97. Bauer and Weinitschke, *Landschaftspflege und Naturschutz*, 3rd ed. (Jena: VEB Gustav Fischer, 1973), 310–311; Gruhn, "Umweltschutz in der DDR," 1043–1045; DeBardeleben, *The Environment and Marxism-Leninism. The Soviet and East German Experience*, 222–224; Nohara-Schnabel, "Zur Entwicklung der Umweltpolitik in der DDR," 812.
98. Reichelt [Environment Minister], comments in "Stenografisches Protokoll der Präsidialratstagung des Deutschen Kulturbunds am 12. Mai 1972 in Berlin," SAPMO-BA, DY 27 952/295. See also DeBardeleben, *The Environment and Marxism-Leninism. The Soviet and East German Experience*, 43; and Schurig, "Politischer Naturschutz," 368.
99. Werner Barm, "Umweltschutz in der DDR," *Deutsche Studien*, no. 38 (June 1972): 198–199; Gruhn, "Umweltschutz in der DDR," 1045; Nohara-Schnabel, "Zur Entwicklung der Umweltpolitik in der DDR," 827–828.
100. For petitions from the early 1970s to the new environment ministry see BArch DK 5/3472, 5/4393.
101. Barm, "Umweltschutz in der DDR," 194, 201; Gruhn, "Umweltschutz in der DDR," 1038, 1050.
102. Barbara Ward and René Dubos, *Only One Earth. The Care and Maintenance of a Small Planet* (New York: W. W. Norton, 1972).
103. Weinitschke's comments in "Stenografisches Protokoll der Präsidialratssitzung des Deutschen Kulturbundes am 12. Mai 1972 in Berlin," 258–259. By 1973 Weinitschke was ILN director and NHF Central Commission chairman.
104. DeBardeleben, *The Environment and Marxism-Leninism. The Soviet and East German Experience*, 176–193, quotes from 178; Willy Brandt, "Notizen des Bundeskanzlers, Brandt, zur Umweltpolitik. 9. Juli 1972," Document no. 70, *Berliner Ausgabe*, vol. 7, *Mehr Demokratie wagen. Innen- und Gesellschaftspolitik, 1966–1974*, ed. Wolther von Kieseritzky (Bonn: Verlag J. H. W. Dietz, 2000), 338.
105. Politbüro, SED Central Committee, "Reinschriftenprotokoll, Nr. 47 vom 6. Nov. 1973," "Anlage 1b," entitled "Erfassung der Interessenlage der DDR zur Aufnahme von Verhandlungen auf dem Gebiet des Umweltschutzes mit der BRD," SAPMO-BArch DY 30/J IV 2/2/1475, 23–30, quote on 25.
106. Ernest D. Plock, *The Basic Treaty and the Evolution of East-West German Relations* (Boulder: Westview Press, 1986), 124–125; Genscher, *Erinnerungen*, 132–134.

107. Udo Kempf, "Bürgerinitiativen—der empirische Befund," in Bernd Guggenberger and Udo Kempf, eds., *Bürgerinitiativen und repräsentatives System*, 2nd rev. ed. (Opladen: Westdeutscher Verlag, 1984), 295–315; Brüggemeier, *Tschernobyl*, 211–220. By 1975 BBU united 100 citizens' initiatives and claimed over 300,000 members. On the BBU's interaction with the interior ministry see Referent für Öffentlichkeitsarbeit, "Betr.: Zusammenarbeit mit dem Bundesverband BBU," 10 July 1972, BAK B 106/63670.
108. Thomasius, Chairman, Central Executive Committee, GNU, to Karl-Heinz Schulmeister, Vice President, DKB, 25 January 1988, SAPMO-DY 27/456; n.a. [Reichelt?], "Konzeption für Gespräche mit Vertretern von Kirchen zur Umweltpolitik der DDR," n.d. [July 1989?] SAPMO-BArch DY 27/465.
109. Behrens et al., *Wurzeln der Umweltbewegung*. See also Peter Wensierski, "Die Gesellschaft für Natur und Umwelt. Kleine Innovation in der politischen Kultur der DDR," in *Umweltprobleme*, 151–168.

Notes on Editors and Contributors

THOMAS LEKAN is an assistant professor of history at the University of South Carolina in Columbia. He received his Ph.D. from the University of Wisconsin–Madison in 1999. He is the author of *Imagining the Nation in Nature: Landscape Preservation and German Identity, 1885–1945* (Harvard, 2004) as well as several essays and book chapters on the history of German nature conservation, regional planning, and popular culture. He is currently working on an environmental history of nature tourism, consumer culture, and ecological activism in twentieth-century Germany.

THOMAS ZELLER is an assistant professor of history at the University of Maryland in College Park. He received his Ph.D. from the Ludwig-Maximilians University in Munich in 1999. He is the author of *Straße, Bahn, Panorama. Verkehrswege und Landschaftsveränderung in Deutschland 1930 bis 1990* (Campus, 2002), a revised version of which will be published as *Driving Germany* in 2006, as well as numerous essays and book chapters on the history of the Autobahn system, landscape architecture, and technology in modern Germany. He is currently at work on a second monograph, *Consuming Landscapes: The View from the Road in the United States and Germany, 1910–1990*. Zeller is also a visiting research fellow at the German Historical Institute in Washington, D.C.

SANDRA CHANEY is an associate professor of history at Erskine College in South Carolina. She received her Ph.D. from the University of North Carolina at Chapel Hill in 1997. She is currently completing a manuscript, *Visions and Revisions of Nature in West Germany, 1945–1975*, and has published several essays and book chapters on post–World War II nature conservation in Germany.

RITA GUDERMANN received her Ph.D. from the Free University in Berlin in 1998, specializing in environmental and agricultural history. She is the author of *Morastwelt und Paradies: Ökonomie und Ökologie in der Landwirtschaft am Beispiel der Meliorationen in Westfalen und Brandenburg (1830–1880)* and a co-editor of the ambitious essay collection *Agrarmodernisierung und ökologische Folgen: Westfalen vom 18. bis zum 20. Jahrhundert* (Ferdinand Schöningh, 2001).

MICHAEL IMORT is an associate professor of geography and environmental studies at Wilfrid Laurier University in Waterloo, Ontario, Canada. He received his Ph.D. in cultural and historical geography in 2000 from Queen's University with a dissertation entitled *Forestopia: The Use of the Forest Landscape in Naturalizing National Socialist Ideologies of Volk, Race, and Lebensraum, 1918–1945*. In 1990, he received his master's degree in forest science from Freiburg University, where he studied fire ecology and conservation forestry.

RUDY KOSHAR received his Ph.D. from the University of Michigan in 1979. He taught at the University of Southern California from 1980 to 1991, then moved to the University of Wisconsin–Madison, where he is a professor of history. His recent publications include *German Travel Cultures* (Berg, 2000); *From Monuments to Traces: Artifacts of German Memory, 1870–1990* (California, 2000); *Germany's Transient Pasts: Preservation and National Memory in Twentieth-Century Germany* (North Carolina, 1998); (with Alon Confino) "Regimes of Consumer Culture," theme volume for *German History* (May 2001); and an edited volume, *Histories of Leisure* (Berg, 2002). He has held Guggenheim, ACLS, German Marshall Fund, Jean Monet, and other fellowships. He is series editor for "Leisure, Consumption, and Culture" with Berg Publishers. His current research is a study of automotive driving practices in Europe and North America from ca. 1900 to the 1960s.

SUSANNE KÖSTERING is the director of the association for museums in the state of Brandenburg in Germany. She received her Ph.D. in history from the Technical University in Berlin in 2001. Her dissertation was published as *Natur zum Anschauen. Das Naturkundemuseum des deutschen Kaiserreichs, 1871–1914* in 2003 by Böhlau Verlag. She is the author of essays on German natural history museums and the history of waste disposal in Berlin and has helped organize museum exhibitions on topics ranging from environmental history to the history of contraception.

JOACHIM RADKAU is a professor of environmental history and the history of technology at the University of Bielefeld, Germany, and one of Europe's leading environmental historians. His global environmental history *Natur und*

Macht (2000, 2nd rev. ed. Munich, 2002) is currently being translated into English. He has also published monographs on the history of nervousness (*Das Zeitalter der Nervosität*, 1998), the history of German technology (*Technik in Deutschland*, 1989), and the rise and fall of the German nuclear industry (*Aufstieg und Krise der deutschen Atomwirtschaft*, 1983). Among his current projects is a biography of Max Weber.

FRIEDEMANN SCHMOLL is a lecturer at the University of Tübingen in Germany, where he received his Ph.D. in 1994. He specializes in cultural history and historical ethnography and is the author of *Erinnerung an die Natur. Die Geschichte des Naturschutzes im deutschen Kaiserreich* (Campus, 2004) as well as numerous essays on German nature conservation, historical preservation, aesthetic reform, and monument culture in the late nineteenth and early twentieth centuries.

THADDEUS SUNSERI is an associate professor of African history at Colorado State University. He is the author of *Vilimani: Labor Migration and Rural Change in Early Colonial Tanzania* (Heinemann, 2001), "Reinterpreting a Colonial Rebellion: Forestry and Social Control in German East Africa, 1874–1915" in *Environmental History* (2003), and other essays and book chapters on the history of colonialism, rural change, and indigenous resistance in colonial Tanzania. He is currently working on a social history of forest use in Tanzania, ca. 1850–1980.

JOHN ALEXANDER WILLIAMS is an associate professor of modern European history at Bradley University in Peoria, Illinois. He received his Ph.D. from the University of Michigan in 1996. His publications include: "'The Chords of the German Soul are Tuned to Nature': The Movement to Preserve the Natural *Heimat* from the Kaiserreich to the Third Reich" in *Central European History* (1996) and "Ecstasies of the Young: Sexuality, the Youth Movement, and Moral Panic in Germany on the Eve of the First World War" in *Central European History* (2001). He is currently completing a book manuscript entitled "Turning to Nature in Germany, 1900–1939: Hiking, Nudism, and Conservation from the Second Empire to the Third Reich."

Index

ABN. *See* Working Association of German Commissioners for Conservation
accidents, 118, 130, 134n24
Adenauer, Konrad, 213
aesthetics: and animal displays, 154; and automobiles, 114; and back-to-nature forestry, 62, 65, 68, 69; and bird protection movement, 162, 163, 164, 168, 169, 170–171, 172, 175, 176, 177, 178, 179, 180; and conservation, 3, 210; Düesberg on, 68; and forests, 61; and Germanization of forest, 57; and Heimat ideal, 4; inheritance of, 6; and nature, 179; and occupied Germany, 214; and water management, 37–38; and Weimar conservation movement, 195; Worster on, 13n21
age-class forest (*Schlagwald*), 61, 67–68, 74, 83, 92
agricultural production cooperatives (Landwirtschaftliche Produktionsgenossenschaften, LPGs), 221, 224
agriculture: and bird protection movement, 162, 166, 172, 175; British schemes for, 98; and China, 29; and colonial forestry, 85; and environmentalism, 27; in Federal Republic of Germany, 218; and forests, 60, 64, 77n21; in German Democratic Republic, 220, 221, 230; in German East Africa, 81, 87, 88, 89, 91, 94, 95, 96–99, 101; and nature, 178; and nature parks, 216; and postwar conservation, 212; rationalized, 33; and water management, 33, 34, 35, 36, 37, 38, 41, 42, 43, 44, 45, 47, 48; in Weimar period, 184, 194, 195. *See also* pastoralism
Alpers, Friedrich, 73
Alps, 36
alternative medicine movement (*Naturheilbewegung*), 22
Altona Museum, 141, 145, 155n8
animals: beneficial vs. harmful, 167, 172; in British Tanganyika, 82–83; display of, 141, 142, 143, 144, 145, 147–154; and forest conservation, 25; in German Democratic Republic, 220; in German East Africa, 85, 86–87; grazing lands for, 33; indigenous species of, 5; and industry, 141; and monoculture forests, 62; moral obligation toward, 169–170; and national parks, 218; political classification of, 143–144; and postwar conservation, 212; protection of, 17, 169; taxonomic classification of, 141, 142, 143, 145, 146, 148, 151, 154; vivisection of, 17; and water management, 44
anthropocentrism, 161, 167, 190
anthropogenic terrain, 3, 5
anthropomorphism, 36, 122, 126, 151, 152, 163, 176, 178. *See also* organic machine
antimodernism, 7, 64, 65, 177, 183, 184, 214

249

Applegate, Celia, 7
Arbeitsgemeinschaft Deutscher Beauftragter für Naturschutz und Landschaftspflege. *See* Working Association of German Commissioners for Conservation
ARCHE (ecology group), 234
aristocracy: and animal display, 150; and automobiles, 116, 118; and bird protection movement, 162, 175; and Weimar conservation movement, 185. *See also* class
Arminius (known as Hermann). *See* Hermann the German
Arndt, Ernst Moritz, 57, 58, 60
Asia, 20, 24–25, 85, 86
Autobahn, 114, 117, 127–128, 129, 130, 199
automobiles, 9, 111–132

back-to-nature movement, 62–71, 72, 74, 78n22, 183, 187, 190, 192, 200
Bad Reichenhall, 23
Baedeker handbook, 122
BANL. *See* Federal Institute for Conservation
Bannwald (forest-shield), 60
Bantu-speaking peoples, 93
barren lands, 33, 34
Barton, Geoffrey, 82
Bavaria, 185–196, 198, 218
Bavarian State Advisory Committee for the Care of Nature, 187
BBU. *See* Federal Association of Citizens' Initiatives for Environmental Protection
beautiful village competition, 224
Bechstein, Johann Matthäus, 165–166
Belgium, 41
Benz, Carl, 112, 121
Bergmann, Klaus, 7
Berlepsch, Hans von, 161; *The Complete Bird Protection (Der gesamte Vogelschutz)*, 167
Berlin, 117, 143
Berlin, University of: Natural Science Museum, 141, 144; Zoological Museum, 148
Bernadotte, Duke Lennart, 217
Bierbaum, Otto Julius, 117–118, 119, 120, 122–123, 124, 126, 129, 130, 131; *Mit der Kraft,* 120
Bing-Höhle, 198
biocoenotic principles, 68, 79n45, 146
biological holism, 149
biology: and bird protection movement, 171, 172, 178; and ecology, 151; and Martin, 149; and natural history museums, 140–141, 143, 144, 146, 147, 153–154; Mammen on, 71; and Weimar conservation movement, 192. *See also* science
bird protection movement, 10, 17–18, 161–180, 221. *See also* nature conservation
Bizonia, 213
Blackbourn, David, 18
Black Forest, 56
Black Forest Society (Baden-Wurttemberg), 213
Brahms, Johannes, 58
Brandenburg, 44, 49, 50
Brandt, Willy, 216
Brehm, Christian Ludwig, 151–152, 156n32, 168, 175
Bremen, Museum for Natural Science, Ethnology, and Commerce, 144–145
Breslau, University of, Zoological Museum, 143
Buch, Richard, 146
Buchwald, Konrad, 217–218
Buddhism, 24
Bundesanstalt für Naturschutz und Landschaftspflege. *See* Federal Institute for Conservatism
Bundesverband Bürgerinitiativen Umweltschutz. *See* Federal Association of Citizens' Initiatives for Environmental Protection
Bund für Vogelschutz. *See* Society for Bird Protection
bureaucracy: in China, 20; and environmental concern, 19, 20; and

environmental politics, 29; and environmental regulation, 30; in German Democratic Republic, 228, 229; and water management, 47. *See also* civil service; government
Burma, 85

canals, 38, 41–42, 214, 226. *See also* water management
capitalism, 69, 113, 122, 183, 185, 208, 223, 226, 232. *See also* economy
Carson, Rachel, *Silent Spring*, 182n46
Central Europe, 5, 19
China, 20, 24–25, 29
Cioc, Mark, 3
civilization, 87, 88, 163, 170, 173, 178, 179, 215. *See also* culture
civil service, 41, 43, 44, 46, 47, 48, 65, 210–211. *See also* bureaucracy; commissioner; foresters; government
class: and animal displays, 154; and bird protection movement, 175; and environmental history, 5; and natural history museums, 146; and Weimar conservation movement, 186, 187, 191. *See also* aristocracy; middle class; peasantry; society
Cleghorn, Hugh, 27
close-to-nature forestry, 72, 73, 74, 75
Club of Rome, 232
Colbert, Jean-Baptiste, 20, 25
collective: and back-to-nature forestry, 68, 69; and environmental concern, 18; and Nazis, 72, 127. *See also* community; society
colonialism, 23, 25, 26–28, 81–102
Columbia River, 116
commerce: in German East Africa, 83, 86, 88, 89, 90, 91, 92, 94, 99, 101–102; and *Neue Sachlichkeit*, 125; and *Wandervogel* movement, 122; and water management, 42
commissioner: in Federal Republic of Germany, 210–212, 214, 218, 221; in German Democratic Republic, 221–223, 226–227, 229. *See also* civil service; government

community: and automobiles, 122, 131; Düesberg on, 69; and landscape, 3; national, 72–73, 191, 201; and Nazis, 72–73, 131; and Weimar conservation movement, 191, 201. *See also* collective; society
Confino, Alon, 7
conservation. *See* nature conservation
Cook's Tours, 122
cooperative, 48, 67, 68
Cottbus, 229
cotton production, 95, 96
Council of Europe, 219
countryside, 171, 184–185, 209, 210. *See also* landscape
Crosby, Alfred W., *Ecological Imperialism*, 26, 27
Crosse-Upcott, A.R.W., 94
cultural landscape (*Kulturlandschaft*), 3, 4, 6, 7, 9, 11, 18, 141, 168, 171, 177, 178
culture: and accidents, 130; aeronautic, 118, 126; and animal displays, 141; and anthropomorphization of rivers, 36; and automobile, 119, 122; Bierbaum on, 119, 120, 122; and bird protection movement, 163, 173–174, 175, 176, 177, 178; equestrian, 126; and forests, 55, 56–61, 64; in German Democratic Republic, 224; and Heimat movement, 120; and industry, 123; and landscape tradition, 2; and natural history museums, 142; and nature, 140, 143, 147–148, 149, 161, 162, 169, 171; and Nazis, 131; and occupied Germany, 214; and postwar forest conservation, 213; Rathenau on, 123; reactionary tendencies in, 7; as rooted in soil, 4; shaping power of, 5; and walking, 122; and Weimar conservation movement, 186, 191, 192

Dahl, Friedrich, 146
Daimler, Gottlieb, 112
DAL. *See* German Academy of Agricultural Sciences

Dar es Salaam, 86, 90
Darmstadt: Grand-Ducal State Museum, later Hessian State Museum, 141, 142, 143
Darwin, Charles, 141, 149
Dauerwald (journal), 75
Dauerwald (sustainable forestry). *See under* forest(s)
democracy, 7, 69, 74, 186, 197, 208, 233. *See also* government; politics
Denmark, 20
dermoplasty, 147, 149, 151, 152, 154, 157n40. *See also* animals: display of
Dernburg, Bernhard, 98
Deutsche Akademie der Landwirtschaftswissenschaften. *See* German Academy of Agricultural Sciences
Deutscher Bund für Heimatschutz. *See* German League for Heimat Protection
Deutscher Kulturbund. *See* League of Culture for the Democratic Renewal of Germany
Deutscher Naturschutzring. *See* German Conservation Ring
Deutscher Rat für Landespflege. *See* German Council for Landscape Planning
Deutscher Verein zum Schutze der Vogelwelt. *See* German Organization for the Protection of the Bird Kingdom
Dieck, 40
disease. *See* health
Ditt, Karl, 184
diversity: and age-class forestry, 83; and back-to-nature forestry, 64, 67; and cultural landscape, 3; of forests, 22, 26, 71; and monoculture forests, 62; and sustainable forestry, 71; and water management, 36; and Weimar conservation movement, 193
division of labor, by gender, 152, 153
DKB. *See* League of Culture for the Democratic Renewal of Germany
DNR. *See* German Conservation Ring

Domaszewski, Victor von, 36
Dominick, Raymond, 184, 208
Dortmund, 117
drainage, 34, 35, 37, 38, 41, 43, 45, 46, 49. *See also* water management
Dresden, 143
drivers, 114, 116, 119, 124, 126. *See also* automobiles
Droste-Hülshoff, Ferdinand von, 168–169
Düesberg, Rudolf, *The Forest as Educator (Der Wald als Erzieher)*, 67–68, 69

Earth Summit, 23
East Germany. *See* German Democratic Republic
Eckert, Otto, 92
ecology: and back-to-nature forestry, 64, 67, 74; and biology, 151; and bird protection movement, 162, 163, 164, 168, 171, 176, 177, 178, 179, 180; and China, 29; in Federal Republic of Germany, 218; of forests, 165; and industrialism, 28; and monoculture forests, 62; and natural history museums, 140–141, 145–146; and Nazis, 72, 73; and sustainable forestry, 71; and sustainable vs. scientific forestry, 55–56; and taxidermy, 147
economy: and back-to-nature forestry, 64, 74; and back-to-nature movements, 183; and bird protection movement, 169, 175; centralized, 18; and colonial forestry, 85, 86; and conservation, 207, 208; in Federal Republic of Germany, 207, 208; and forest management, 83, 84; and Friends of Nature, 189; of German Democratic Republic, 207, 208, 222, 225, 226, 227, 229, 230, 231, 233, 234; in German East Africa, 86, 88, 89, 90, 91, 92, 93, 94, 96, 97, 99, 100, 101–102; and landscape tradition, 2; liberal approach to, 33, 64; and monoculture forests, 62; and national parks, 218, 219; and nature, 179; and nature parks, 216; and Nazi

conservationism, 199; of occupied Germany, 214, 215; and postwar conservation, 212, 232; and sustainability, 23, 71, 72, 73; and water management, 33, 37, 39, 41, 42, 43, 44, 45, 46, 48, 49; and Weimar conservation movement, 186, 193, 194. See also capitalism; socialism

ecosystem, 79n55; in Federal Republic of Germany, 218; human modification of, 5; and sustainability, 23, 74. See also ecology; environment; nature

education: and affinity for forest, 55, 67; and back-to-nature forestry, 64, 67, 70; and bird protection movement, 164, 175; in Federal Republic of Germany, 212, 219; in Hauser, 124; and middle-class conservation movement, 184; in scientific forestry, 61; through museums, 140, 144; and Weimar conservation movement, 185, 187, 188, 189, 197, 200

Eichendorff, Joseph von, 58
Elbe River, 228, 232
elites, 19, 47, 49, 85, 188, 191
Elvin, Mark, 29
emotion: and animal displays, 151, 152; and bird protection movement, 162, 164, 166, 172, 175, 176, 180; and Heimat, 189; and occupied Germany, 214

Ems River, 44
Engler, Eduard, 119
Enlightenment, 7, 57, 62, 165, 209
environment: and Autobahn, 128, 129; and automobiles, 114, 115, 117, 128; and bird protection movement, 172; and bureaucratic regulation, 30; and developing world, 23; in East Germany, 231; in Federal Republic of Germany, 217, 218, 220, 232; and forests, 83–84; French indifference to, 18; in German Democratic Republic, 232; in German East Africa, 82, 83, 88, 92; in Hauser, 128, 129; and history, 6; and landscape tradition, 2; and museums, 140; and national identity, 17, 18, 19, 20, 21, 23, 25, 29–30; and natural history museums, 142; Prussian-German policy toward, 29; stewardship of, 3; and sustainability, 23; and technology, 114; and water management, 43, 50; and Weimar conservation movement, 201. See also forest(s); landscape; nature; nature conservation

Ernst Thälman Pioneers, 224
Eternal Forest, The (Der ewige Wald) (film), 22
ethnicity/ethnic group, 2, 22. See also race
Europe, 18–19, 20, 23, 41, 83
European Conservation Year (ECY), 219
Evelyn, John, 25
evolution, 141, 142

family: and animal displays, 143, 147, 149–151, 152–153, 154, 157nn38, 40; and bird protection movement, 163; and Weimar conservation movement, 192

farming. See agriculture
fascism, 7, 113. See also Nazis
fashion, 164, 172, 173–174
fauna. See animals
feather industry, 164
Federal Association of Citizens' Initiatives for Environmental Protection (Bundesverband Bürgerinitiativen Umweltschutz, BBU), 233
Federal Forest Law (1975), 74
Federal Institute for Conservation (Bundesanstalt für Naturschutz und Landschaftspflege, BANL), 211, 216, 219, 221
Federal Republic of Germany, 10, 66, 74, 207–220; Basic Law, 217; Federal Environmental Office, 232; Federal Ministry of the Interior, 233; and German Democratic Republic, 221, 223–224, 228, 229, 230, 232; Ministry of Health Affairs, 217, 219

Fischer, Eugen, 192
Flink, James J., 111
Floericke, Curt, 163
flood protection, 4, 34, 35, 36, 42, 45. See also water management
food: and bird protection movement, 164, 172, 174–176; in German Democratic Republic, 220
foresters, 55, 61, 62, 63–64, 66–67, 102
forest reserve, 85, 86, 93, 96, 97, 98, 100, 102, 107n103. See also nature reserve
forest(s): affinity for, 55; age-class, 61, 67–68, 74, 83, 92; and agriculture, 60, 64, 77n21; back-to-nature, 63–64, 71, 74, 75, 78n22; and back-to-nature movement, 62–71, 72, 74, 75; and bird protection movement, 162, 165, 166, 168, 172, 175; Buchwald on, 218; in China and Japan, 20; close-to-nature, 72, 73, 74, 75; collective past in, 66–67; and colonialism, 25, 28; and common good, 84; as commons, 85; communally owned, 62; for construction, 88; and culture, 55, 56–61, 64; cutting of, 71, 72, 78n22; destruction of, 26, 92, 96, 97; diversity of, 22, 26, 62, 64, 67, 71, 83; ecology of, 165; and economy, 83, 84; and education, 55, 61, 64, 67, 70; and ethnic groups, 22; in Europe, 20; in Europe vs. Asia, 24–25; in France, 20; in German Democratic Republic, 222, 224, 230; as Germandom symbol, 55–75; in German East Africa, 81–102; Germanization of, 57–61; and *Gruppen-Plenterwald*, 68, 79n45, 146; hunting in, 25; and industry, 60, 62, 64, 65, 67, 83, 86, 90, 213; institutional relationship to, 20–21, 24–26; and landscape tradition, 2; and law, 19, 74, 84–85, 86–87, 88–89, 95, 96, 97, 99–100; mangrove, 83, 86, 89, 90, 91–92, 99, 102; medieval, 24, 77n17; and miombo woodland, 86, 87, 89, 93, 102; and *Mischwald*, 63; monoculture, 62, 63, 68, 75, 78n22, 165; monotonization of, 171; montane, 82, 102; and *Nachhaltigkeit*, 23; and nation, 68, 69, 72; and national identity, 8–9, 59, 60; and natural environment, 83–84; and nature, 56, 178; and Nazis, 55–56, 65, 69, 70, 72–74, 75; people of, 21–22; and postwar conservation, 212, 213, 214, 215, 218, 222, 224, 230; private, 61, 62; and professorships, 22; and race, 60, 61; and reforestation movement, 22; Riehl on, 21–22; and Romanticism, 22; saline, 23, 24, 26, 31n17; and scientific forestry, 55, 61, 62, 63, 65, 67–68, 71, 72, 77n16, 81–82, 83, 85, 87, 88, 99, 101–102; as shield, 60; and state, 61–62, 65, 68, 83, 84, 85, 86; sustainable, 23, 24, 55–56, 71–75, 78n22; in Tacitus, 57–58; and technology, 56; timber from, 61–62; and urbanization, 60; as *Urwälder*, 87; völkisch perspective on, 55, 56, 61, 63, 65, 66–67, 70, 74; and Weimar conservation movement, 195; wildness vs. order in, 56, 57
fowlers, 167–168, 175. See also hunting
France: automobiles in, 112, 116, 130; and bird protection movement, 17, 173–174; forests in, 20; and garden theory, 21; and Germanization of forest, 57, 58, 60, 61; indifference of, 18; institutions of, 20; occupation of Rhineland by, 190; and postwar forest conservation, 213; shipbuilding in, 25; and water management, 41
Franco-Prussian War, 130
Frankfurt am Main, 142; Museum of the Senckenberg Nature Research Society, 145
Frederick the Great, 45, 65
freedom, 59, 60, 125
Free German Youth, 224
Friedrich, Caspar David, 57
Friends of Nature (Naturfreunde), 162, 189, 223
Frömbling, Friedrich Wilhelm, 60, 61
Fuchs, Carl, 195

garden theory, 21
gender: and automobiles, 123, 124; in Brehm, 152; and division of labor, 152, 153; and environmental history, 5; and natural history museums, 143, 147, 150, 151, 152–153, 156n25; and nature, 178
Genscher, Hans-Dietrich, 219
geography: and natural history museums, 141, 143, 144, 150, 151; and water management, 36. *See also* landscape
German Academy of Agricultural Sciences (Deutsche Akademie der Landwirtschaftswissenschaften, DAL), 221, 229
German Conservation Ring (Deutscher Naturschutzring, DNR), 212, 213, 216, 218, 219
German Council for Landscape Planning (Deutscher Rat für Landespflege), 237n36
German Democratic Republic, 10, 74, 207, 208, 220–232; and Federal Republic of Germany, 221, 223–224, 228, 229, 230, 232; Ministry of Agriculture, 221, 222, 223, 226–227, 229
German East Africa, 81–102
German East Africa Corporation (DOAG), 90
German Horticulture Society, Green Charter of Mainau, 217
German League for Heimat Protection (Deutscher Bund für Heimatschutz), 186, 196
German Museum Conference, 145
German Organization for the Protection of the Bird Kingdom (Deutscher Verein zum Schutze der Vogelwelt), 164
German Reich, 41, 140, 143, 147
Germany: post–World War II era, 207–234. *See also* Federal Republic of Germany; German Democratic Republic; German Reich; Nazis; Third Reich; Weimar period; Wilhelmine era

Gesellschaft für Natur und Umwelt. *See* Society for Nature and the Environment
Giannoni, Karl, 191–192
Gilsenbach, Reimar, 227–228
Gleichschaltung (synchronization), 184, 196
Gloger, Constantin Wilhelm Lambert, 166
GNU. *See* Society for Nature and the Environment
Göring, Hermann, 72, 73, 199, 210, 226
Götzen, Gustav Adolf Graf von, 88, 91
government: authoritarian, 74; and bird protection movement, 169; centralized, 19, 20, 77n16, 189, 196, 199, 208, 220, 221, 227; in China, 20; decentralized, 19; and environmental concern, 18, 19, 20, 24–26, 28, 29, 30; in Federal Republic of Germany, 210–212, 214, 216–217, 218, 219–220, 233; in German Democratic Republic, 208, 220, 221, 222, 223, 226–227, 228, 229, 231, 233; in German East Africa, 90, 91, 92, 93, 97, 98, 99; and Heimat, 188–189; local, 20; polycentric, 19, 20, 21; and postwar conservation, 212; Prussian-German, 19; and scientific forestry, 77n16; and water management, 41, 42–43, 46, 47, 48, 49. *See also* politics; state
Grass, Karl, 90
Great Britain, 19; automobiles in, 112, 116, 121, 127; and bird protection movement, 17; colonies of, 28, 81, 82, 85, 98, 102; conservation movement in, 184; shipbuilding in, 25; and water management, 41
Great Depression, 125
Green Charter of Mainau (German Horticulture Society), 217, 237n36
Green movement, 66
Green Party, 2–3, 10, 234
Grimm, Jacob, 58, 59
Grimm, Wilhelm, 58, 59
Grove, Richard H., 26, 27, 31n29

Gruppen-Plenterwald (biocoenotic principles), 68, 79n45, 146
Grzimek, Bernhard, 218
Guenther, Konrad, 193–194, 197; *Nature Conservation*, 187

Haeckel, Ernst, 142
Hähnle, Lisa, 164
Halle, 229
Hamburg, Natural History Museum, 142, 153
Hanover School Museum, 157n38
harbors, 35
Hartert, Ernst, 162
Hartig's forestry rules, 21
Hassinger, Heinrich, 190
Haßler, Friedrich, 195
Hauser, Heinrich, 123–124, 125–127, 128–129, 130, 131–132; *Brackwasser*, 124; *Friede mit Maschinen*, 124
Haushofer, Karl, 3
Havel River, 47, 49
health: and alternative medicine movement, 22; and back-to-nature forestry, 63, 64; Buchwald on, 217–218; and colonial forestry, 85; Düesberg on, 68; in German East Africa, 87, 95–96; and postwar conservation, 208, 215, 225, 228, 233; and water management, 37, 44, 46
heathland, 33, 38, 42, 44, 47
hedges, 5, 168, 171, 172, 212, 221, 230
Heimat discourse: and animal displays, 145, 154; and Autobahn, 128; and back-to-nature forestry, 68; in Bierbaum, 120; and ecology, 146; and German East Africa, 82, 83; and landscape, 3–4; and modernity, 7; and natural history museums, 143, 144; and nature, 4; and Nazis, 197–198; as reactionary, 7; and urban entity, 146; and Weimar conservation movement, 184, 185, 186, 188–190, 191, 192–193, 195–196, 200
Heimat League, 223
Heimat protection movement: and anthropogenic terrain, 3; and bird protection movement, 170, 171, 176–177, 178; and forests, 58, 64, 65; and landscape, 177; and modernity, 119–120; and natural history museums, 146; and nature, 193; organizations, 196; in postwar Germany, 214, 223–224; and regionalism, 189, 205n57; rhetoric of, 178; and Weimar conservation movement, 185, 186
Helbok, Anton, 192
Hennicke, Carl R., *Book of Bird Protection (Vogelschutzbuch)*, 173
Herder, Johann Gottfried, 2, 3, 58
Herf, Jeffrey, 113, 202n10
Hermand, Jost, 7
Hermann the German (Arminius), 58, 59, 60, 76n4; Hermann's Battle (*Hermannsschlacht*), 58, 59, 60
highways, 214, 215, 216. See also automobiles
Hirschfeld, Christian Cay Lorenz, 21
Hitler, Adolf, 22, 117, 128, 129–130, 196
Hofstadter, Richard, 2
Hölderlin, Friedrich, 57
Holy Roman Empire, 19, 21
homeland. See Heimat discourse; Heimat protection movement
Honecker, Eric, 231
Huch, Ricarda, 17
Humboldt, Alexander von, 27, 149, 150
Humboldt University, Institute of Horticulture and Resource Planning, 225
hunting: and bird protection movement, 17–18, 167–168, 172, 174–176; in European forests, 25; in German Democratic Republic, 220; in German East Africa, 81, 86–87, 95, 100; and Göring, 199; and national parks, 218; and water management, 44
hydraulic engineers, 33, 34, 37, 38–39, 40, 41, 44, 45, 47, 48, 49, 50

ILN. *See* Institute for Land Research and Conservation

Imperial Bird Protection Law, 172, 176
Imperial Germany, 7, 117
import/export trade, 86, 88, 89, 90, 91, 92, 99. *See also* commerce; economy
India, 24, 25, 85
Indian Ocean, 89, 90
individual, 19, 69, 72, 114, 115, 122, 127. *See also* liberalism
Indonesia, 85
industry/industrialism: and animal displays, 153, 154; automobile, 115, 116, 127; and back-to-nature movements, 64, 65, 67, 183; and bird protection movement, 163, 171, 176, 178, 179; Buchwald on, 218; and countryside preservation, 185; and culture, 123; and ecology, 28; in Federal Republic of Germany, 208, 213, 215, 218; and forests, 60, 62, 64, 65, 67, 83, 86, 90, 213; French, 213; and Friends of Nature, 189; in German Democratic Republic, 208, 221, 227, 228, 229, 230, 233; and German East Africa, 86, 90; in Hauser, 124, 125; and Heimat movement, 120; and Martin, 148; Marx on, 113; and modernity, 119; and natural history museums, 140–141, 147; and nature, 123, 141, 178; Rathenau on, 123; reactions against, 7; and state supervision of forests, 62; study of, 6; and water management, 36, 39, 41–42; and Weimar conservation movement, 184, 185, 186, 187, 191, 194, 195, 196, 201
Inner-German Frontier Commission, 232
Institute for Land Research and Conservation (Institut für Landesforschung und Naturschutz, ILN), 221, 222, 226, 227, 228, 229, 230
Institute of Horticulture and Resource Planning, Humboldt University, 225
Institut für Landesforschung und Naturschutz. *See* Institute for Land Research and Conservation institutions, 20–21, 23, 24–26, 28, 140, 162, 166, 176; defined, 18. *See also* government
International Conventions of Farmers and Foresters (Vienna, 1873), 166
Internationaler Frauenbund für Vogelschutz. *See* International Women's Association for the Protection of Birds
International Union for the Conservation of Nature (IUCN), 226
International Women's Association for the Protection of Birds (Internationaler Frauenbund für Vogelschutz), 173
irrigation, 34, 37, 39, 41, 43, 45. *See also* water management
Italians, 17, 176
IUCN. *See* International Union for the Conservation of Nature

Jackson, John Brinckerhoff, 134n30
Jaeger, Gustav, 149
Jäger, Helmut, 23
Jahn, Friedrich Ludwig, 57, 58, 60
Japan, 20
Jews, 68, 69–70, 122, 191, 192, 197, 198. *See also* race
John, Owen, 127
Jones, Eric L., 28
Jünger, Ernst, 124–125, 127, 129
Juvenal, 76n4

Keaton, Buster, 127
Kersting, Georg Friedrich, 57
Keudell, Walter Graf von, 72, 73, 74, 200
Khutu-Khutu forest reserve, 97, 98
Kiel, University of, Zoological Museum, 141
Kilwa region, 81, 95, 96, 98
Kipo forest, 100
Klages, Ludwig, *Mensch und Erde*, 173
Klamroth, Martin, 93, 95
Kleist, Heinrich von, 57, 58
Klemperer, Victor, 133n8

Klenke, Dietmar, 199
Klopstock, Friedrich Gottlieb, 58
Klose, Hans, 196, 210, 211, 212
Knaut, Andreas, 7
Kneschke, Karl, 223
Koehler, inspector, 47
Kohrener Land, 228
Kragh, Gert, 209
Kretschmann, Kurt, 226
Krieg, Hans, 216–217
Kulturbund zur demokratischen Erneuerung. See League of Culture for the Democratic Renewal of Germany
Kulturland/Kulturmenschen (cultivated land/cultured peoples), 87
Küster, Hansjörg, 5

labor, 121, 186, 226. See also workers
Lagarde, Paul de, 66
landowners, 47, 48, 208
landscape: and affinity for forest, 61; and animal displays, 141, 145; anthropogenic, 3; and automobiles, 114, 119, 120, 123; in Bierbaum, 120, 123; and bird protection movement, 171, 177; in German Democratic Republic, 225, 229; in German East Africa, 102; and Heimat movement, 120, 177; idealized, 9; as *Kulturland*, 87; and Nazis, 2, 3, 198–199, 210; pristine, 214; protected, 216, 225; and regionalism, 184; and tourism, 178; tradition of, 2; and water management, 34, 38; as way of seeing, 5; and Weimar period, 186, 193–196, 198–199, 201. See also countryside; geography
landscape planning (*Landschaftspflege*), 10, 184–185, 194, 205n57, 209, 214, 217
landscape view (*Landschaftsanschauung*), 6
land use, 210; and bird protection movement, 172; and common property, 33, 34; in Federal Republic of Germany, 233; in German Democratic Republic, 220, 231; and nature parks, 216; and Nazis, 198;

and occupied Germany, 215; planning for, 3; and postwar conservation, 212; and private property, 33–34; and reclamation, 42; and water management, 50
Landwirtschaftliche Produktionsgenossenschaften. See agricultural production cooperatives
Langbehn, Julius, 66; *Rembrandt as Educator*, 67
law: and bird protection movement, 169, 171–172; and customs and interests, 26; decentralized regulation by, 19; and elites, 19; and environmental concern, 18; European, 19; in Federal Republic of Germany, 219, 225; and forests, 19, 74, 84–85, 86–87, 88–89, 95, 96, 97, 99–100; in German Democratic Republic, 221, 225, 230; in German East Africa, 81, 86–87, 88–89, 95, 96, 97, 98, 99–100; and individual, 19; Nazi, 199–200, 210, 214, 221, 225, 232; and occupied Germany, 210, 214; and postwar conservation, 212; and water management, 33–34, 38; and Weimar conservation movement, 185, 187–188, 197
League of Culture for the Democratic Renewal of Germany (Kulturbund zur demokratischen Erneuerung Deutschlands/Deutscher Kulturbund, DKB), 223, 224, 226, 227, 228, 234
Lebensraum (living space) 2, 217
Lebensreform (life reform), 64, 183
Leed, Eric, 120
Leipzig, 229; Local Natural History Museum, 141, 146
Lekan, Thomas, 184
Lenz, Harald Othmar, 166
Leopold Friedrich Franz, prince of Anhalt-Dessau, 21
Lethen, Helmut, 125
liberalism, 7, 113, 115, 163, 183. See also individual
Lichtenstein, Hinrich, 148
Liebe, Karl Theodor, 169–170

Liebert, Eduard von, 97
Liebieg, Theodor von, 119, 126
Liebig, Justus von, 29
life reform (*Lebensreform*), 64, 183
Lindner, Werner, 204n37
Linnaean classification system, 141, 143, 146. *See also* animals; biology; science
Liwale Forest Reserve, 93
local government, 20
local identity, 3, 146
local self-sufficiency, 23
Lomborg, Bjorn, 115
Löns, Hermann, 188
Lortzing, Albert, 58
Louis XIV, king of France, 20
lowlands, 33, 34, 42
LPGs. *See* agricultural production cooperatives
Lüneburger Heath, 5, 226

machine, 123, 124, 125, 126, 127, 129. *See also* automobiles; industry/industrialism; technology
McClelland, Charles, 41
McNeill, J. R., 1, 115
Maji Maji rebellion, 95, 96, 99, 100
Mammen, Franz Graf von, *Der Wald als Erzieher*, 70–71
mangrove forests, 83, 86, 89, 90, 91–92, 99, 102
Marinetti, F. T., 118
Marshall, Robert, 3
Martin, Philipp Leopold, 147–151, 152, 154, 156n25; *Natural History in Practice*, 147–148, 149
Marx, Karl, 59
Marx, Leo, 113, 114, 115
Marxism, 186, 189
mass society, 186, 187, 188, 197
materialism, 38, 165, 214, 232
Matumbi villagers, 96
meadowland, 43
melioration projects, 42–43, 44, 47, 222
Menzies, Nicholas K., 24
Meusel, Hermann, 221, 222, 225, 226
Meyer, Konrad, 210
Middle Ages, 23, 24, 35, 77n17, 218

middle class: and animal displays, 141, 147, 150, 153, 154; and automobiles, 116, 117; and back-to-nature forestry, 64; and bird protection movement, 162, 163, 164, 175, 179; and conservatism of foresters, 66; in German Democratic Republic, 224, 226, 227; in Hauser, 124; and Nazism, 196; and occupied Germany, 214; and postwar conservation, 212; and tourism, 122; and water management, 33, 45, 47, 50; in Weimar period, 184, 185, 186, 189–190, 196, 200. *See also* class; society
military, 60, 61, 66, 117, 130, 199, 200, 204, 220. *See also* World War I; World War II
mining, 24, 25, 40, 224, 227, 229, 230
miombo woodland, 86, 87, 89, 93, 102
Mischwald (mixed forest), 63
Möbius, Karl, 142, 146
modernism, 7, 113; and Autobahn, 199; and countryside preservation, 184–185; impulses against, 64, 65, 177, 183, 184, 214; and Nazis, 184, 202n10; and *Neue Sachlichkeit*, 125; reactionary, 124–125; and Weimar conservation movement, 194, 201
modernity: and automobile culture, 119; and back-to-nature forestry, 64, 65, 67; Bierbaum on, 123; and bird protection movement, 179; dislocations of, 3; in Hauser, 124; and labor, 121; and Martin, 148; Marx on, 113; and museums, 140; and nature, 7, 119–120, 178, 179; and Nazi back-to-nature ideology, 183; in occupied Germany, 214; Rathenau on, 123; reactions against, 7; and *Wandervogel* movement, 122; and water management, 40; in Weimar period thought, 184, 191, 196. *See also* progress
Mohoro, 90, 99
Möller, Alfred, 71, 79n55
monoculture, 62, 63, 68, 74, 78n22n, 165
moors, 38, 42, 44, 47

morality: and animal displays, 151–152; and anthropomorphized nature, 36; and bird protection movement, 162, 164, 165, 169, 170, 172, 173, 174, 176, 178, 179, 180; and Germanization of forest, 57; and nature, 170; and obligation toward animals, 169–170; and occupied Germany, 214
Moselle River, 213, 226
Mosse, George, 7
movement, pleasure of, 120, 121
Mpanga Forest Reserve, 96
Muir, John, 3
Munich, 117
Munich Soviet Republic, 187
Münster, 142, 143
Müritz, Lake, 228
museums, 9, 140–154, 177. *See also* institutions
Museums Conference (Mannheim, 1903), 152

Nader, Ralph, 2
Naminangu, 92
Napoleon I, emperor of France, 57, 58
nation: and automobiles, 131; and Düesberg, 69; and environmental history, 5; and forests, 68, 69, 72; living space for, 3; socially homogeneous, 185; unity of, 59, 67, 127, 199; in Weimar period thought, 184, 191
National Association of German Foresters, 66
national community, 72–73, 191, 201
national identity: and affinity for forest, 55, 61; and automobiles, 126; and back-to-nature forestry, 64; and bird protection movement, 173–174, 175–176, 180; and environmental concern, 17, 18, 19, 20, 21, 23, 25, 29–30; and environmental politics, 29; and forests, 8–9, 57, 59, 60, 64; and Germanization of forest, 57; and landscape tradition, 2; and nature, 8, 17, 22, 120, 170, 213; and occupied Germany, 214; and order, 56; and Weimar conservation movement, 192

nationalism: and bird protection movement, 178; and forests, 58; inheritance of, 6; and landscape planning, 209; and natural history museums, 143; and nature, 22; and Nazis, 197; and Weimar conservation movement, 188, 191, 192, 200–201
national park, 218–219, 228, 233
natural history museums, 9, 140–154
natural monument, 4, 178, 186, 187, 195, 209, 210, 216
nature: and aesthetics, 179; and agriculture, 178; alienation from, 154; as ambiguous concept, 184; and animal displays, 141; anthropomorphized, 36, 122, 126, 151, 152, 163, 176, 178; and Autobahn, 128; and automobiles, 114–123; and back-to-nature forestry, 64, 65; balance in, 165; in Bierbaum, 120, 123; and bird protection movement, 166, 177; and Black Forest, 56; changing definitions of, 5; and civilization, 170, 179; and culture, 140, 143, 147–148, 149, 161, 162, 169, 171; and dermoplasty, 152; design in, 170; domestication of, 154; ecologically defined, 146; and economy, 179; in Federal Republic of Germany, 215, 232; and forests, 56, 64, 65, 178; and gender, 178; in German Democratic Republic, 224, 227, 228, 233, 234; German feeling for, 17–18; as harmonious order, 169; in Hauser, 124; and health, 22; as Heimat, 190; and Heimat movement, 119–120; homogeneity vs. diversity in, 193–194; human harmony with, 4; human impact on, 3, 171; human use of, 23, 37; idealization of, 7, 179; and ideology, 21–22; improvement on, 4; and industry, 123, 178; instrumental approach to, 161, 162, 165, 166, 170, 172, 178, 179; integrity of, 169, 178; intrinsic value of, 178; and Jünger, 125; and landscape planning, 209; liberation from, 36–37; in Martin, 149; and

modernity, 7, 119–120, 178, 179; as moralizing, 170; and national identity, 8, 17, 22, 120, 170, 213; and nationalism, 22; and natural history museums, 147; and nature parks, 216; and Nazis, 131, 201; and occupied Germany, 214; order imposed on, 37; paternalism toward, 165–166; as pristine, 170; protection of, 38; as pseudo-mechanical, 161; and race, 120; as self-regulating, 170; and socialism, 226, 234; and society, 18–19, 190, 193–194; stewardship of, 165–166; and sustainability, 22–24; and technology, 113, 179, 201; vanishing world of, 154; walking through, 122; and water management, 45; in Weimar period thought, 184, 190, 193, 201; wildness vs. order in, 56, 57. *See also* ecology; ecosystem; environment

Nature and Heimat Friends (NHF), 221, 223, 224, 225, 227, 229, 230, 234

nature conservation, 3, 10; and bird protection movement, 162, 170, 171, 172, 176–177, 178; in German East Africa, 82, 86, 101; and landscape planning, 209; in post–World War II era, 207–234; in Weimar era, 183–201. *See also* bird protection movement; forest reserve; national park; nature park; nature reserve

nature park, 215–216, 229, 233

Nature Park Society (Verein Naturschutzpark, VNP), 215–216

nature reserve, 164, 188, 196, 210, 216, 220, 222. *See also* forest reserve

Naturfreunde. *See* Friends of Nature

naturgemäße Waldwirtschaft (close-to-nature forestry), 73, 74. *See also under* forest(s)

naturnahe Waldwirtschaft (close-to-nature forestry), 72

Natur und Landschaft (Nature and Landscape) (BANL journal), 211

Naumann, Johann Andreas, 168, 175

Naumann, Johann Friedrich, 165, 168

Nazis: and Autobahn, 128, 129; and automobiles, 117, 131, 132; and back-to-nature forestry, 65, 70; and back-to-nature movements, 183; civil service ties to, 210–211; and collectivity, 127; and community, 131; and culture, 131; and ecology, 72, 73; and forest-state analogy, 69; and *Gleichschaltung*, 184, 196; in Hauser, 127; and Heimat, 197–198; and Huch, 17; and identity based on nature, 22; inheritance from, 6; and Jews, 197, 198; and landscape, 2, 3, 198–199, 210; and machines, 127; and Mammen, 70; and mass society, 197; and modernism, 184, 202n10; and modernity, 7; and national community, 72–73; and nationalism, 197; and national unity, 127; and nature, 22, 131, 201; and nature preservation, 210; and occupied Germany, 214; and race, 6, 127, 131, 197; and socialism, 197; and society, 70–71, 198, 201; and state, 199–200, 201; and sustainable forestry, 55–56, 72–74, 75; and völkisch perspective, 72, 183; and Weimar conservation movement, 184, 185, 196–200, 201. *See also* Third Reich

Netherlands, 20; colonies of, 81, 85, 90

Neue Sachlichkeit (New Objectivity), 125

Neumann, Roderick, 82

New Brandenburg, 224

newness, 143

NHF. *See* Nature and Heimat Friends

nomadism, 68, 69

Norway spruce, 62

nostalgia, 56, 184

Novalis, 57

nuclear technology, 18

Nye, David, 7

objectivity, 125, 152

order, 37, 56, 57, 66, 169

organic machine, 9, 115, 116, 118, 121, 125, 128, 131
ornithology, 167–168. *See also* bird protection movement
Ostpolitik, 219
Otto, Heinrich, 97–98

Paris International Agreement on the Protection of Birds Useful to Agriculture, 166
particularism (*Zersplitterung*), 21, 200
pastoralism, 3, 7, 81, 85, 86, 87, 101, 113, 185. *See also* agriculture
peasantry: and bird protection movement, 175; and colonial forestry, 85; and forests, 26; in German East Africa, 81, 86, 87, 93, 95, 96–97, 98–100, 101; and water management, 44, 48, 49. *See also* class; society
Pechmann, H. von, 37
People's League for Nature Conservation (Volksbund für Naturschutz), 186, 188
Pfeil, Wilhelm, 21
Pinchot, Gifford, 2
Pirna, 224
planning: centralized, 208; in Federal Republic of Germany, 233; in German Democratic Republic, 225, 226, 231, 234; landscape, 194, 205n57, 209, 214, 217; land use, 3; and occupied Germany, 214; regional conservation, 208, 219; spatial, 216
Plattdeutsch writers, 188
Pleschberger, Werner, 66
Ploucquet, Hermann, 149
Pniower, Georg, 225, 226
Polish people, 68
politics: and animal classification, 143–144; and animal displays, 145; and back-to-nature forestry, 64, 67; and back-to-nature movements, 183; and bird protection movement, 164, 177; and changing definitions of nature, 5; environmental, 29; in Federal Republic of Germany, 219, 234; and forest identity, 60, 61; in German Democratic Republic, 226, 227; and Germanization of forest, 57; and Heimat, 188–189; and hydraulic engineers, 39; and landscape tradition, 2–3; and occupied Germany, 214; Ratzel on, 60; and water management, 39, 45; and Weimar conservation movement, 192, 194, 200. *See also* government; state
pollution: and automobiles, 115; and back-to-nature forestry, 75; in Federal Republic of Germany, 215, 219, 220; in German Democratic Republic, 225, 228, 229, 230, 231; and nature parks, 216; and postwar conservation, 212; and postwar forest conservation, 213; study of, 6; in United States vs. in Germany, 29–30; and urban health movement, 208; and Weimar conservation movement, 185
polycracy, 19, 20, 21
Popp, Max, 38
population, 62, 85, 87, 96, 215, 216
populism, 122
poverty, 33, 37, 44, 45
private property, 33–34, 35, 42, 45, 64, 185, 187, 199, 208, 220
progress, 43, 148, 177, 184, 194, 207, 225. *See also* modernity
Protective Association of the German Forest (Schutzgemeinschaft Deutscher Wald, SDW), 213
protest movements, 6, 18, 213–214
Prussia, 19, 21, 33, 39, 41, 42, 43, 50, 143, 186, 211

Rabinbach, Anson, 121
race: and automobiles, 131; and back-to-nature forestry, 68, 69–70; Düesberg on, 68; eugenic approach to, 69; and forests, 60, 61; and German East Africa, 86, 96; and landscape tradition, 2; Mammen on, 71; and nature, 120; and Nazis, 6, 127, 131, 197; and *Wandervogel* movement, 122; in Weimar period thought, 184,

191, 192, 196, 200–201. See also
    ethnicity/ethnic group; Jews
Rackham, Oliver, 26
Radkau, Joachim, 6, 49
railroads, 40, 41, 43, 86, 93, 98, 117,
    119, 120, 122, 123, 214
Rathenau, Walter, 123
Ratzeburg, Julius Theodor, 166
Ratzel, Friedrich, 3, 119; *Anthropogeographie*,
    60
recreation, 146, 216, 222, 225, 228–229,
    233
regionalism: and conservation planning,
    208, 219; and cultural landscape, 3; in
    Federal Republic of Germany, 233; and
    Heimat, 188–189; and Heimat
    movement, 120; and Heimat organizations, 196; and Heimat protection,
    205n57; and landscape, 184; and
    natural history museums, 143, 144,
    146; and occupied Germany, 208, 214,
    219, 221; in Weimar era, 184, 200
Reich Conservation Law
    (Reichsnaturschutzgesetz, RNG),
    199–200, 210, 214, 221, 225, 232
Reichelt, Hans, 226
Reich League for Nationhood and
    Homeland (Reichsbund für Volkstum
    und Heimat), 196, 200
Reichsnaturschutzgesetz. See Reich
    Conservation Law
Reinhardt, Ludwig, 175
religion, African, 93–94, 95, 96
revolution, 84, 186, 187, 200
Rheinland Heimat League, 188, 190
Rhine River, 35–36, 58
Riehl, Wilhelm Heinrich, 3, 5–6, 66,
    119, 120, 170; *Natural History of the
    German People*, 21–22, 59, 149
river(s): anthropomorphized, 36; in
    Federal Republic of Germany, 232;
    in German Democratic Republic,
    230, 232; regulation of, 171; and
    water management, 34, 35–36, 37,
    38, 39–41, 42. See also specific rivers
RNG. See Reich Conservation Law
road trip literature, 117–118, 119–123

Rohkrämer, Thomas, 7, 65
Rollins, William, 5, 7, 65, 82, 129, 184
Roman Empire, 35, 57, 60, 76n4
romanticism: and back-to-nature
    forestry, 64, 65; and bird protection
    movement, 170; and forests, 22; and
    Germanization of forest, 57–58; in
    Hauser, 124; inheritance of, 6; and
    landscape tradition, 2; as reactionary,
    7; technology and nature in, 113; and
    Weimar conservation movement,
    195, 196
Rosenberg, Alfred, 3
Roßmäßler, Emil Adolf, 63, 166
Rostock District, 229
Ruanda-Urundi, 85
Rubner, Heinrich, 73
Rudorff, Ernst, 50, 58
Rufiji Industrial Corporation, 90
Rufiji River region, 81, 82, 83, 89, 92,
    93, 95, 96, 99, 100
Ruhr River, 216
rural areas: in German East Africa, 91;
    and national parks, 218; and nature
    parks, 216; and occupied Germany,
    215; and water management, 33, 34,
    44–45, 48, 50; in Weimar period
    thought, 184, 185, 194
rural cooperative, 42–43
Ruttmann, Walther, 126

Saar River, 213
Sächsische Schweiz, 224, 228
salt-works, 24, 26, 31n17
Saxony, 224
Schabel, Hans, 82
Schama, Simon, 4–5, 76n4
Schele, Friedrich Radbod Freiher Graf
    von, 90
Schillings, Carl Georg, 173, 174
Schlagwald (age-class forest), 61, 67–68,
    74, 83, 92
Schlegel, August Wilhelm, 57
Schlegel, Friedrich, 57
Schleifenbaum, Peter-Christoph, 66
Schlichting, J., 40–41
Schnee, Albert Heinrich, 101–102

Schnurre, Otto, 171
Schoenichen, Walther, 194–195, 196, 197, 198, 199, 200, 204n37
Schubert, Franz, 58
Schultze-Naumburg, Paul, 3, 194, 204n37
Schumann, Robert, 58
Schutzgemeinschaft Deutscher Wald. *See* Protective Association of the German Forest
science: amateur, 178; and animal displays, 152, 153–154; and bird protection movement, 164, 178, 179, 180; in Federal Republic of Germany, 218; in German Democratic Republic, 226, 227; and Martin, 148, 149; and occupied Germany, 214. *See also* biology
scientific forestry: and back-to-nature forestry, 63, 65, 67–68, 71, 72; and centralized state-making, 77n16; and colonialism, 85; development of, 61; in German East Africa, 81–82, 83, 87, 88, 99, 101–102; and property reform, 62; and sustainable forestry, 55, 72
Scotch pine, 62
Scott, James C., 8, 92
SDW. *See* Protective Association of the German Forest
Second Budapest International Convention of Ornithologists (1891), 166, 169
SED. *See* Socialist Unity Party
Segeberg, Harro, 118
Seifert, Alwin, 128, 210
sexuality, 21, 163, 167
shipbuilding, 25, 86, 90
Sieferle, Rolf Peter, 64, 202n10
slavery, 94, 95
Slavic peoples, 69
social Darwinism, 3
social-democratic state, 68
Social Democrats, 189
socialism, 65, 208; in German Democratic Republic, 223, 224, 226, 227, 228, 231, 232, 233, 234; and nature, 226, 234; and Nazis, 197. *See also* economy; labor; workers

Socialist Unity Party (SED), 226, 227
society: and animal displays, 141, 149, 150, 154; and automobiles, 115, 127, 131; and back-to-nature forestry, 64, 65, 67–68, 69, 70; and bird protection movement, 162, 163, 164, 172, 174, 176, 177; common good of, 84, 97, 101; and environmental history, 5; and forest-state analogy, 69; and Friends of Nature, 189; in German Democratic Republic, 226, 227; in German East Africa, 89, 93, 94, 95, 96; homogeneity of, 185; Mammen on, 70–71; Marx on, 113; and natural history museums, 142, 143, 145, 146, 147; and nature, 18–19, 190, 193–194; Nazi view of, 70–71, 198, 201; and *Neue Sachlichkeit*, 125; and water management, 39, 42, 44–45; and Weimar conservation movement, 184, 185, 187, 188, 190, 191, 193–194; welfare of, 164, 172, 176. *See also* class; collective
Society for Bird Protection (Bund für Vogelschutz), 164, 173, 174
Society for Nature and the Environment (Gesellschaft für Natur und Umwelt, GNU), 234
Society of German Engineers (Verband Deutscher Ingenieure, VDI), 41
Society of the Friends of German Heimat Protection, 204n37; *Der deutsche Heimatschutz*, 190–192
Soviet Union, 74, 208, 220, 227, 229
spatial planning, 216
special path (*Sonderweg*), 7, 183–184
Spengler, Oswald, 127
state: and back-to-nature forestry, 68; and bird protection movement, 169, 172; and colonial forestry, 85, 86; and Düesberg, 68, 69; European, 83; in Federal Republic of Germany, 218; and forests, 61–62, 83, 84; in German East Africa, 98; and Heimat, 188–189; and natural history museums, 141, 144; and Nazi conservationism, 199–200, 201; Prussian-German vs.

Anglo-American, 19; social-democratic, 68; and water management, 33–34, 38–39; and Weimar conservation movement, 185–186, 187–188, 195; and Wilhelmine foresters, 65. See also government; politics
Stechlin Lake, 222
Stern, Fritz, 7
Stine, Jeffrey, 114
Stubbe, Hans, 229
Stuttgart, 143
surveying, 43, 44
sustainability, 22–24, 28, 29, 31n17, 55–56, 62, 71–75, 78n22
swampland, 33, 35, 37–38, 39
Sweden, 142
Switzerland, 75
synchronization (*Gleichschaltung*), 184

Tacitus, *Germania*, 57, 76n4
Tamburu Forest Reserve, 93
Tanganyika, 82–83, 98, 102
Tanzania, 81–102, 107n103
Tarr, Joel, 114
taxation, 87, 89, 90, 91, 95
taxidermy, 147, 154. *See also* animals: display of
taxonomy. *See under* animals
technology: and Autobahn, 128, 129; and automobiles, 117; and back-to-nature forestry, 64, 65; in Bierbaum, 118, 120, 123; and countryside preservation, 185; and environment, 114; and environmental history, 5; in Federal Republic of Germany, 232; and forests, 56; in German Democratic Republic, 225, 233; in Hauser, 124; and Heimat movement, 120; in Jünger, 124, 125, 129; Marx on, 113; and modernity, 119; and nature, 113, 179, 201; and Nazi back-to-nature ideology, 184; and *Neue Sachlichkeit*, 125; Rathenau on, 123; and water management, 46; and Weimar conservation movement, 194, 195, 201, 202n10

Teutoburg Forest, 58, 59, 60
textiles, 95
Third Reich, 7, 69, 70, 73, 212. *See also* Nazis
Tieck, Ludwig, 57
Todt, Fritz, 128
Toepfer, Alfred, 215
Tongamba forest reserve, 98
tourism, 117, 122, 178, 216, 218, 224, 228, 229
town, 23, 86. *See also* urban location; village
traffic, 125
Tulla, Johann Gottfried, 4, 35–36

Uekötter, Frank, 29
Ulbricht, Walter, 226
UNESCO, Man and the Biosphere conference (1968), 218
United Nations, Conference on the Human Environment (1972), 231
United States: ad hoc environmental policy in, 29; automobiles in, 111, 112, 115, 116, 117, 121, 127, 130; conservation movement in, 184; Marx on, 113; natural history museums in, 142; pollution control in, 29–30; state in, 19; and sustainability, 23; and water management, 41, 50; and wilderness, 3
urbanization: and animal displays, 153, 154; and back-to-nature forestry, 64, 65; and back-to-nature movements, 183; and environmental history, 5; and forests, 60; and Heimat movement, 120; study of, 6; in Weimar period thought, 184, 201
urban location: Buchwald on, 218; and growth, 216; and health, 208; and natural history museums, 144, 146; and nature parks, 216; and occupied Germany, 215; traffic in, 125; and water management, 39; and Weimar conservation movement, 186
utopianism, 42, 68

Verband Deutscher Ingenieure. *See* Society of German Engineers

Verein Naturschutzpark. *See* Nature Park Society
Vermeer, Eduard B., 24–25
village, 23, 43, 84, 86, 87, 98–99, 224. *See also* town; urban location
VNP. *See* Nature Park Society
völkisch perspective: and forests, 55, 56, 61, 63, 65, 66–67, 70; and Göring, 72; inheritance of, 6; and Nazis, 72, 183; and occupied Germany, 209; as reactionary, 7; and sustainable forestry, 74
Volksbund für Naturschutz. *See* People's League for Nature Conservation
Volti, Rudi, 111

Wackenroder, Wilhelm Heinrich, 57
Wagenfeld, Karl, 191
Wagner, Richard, 17, 58
Waldbewußtsein, *Waldgesinnung* (forest-mindedness), 55
Waldweben (sough of the forest), 58
Wandervogel movement, 122
water management, 6, 8, 33–50, 84; in Federal Republic of Germany, 219, 220, 232; in German Democratic Republic, 220, 226, 230, 232; in German East Africa, 98; and institutional distance, 20; and Nazis, 198; and occupied Germany, 214, 215; and postwar conservation, 212, 213–214
Weber, Carl Maria von, 58
Weber, Max, 19
Wehrwald (defensive forest), 60
Weimar constitution, 186
Weimar period, 7, 66, 69, 117, 183–201
Weinitschke, Hugo, 224, 231
Weinmann, Richard, 205n64
Weisbrod, Bernd, 186
Wendler, C. F., 49–50
Werra River, 232
Weser River, 232
West Germany. *See* Federal Republic of Germany
Westphalia, 44, 48, 49, 50, 191
White, Lynn, 24, 28
White, Richard, 115–116, 134n28
whites, 86, 96. *See also* race
Wiepking, Heinrich, 210
wilderness, 3, 4, 56, 57, 185
wildlife. *See* animals
wildness, 56
Wilhelm I, emperor of Germany, 59
Wilhelmine era, 7, 10, 65, 153, 176, 184
Wittfogel, Karl August, 49
Wodziki, Casimir Graf von, 166
women, 123, 124, 152, 153, 173–174
workers: East German, 220, 222, 228; and Friends of Nature, 189; in German East Africa, 85, 90, 91, 92, 95, 99; and nature parks, 215; study of, 6; and Weimar conservation movement, 187, 200. *See also* labor; socialism
Working Association of German Commissioners for Conservation (Arbeitsgemeinschaft Deutscher Beauftragter für Naturschutz und Landschaftspflege, ABN), 211, 219
working class, 154, 184, 226. *See also* class; society
Working Group for Close-to-Nature Forestry, 74, 80n67
World War I, 61, 72, 117, 123, 124, 154, 195
World War II, 73, 102, 112, 201, 209
Worster, Donald, 13n21, 22
Wurffbain, Rasch, 46, 47, 48
Wutach Gorge, 213
Wutach River, 213–214

youth, 122, 183, 184, 185, 186, 187, 188, 190, 200

Zanzibar, 89
Zaramo people, 93
Zeller, Thomas, 128
zoogeographical regions. *See* geography